Environmental Molecular Microbiology

Edited by

Wen-Tso Liu

University of Illinois at Urbana-Champaign
Urbana, IL
USA

and

Janet K. Jansson

Lawrence Berkeley National Laboratory
Berkeley, CA
USA

Caister Academic Press

Caister Academic Press
Norfolk, UK

www.caister.com

British Library Cataloguing-in-Publication Data
A catalogue record for this book is available from the British Library

ISBN: 978-1-904455-52-3

Description or mention of instrumentation, software, or other products in this book does not imply endorsement by the author or publisher. The author and publisher do not assume responsibility for the validity of any products or procedures mentioned or described in this book or for the consequences of their use.

Cover image taken from Fig. 10.7.

Printed and bound in Great Britain

Contents

Contributors

Shuwen An
Institute of Molecular and Cell Biology
Singapore

g0402975@nus.edu.sg

Gary L. Andersen
Center for Environmental Biotechnology
Lawrence Berkeley National Laboratory
Berkeley, CA
USA

GLAndersen@lbl.gov

Eoin L. Brodie
Center for Environmental Biotechnology
Lawrence Berkeley National Laboratory
Berkeley, CA
USA

elbrodie@lbl.gov

Mark Bailey
Natural Environment Research Council
Centre for Ecology and Hydrology
Wallingford
UK

mbailey@ceh.ac.uk

Peter A. Bryan
Howard University
Department of Biology
Washington, DC
USA

peterabryan@gmail.com

Patricia A. Carey
Department of Electrical Engineering and Computer Science
University of Tennessee/Oak Ridge National Laboratory
Knoxville/Oak Ridge, TN
USA

careypa@ornl.gov

James R. Cole
Center for Microbial Ecology
Michigan State University
East Lansing, MI
USA

colej@msu.edu

Todd Z. DeSantis
Center for Environmental Biotechnology
Lawrence Berkeley National Laboratory
Berkeley, CA
USA

tdesantis@lbl.gov

Johan Dicksved
Department of Ecology
Earth Science Division
Lawrence Berkeley National Laboratory
Berkeley, CA
USA

johan.dicksved@mikrob.slu.se

Brian D. Dill
Oak Ridge National Laboratory
Oak Ridge, TN
USA

dillbd@ornl.gov

Yi-Hu Dong
Institute of Molecular and Cell Biology
Singapore

dongh@imcb.a-star.edu.sg

Jan Dirk van Elsas
Dep. Microbial Ecology
University of Groningen
Haren
The Netherlands

j.d.van.elsas@rug.nl

Ryan J. Farris
Center for Microbial Ecology
Michigan State University
East Lansing, MI
USA

farrisry@msu.edu

Stephen J. Giovannoni
Department of Microbiology
Oregon State University
Corvallis, OR
USA

steve.giovannoni@oregonstate.edu

Zhili He
Institute for Environmental Genomics (IEG)
Stephenson Research and Technology Center
University of Oklahoma
Norman, OK
USA

zhili.he@ou.edu

Janet K. Jansson
Department of Ecology
Earth Science Division
Lawrence Berkeley National Laboratory
Berkeley, CA
USA

janet.jansson@lbl.gov

Yoichi Kamagata
Research Institute of Genome-Based Biofactory
National Institute of Advanced Industrial Science and Technology (AIST)
Toyohira
Sapporo
Japan

y.kamagata@aist.go.jp

Kostas Konstantinidis
Department of Civil and Environmental Engineering
Georgia Institute of Technology
Atlanta, GA
USA

kostas.konstantinidis@ce.goatech.edu

George A. Kowalchuk
Netherlands Institute of Ecology (NIOO-KNAW)
Centre for Terrestrial Ecology
Heteren
The Netherlands

G.Kowalchuk@nioo.knaw.nl

Johan H.J. Leveau
University of California, Davis
Department of Plant Pathology
Davis, CA
USA

jleveau@ucdavis.edu

Wen-Tso Liu
Department of Civil and Environmental Engineering
University of Illinois at Urbana-Champaign
Newmark Civil Engineering Laboratory
Urbana, IL
USA

wtliu@illinois.edu

Karen E. Nelson
Director of Human Microbiology
Department of Human Genomic Medicine
The J. Craig Venter Institute
Rockville, MD
USA

kenelson@jcvi.org

Satoshi Okabe
Department of Urban and Environmental Engineering
Graduate School of Engineering
Hokkaido University
Kita-ku
Sapporo
Japan

sokabe@eng.hokudai.ac.jp

James Prosser
Institute of Environmental and Biological Sciences
University of Aberdeen
Aberdeen
UK

j.prosser@abdn.ac.uk

Angela Sessitsch
Austrian Research Centers GmbH – ARC
Department of Bioresources
Seibersdorf
Austria

angela.sessitsch@arcs.ac.at

Kornelia Smalla
Julius Kühn-Institut Bundesforschungsinstitut für Kulturpflanzen
Institut für Epidemiologie und Pathogendiagnostik
Braunschweig
Germany

kornelia.smalla@jki.bund.de

Ulrich Stingl
Department of Microbiology
Oregon State University
Corvallis, OR
USA

ustingl@syntheticgenomics.com

James M. Tiedje
Center for Microbial Ecology
Michigan State University
East Lansing, MI
USA

tiedjej@msu.edu

Alexander H. Treusch
Department of Microbiology
Oregon State University
Corvallis, OR
USA

alexander.treusch@science.oregonstate.edu

Bryan A. White
Departments of Animal Sciences and Pathobiology
Division of Nutritional Sciences
The Institute for Genomic Biology
University of Illinois at Urbana-Champaign
Urbana, IL
USA

bwhite44@illinois.edu

Nathan C. VerBerkmoes
Oak Ridge National Laboratory
Oak Ridge, TN
USA

verberkmoesn@ornl.gov

Etienne Yergeau
Biotechnology Research Institute
National Research Council of Canada
Montreal
Canada

Etienne.Yergeau@cnrc-nrc.gc.ca

Jacque C. Young
Graduate School of Genome Science and Technology
University of Tennessee/Oak Ridge National Laboratory
Knoxville/Oak Ridge, TN
USA

youngjc@ornl.gov

Lian-Hui Zhang
Institute of Molecular and Cell Biology
Singapore

lianhui@imcb.a-star.edu.sg

Liping Zhao
School of Life Science and Biotechnology
Shanghai Jiao Tong University
Shanghai
China

lpzhao@sjtu.edu.cn

Jizhong Zhou
Institute for Environmental Genomics (IEG)
Stephenson Research and Technology Center
University of Oklahoma
Norman, OK
USA

jzhou@ou.edu

Preface

Microorganisms inhabited Earth long before any higher organisms. During the course of evolution they have occupied diverse environments such as soil and fresh and marine water, and different ecological niches including symbiotic relationships with plants, animals and humans. Until recently the main source of information we had about microorganisms in the environment arose from what we now refer to as 'conventional' or 'traditional' approaches. These approaches primarily relied on cultivation of selected microbes in broth or agar, whereby we could then identify them and study their properties in isolation or in defined communities. Now we live in an era where 'modern' tools can be utilized to study microorganisms in environmental samples without the necessity to cultivate them. The current molecular toolbox encompasses a range of molecular approaches. The majority of these are based on analyses of DNA, but new approaches for study of RNA and proteins extracted from environmental samples are rapidly becoming available. Currently, there is a major emphasis on the application of 'omics' approaches to directly study molecules extracted from samples of interest and to use these approaches to determine the identities and functions of the microbes inhabiting different environments.

This concept for this book originated from a series of workshops since the first meeting at Wageningen, the Netherlands in 2004 sponsored by The Scientific Committee on Problems of the Environment (SCOPE) and the International Council of Scientific Union Workshop on Microbial Environmental Genomics (MicroEnGen), with participation from several leading, international microbial ecologists and molecular biologists, many of whom are authors of chapters in this book. In addition, several of the authors are involved in the newly launched TerraGenome initiative: the international consortium dedicated to sequencing the soil microbiome (http://terragenome.org) and other ongoing national and international collaborative projects. These initiatives highlight the need for collaboration between scientists and between countries for accomplishing ambitious goals that are currently too challenging for individual laboratories to pursue on their own. The collaborative nature that is now becoming more commonplace is illustrated by the shared authorship of many of these chapters by authors from more than one laboratory and country.

The first half of this book focuses on the microbial diversity and phylogeny of microorganisms on Earth and describes the molecular toolbox now in hand for the study of microbial diversity. Chapter 1 describes the use of the 16S rRNA gene as a phylogenetic marker for identification and classification of Bacteria and Archaea, and then compares this approach with recent advances in knowledge gleaned from sequenced microbial genomes. The subsequent chapters explain the basis of different molecular approaches, including metagenomics, metaproteomics, microarrays, and molecular fingerprinting approaches for the study of the composition and diversity of microbial communities and their functions.

The second half of this book describes how these approaches have been used to study microbes in different environments, such as soil, marine water, plants, humans and wastewater treatment plants. Each of these environments contains a specific type of microbial community that is responsible for key functions. The last chapter of the book describes the formation of microbial biofilms, the genetics behind this process and environmental implications.

What we have learned from application of these tools is that microbial diversity on Earth is enormous and much greater than originally thought. We also know that most of the microbes that we have identified to be present are unknown and their functions remain hidden. Therefore, there is a continuous need for better tools and more information as we continue to explore the microbial world. The importance of microbial functions for our health and environment has also gained increasing recognition. For example, their roles in carbon and nutrient cycling and how these functions are impacted by climate change are key areas of interest. There has also been a shift away from a focus on pathogens, towards understanding the beneficial functions of microorganisms for plant growth and for human health that are discussed in the chapters devoted to those topics.

In conclusion, the field of environmental microbiology is continuously evolving and at such a rapid pace that it is difficult for even experts in the field to keep up with the latest methods and the increase in information that is obtained. This book lays the platform with respect to the current state-of-the-art in the field of Environmental Molecular Microbiology and our hope is that it can serve as an aid to students, research scientists, environmental agencies and to industries that can benefit from application of these tools or this knowledge.

Wen-Tso Liu and Janet K. Jansson

Microbial Diversity and Phylogeny: Extending from rRNAs to Genomes

1

James R. Cole, Kostas Konstantinidis, Ryan J. Farris and James M. Tiedje

Abstract

The small subunit ribosomal RNA gene (SSU rRNA) has been the cornerstone of microbial ecology studies over the last 15 years, and has provided much of what we know about Bacterial and Archaeal diversity and community structure, and has greatly aided microbial taxonomy. This chapter provides information on the development of this gene as a molecular marker and tools for its analysis, but also addresses challenges for the future including advancing the species definition, linking 16S rRNA information to taxonomy, the need for finer-grained resolution of community members and how genomics is aiding our understanding of the relationships among closely related organisms, and ultimately of natural populations. When the available 16S rRNA genes from species type strains were examined, the most distant sequences in the median genus and family were about 4.4% and 14% different, respectively. The largest dissimilarity between a sequence and its closest relative in the same taxa (similar to single-linkage clustering distance) was 3.5% and 10% for the median genus and family. The ratio of the two values, a measure of the 'shape' of the taxa averaged less than 1.5 for all ranks, indicating that most taxa are not elongated, but are fairly spherical. When the near-full-length 16S rRNA gene sequences in the public databases in 2006 were clustered into groups at proxy distances for species, genus, family and order, the number of clusters with time increased exponentially for all ranks documenting the enormous diversity of the microbial world and that much remains to be discovered.

Introduction

Phylogenetic analysis of the small subunit ribosomal RNA gene (SSU rRNA) (Woese *et al.*, 1990) has revolutionized the way that microbial ecology, diversity and phylogeny studies have been exercised for the past three decades. In fact, our view of the microbial world to date is rooted, one way or another, in the analyses of the SSU rRNA genes.

For analysis of environmental populations the small-subunit rRNA gene is, by far, the most commonly used phylogenetic marker. The adoption of the small-subunit rRNA gene as the pre-eminent molecule for microbial phylogenetic analysis predates its use in environmental microbiology. In fact, our view of the evolutionary relationships among life forms on Earth was transformed by the comparative analysis of ribosomal RNA gene sequences (Woese *et al.*, 1990). It was the resulting bacterial phylogeny, along with the availability of large numbers of rRNA gene sequences and the development of a variety of molecular methods that made the rRNA gene the first practical molecular marker for environmental microbial analysis. Hence, for the first time, researchers had the ability to comprehensively determine the composition and structure of microbial communities. Its use revealed that most communities were far more complex than previously imagined, thereby stimulating even more work in this field and spawning the new field of metagenomics (National Research Council, 2007). The first large (for the time) collection of rRNA gene sequences was released by Carl Woese and colleagues in 1992. These 708 sequences formed the initial backbone of what came to be known as the Ribosomal Database Project (RDP; Olsen *et al.*, 1991; Cole *et al.*, 2009). The current Ribosomal Database Project maintained, as of October 2008, over 670,000 aligned and annotated bacterial small-subunit rRNA gene sequences, and 34,000 archaeal sequences (Fig. 1.1). Furthermore, there are over 1.5 million privately held sequences in *my*RDP undergoing users' analyses.

The RDP sequences are available in several subsets, including sequences that are of higher quality and near-full-length sequences (286,680), from environmental samples (559,135), from in-culture bacterial isolates (146,968), and from bacterial species type strains (5534). These type-strain sequences cover about 60% of the validly named bacterial species. Type-strain sequences are of special interest, because species type-strains serve as archetypes and link rRNA-based phylogeny with bacterial taxonomy. For higher taxa, the species type-strains serve to delineate the formal membership in the taxa and can provide information about the formal 'breadth' of the taxon and how well the members form a coherent cluster. The RDP also provides a suite of tools to allow users to place their own sequences in the context of the RDP sequence collection. This includes tools for taxonomic placement (RDP Classifier), nearest-neighbour finding (Sequence Match), primer and probe testing (Probe Match), phylogenetic analysis (Tree Builder), sample comparison (Library Compare) and taxonomic visualization (Taxomatic). The RDP offers a user account system (*my*RDP) for automated alignment and classification of user sequences, and a pipeline for processing pyrosequencing data (Pyrosequencing Pipeline). In addition, the RDP offers a classroom assignment generator (Assignment Generator) for instructors, extensive help files, and a set of short video tutorials demonstrating the RDP tools.

One limitation of the 16S rRNA gene is that it is rather conserved and hence is not reliable for taxonomic identities at the species level (see later sections). Other phylogenetic marker genes can provide more fine-grained resolution, and a number of these are used regularly in pure culture work, e.g. *recA, gyrB, rpoB* (also see later section on MLST/A). Some of these are beginning to find their way into ecology work. They offer the potential advantage of providing resolution that may better relate to function.

While the 16S rRNA gene has revolutionized bacterial and archaeal ecology and taxonomy, the 18S and ITS (internally transcribed spacer)

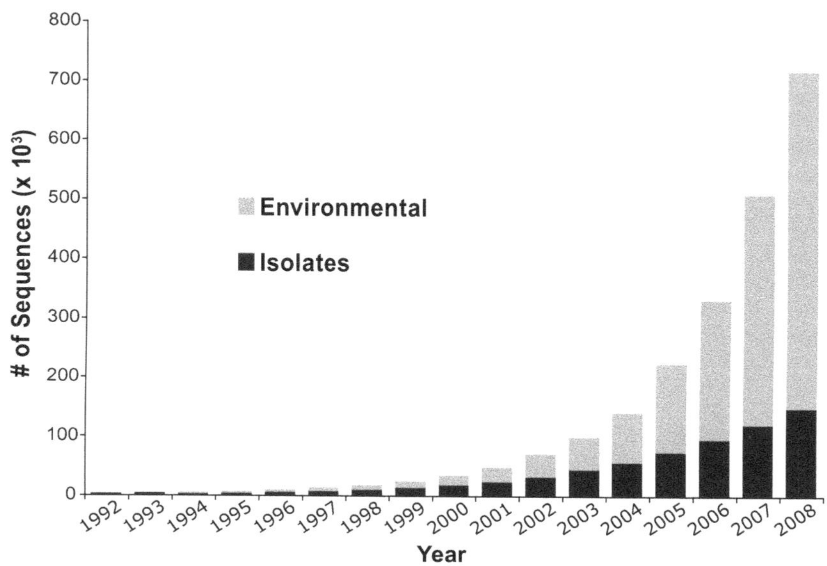

Figure 1.1 Growth in number of 16S rRNA gene sequences in the public databases that are from isolates and from environmental samples.

regions provide analogous methods for fungi and other microbial eukaryote ecology studies (Horton and Bruns, 2001). The databases for these genes or regions, however, are not yet as well developed which continues to limit the use of these markers.

A word about what a species is

Species represents the most fundamental unit of microbial taxonomy and the cornerstone concept for studying and cataloguing the microbial diversity on the planet. Yet the definition of bacterial and archaeal species remains a highly controversial and unsettled issue. An improved understanding of species and its definition has important broader impacts as species is tightly linked to what is regulated for transport, possession, bioterrorism agent oversight, and for diagnostic/clinical purposes. We briefly describe this issue here and the recent developments based on genomics as it relates to several of the sections in this book; the reader is directed to several excellent reviews for further details (Achtman and Wagner, 2008; Konstantinidis *et al.*, 2006a; Rossello-Mora and Amann, 2001; Stackebrandt *et al.*, 2002; Staley, 2006; Ward, 2006).

The current species concept for Bacteria and Archaea (Rossello-Mora and Amann, 2001; Stackebrandt *et al.*, 2002) has been an enormously difficult subject to advance for several reasons. Although Bacteria and Archaea are asexual organisms, they do exchange genes in their unique and unusual ways, e.g. through bacteriophages or plasmid DNA, and the exact rate and magnitude of these processes remains undetermined and probably depends upon the typically fluctuating environmental selection pressure(s). Accordingly, it remains unclear whether or not the processes driving diversification of these microorganisms produce sufficiently discrete groups of individuals (species). While recent environmental surveys (Konstantinidis and Delong, 2008; Tyson *et al.*, 2004) as well as computer simulations (Fraser *et al.*, 2007) have indicated that discrete populations may be identifiable, at least within stable environmental settings, more research is required to advance these previous findings and interpretations for a species concept. Furthermore, it is not always feasible, due to technological limitations related to the tiny size of bacterial and archaeal cells and the poor understanding of the functionality of the cell, to delineate unique phenotypic characteristics that constitute a prerequisite for species description.

The inability to devise a solid theoretical framework for species description has led the microbial taxonomists to agree upon a functional species definition that is based upon a combination of methods and recommended standards (Stackebrandt *et al.*, 2002; Wayne *et al.*, 1987). Recent high-resolution studies with sequenced genomes of several other bacterial families, however, have convincingly demonstrated that the 70% DDH standard does not correspond to coherent groups but rather to a continuum of genetic diversity (Goris *et al.*, 2007). Besides, the 70% DDH or similar standards are simply manmade cut-offs to define evolutionary relatedness and do not necessarily emerge from natural groupings of organisms and natural processes. Lastly, the current species definition for bacteria does not translate well to eukaryotes; e.g. none of the 22 species concepts existing for eukaryotic organisms encompasses the current species definition for bacteria. In fact, several studies have clearly shown that the bacterial species definition may be too broad and encompass organisms that are too heterogeneous under the same 'named' species (Cohan, 2002; Goris *et al.*, 2007; Konstantinidis and Tiedje, 2005a; Sibley *et al.*, 1990; Staley, 1997; Ward, 2006).

Searching for natural groupings of organisms at different levels from the 70% DDH offers a very promising approach for advancing the species definition. Although several studies have shown that coherent groups may emerge due to geographic isolation (Whitaker *et al.*, 2003) and for symbiotic organisms that evolve in close association with their host and, hence, follow the speciation of their host (Staley, 2006), it has not been possible to generalize these findings to the whole microbial world. Indeed, the previous examples are probably exceptions with respect to the continuous gene flow that is observed in most natural (and more complex) environments. Further, several genetic surveys have frequently recovered distinguishable sequence-based groups from complex environments (Acinas *et al.*, 2004a; Rappe and Giovannoni, 2003; Rusch *et al.*, 2007). Most of these surveys, however, were limited in the amount of total natural diversity sampled and/or included organisms that were typically recovered from different populations and habitats further complicating population analysis and comparisons (Konstantinidis and Delong, 2008). With the advent of more high-throughput metagenomic techniques, which provide access to the diversity of naturally occurring populations (Konstantinidis and Delong, 2008; Tringe *et al.*, 2005; Tyson *et al.*, 2004; Venter *et al.*, 2004), more robust conclusions may soon emerge. Until then, however, the current evolutionary- and phenotype-based standards described previously remain the most pragmatic and efficient approach to the species problem.

Characteristics of formal bacterial taxa

In bacteriology, names are controlled by a formal set of rules for the ranks subspecies through order (Lapage, 1992). (The ranks of phylum and above are not formally covered by the Code of Nomenclature.) On the other hand, bacterial taxonomies are effectively a statement of opinion and there is no single 'official' bacterial taxonomy. Until recently, higher-order bacterial taxonomies either failed to follow the rules of nomenclature or were inconsistent with rRNA-based phylogeny. In 2002, Garrity and co-workers published a new bacterial taxonomy that attempted to reconcile taxonomy with rRNA-based phylogeny, while following the rules of bacterial nomenclature (Garrity *et al.*, 2002). The RDP has adopted this taxonomy as its main taxonomy. This taxonomy is still evolving as more is learned and errors are corrected. The RDP is currently using version 7.7 of this 'nomenclatural taxonomy' (Garrity *et al.*, 2007), but modified by the inclusion of recently proposed rearrangements of the Firmicutes (Ludwig *et al.*, 2008) and the Cyanobacteria (Wilmotte and Herdman, 2001), along with published informal classifications where authoritative formal classifications do not exist, such as the *Acidobacteria*, *Verrucomicrobia* and the OP11 candidate phylum (Barns *et al.*, 2007; Harris *et al.*, 2004; Sangwan *et al.*, 2004). Additional information for the classification of chloroplast sequences and the archaeal phyla Korarchaeota, and Nanoarchaeum were taken from the NCBI taxonomy (Wheeler *et al.*, 2000).

We examined the relatedness of 5176 species type-strain sequences from RDP release 9.44 (October 2007) for each genus, family, class and order from the nomenclatural taxonomy (Table 1.1; Cole *et al.*, 2007). We first measured the overall breadth of each taxa, equivalent to complete-linkage clustering distance, by computing the maximum pairwise distance between 16S rRNA genes in each genus, family, order, and class for which type-strain sequences were available for at least two different immediate subtaxa (two genera for a family, etc.). The median breadth was 0.044, 0.14, 0.20, and 0.27 for genera, families, orders and classes, respectively (Fig. 1.2a). Few families (< 1%) were smaller than the median genus, and only about 5% of genera were broader than the median family. For higher ranks, there was more overlap. Although only 3% of orders were smaller than the median family, 18% of families were broader than the median order while 12% of orders were larger than the median class and 15% of classes were smaller than the median order.

To get a better idea of the 'shape' of the taxa, we also measured the internal distance necessary to join all members, equivalent to the single-linkage cluster distance (Fig. 1.2b). For genera, the median distance was similar to the total breadth, 0.035. The median family, order and class single-linkage distances were 0.10, 0.16 and 0.20, respectively. The median ratio of single-linkage distance to complete-linkage distance is less than 1.5 for all ranks (Fig. 1.2c). Only a small number (2%) of taxa had a single-linkage distance more than twice as large as the complete-linkage distance. The relatively small ratios between the two distance measures indicate the formal taxa are, for the most part, relatively compact and not elongated or string-like.

Table 1.1 Number of each rank in the RDP 9.44 taxonomy

Rank	Number
Genus	1,226
Family	230
Order	82
Class	33

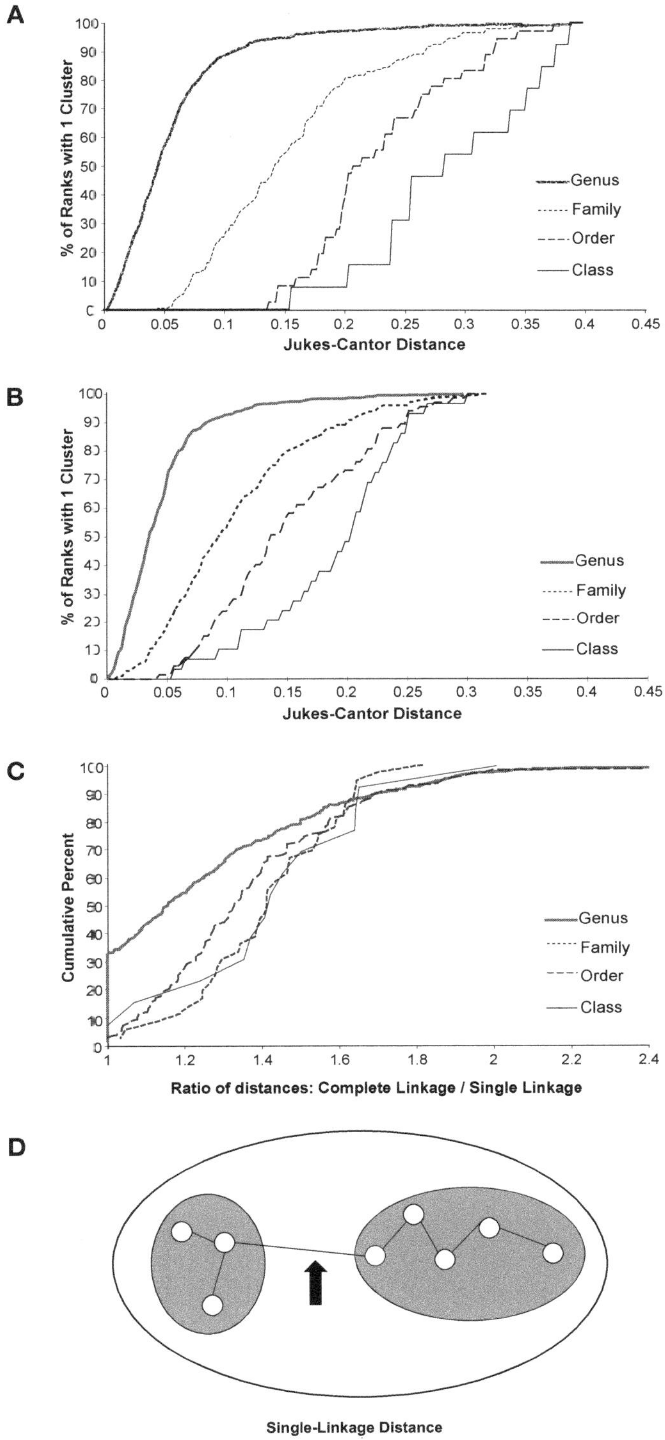

Figure 1.2 **A** Maximum distance between species type-strain sequences in individual taxa, plotted as % of taxa with indicated or shorter distance. **B** Minimum length needed to join all species type-strain sequences in a single graph, plotted as in A. **C** Ratio of distances calculated as in A and B, plotted as % of taxa with indicated or smaller ratio. **D** Graphic representation of minimum edge length (single-linkage-cluster distance).

Diversity represented by the rRNA gene sequence collection

We examined the extent of diversity represented by the bacterial rRNA gene sequence library by clustering 99,949 high-quality near-full-length (> 1200 alignable positions) bacterial small subunit rRNA gene sequences from RDP release 9.44 (October 2007) into operational taxonomic units (OTUs). There are several methods of agglomerative clustering that have been applied to molecular data. We chose to cluster based on single-linkage distance for the following reasons: 1) Unlike other agglomerative clustering methods, single-linkage clusters are independent of the order in which the sequences are clustered. 2) Adding a new member may cause adjacent clusters to consolidate, but unlike other methods, will never cause clusters to separate. We clustered the sequences at minimum intra-cluster distances of 0.01, 0.035, 0.10 and 0.16 differences per position. Although taxonomic concepts do not translate well to specific evolutionary distances, 0.01 was shown to be a relatively conservative species-level proxy (Stackebrandt and Ebers, 2006). The other values represent the median of the type-strain single-linkage distances shown above for genus, family and order, respectively. Clustering produced 28,155 OTUs at the 1% species proxy and 446 OTUs at the 16% order proxy (Table 1.2). At each level, the majority of OTUs contain only a single environmental sequence. It is possible that some of these singletons may be the result of in-vitro artefacts. In a recent study, up to 5% of public rRNA gene sequences were found to contain anomalies (Ashelford *et al.*, 2005). However, all sequences in the RDP are screened for anomalies with the same Pintail anomaly detection program used in the aforementioned study, and only those without detected anomalies were included in these clustering results. These singleton sequences may be unique to habitats that have not yet been well-studied. Or perhaps they represent very minor components of the environment in which they were found, equivalent to the 'rare biosphere' seen in very-deep sequencing studies (Sogin *et al.*, 2006). On the other hand, the majority of sequences are contained in a relatively small number of clusters. For example, for the 3.5% distance OTUs, over half the sequences are contained in the largest 1% of OTUs (not shown).

When these data are arranged in order of submission, a collector's curve is formed (Fig. 1.3). Unlike a collector's curve formed by random selection from a single sample, these collector's curves have been shaped by the purposes of the various researchers collecting rRNA gene sequences at various times. For example, the relatively flat region between *ca* 50,000 and 60,000 in the 1% and 3.5% cluster collector's curve was caused by over 11,000 sequences submitted from human intestinal bacteria from a single study (Eckburg *et al.*, 2005). From the shape of the 1% and 3.5% collector's curves, it is apparent that novel diversity is still accumulating. Following the increase in this discovered diversity by time (Fig. 1.4), the numbers of clusters appear to be increasing exponentially at all four levels of clustering. The exponential fit of these data is quite striking, with r^2 values above 99% for all levels and derived doubling times of between 33 months and 39 months. For comparison, the doubling time for high-quality near-full-length rRNA gene sequences is approximately 27 months. The exponential

Table 1.2 rRNA diversity at each taxonomic rank in RDP Release 9.44 (10/31/06) as measured by clustering

Distance	Total clusters	Clusters with environmental sequences only	Clusters with only one environmental sequence	Clusters with at least one isolate sequence	Clusters with a type sequence
1% (species)	28,155	17,175 (61.0%)	16,005 (56.8%)	10,980 (39.0%)	3337 (11.9%)
3.5% (genus)	13,008	9363 (72.0%)	7900 (60.7%)	3645 (28.0%)	1490 (11.5%)
10% (family)	2394	1858 (77.6%)	1312 (54.8%)	536 (22.4%)	274 (11.4%)
16% (order)	446	367 (82.3%)	230 (51.6%)	79 (17.7%)	45 (10.1%)

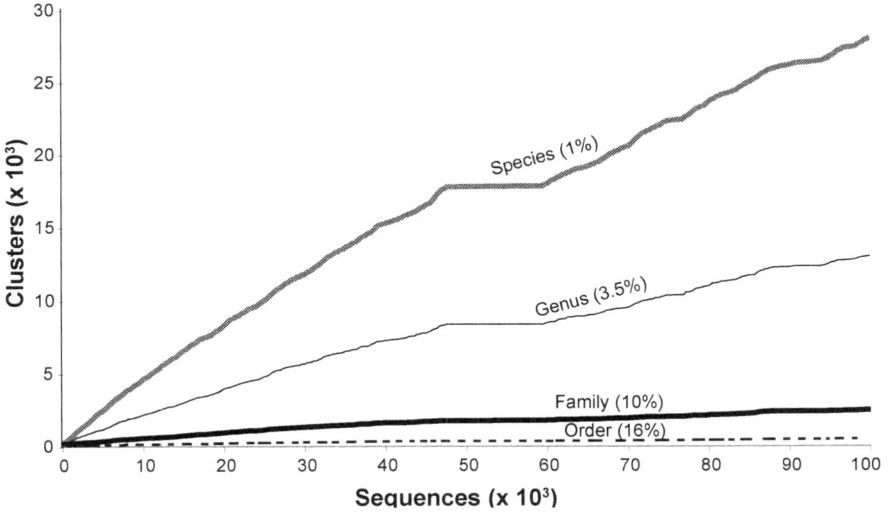

Figure 1.3 Accumulation curves. High-quality, near-full-length sequences from RDP release 9.23 were clustered into OTUs by single-linkage clustering at the given distances. Sequences ordered by initial submission date.

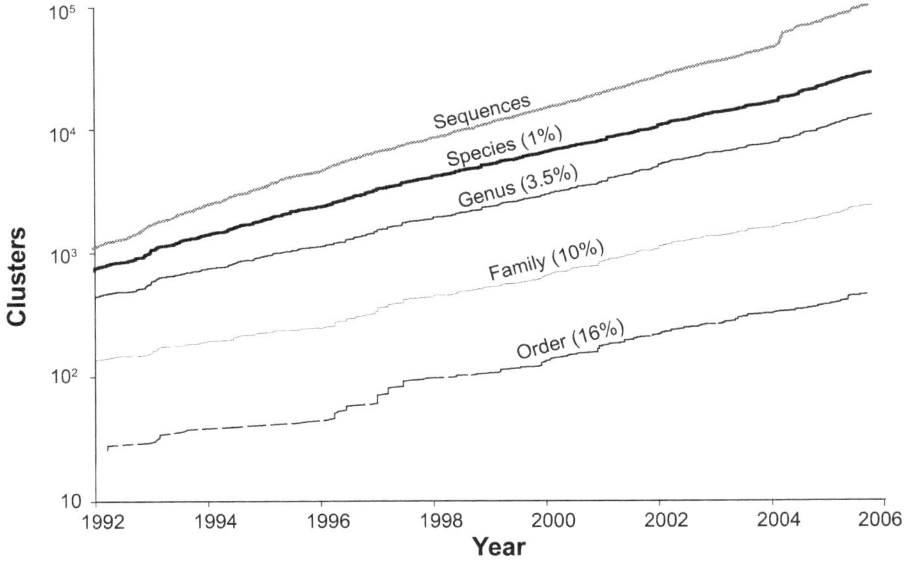

Figure 1.4 Clustering results from Figure 1.2 plotted as number of clusters with one or more member sequences with initial submission date on or before the indicated date. Exponential best-fit gave r^2 values of 0.996, 0.998, 0.998, 0.996, and 0.99, and apparent doubling times of 33, 34, 38, 39 and 27 months for sequences, 1%, 3.5%, 10%, and 16% distance OTUs, respectively.

increase in total sequences has been fuelled, in part, by steady improvements in sequencing technologies. Now, next-generation sequencing technologies, such as pyrosequencing, are now starting to be applied to environmental rRNA gene analysis. These new technologies herald a new era in the exploration of microbial diversity.

Impact of new sequencing technologies

Using the first-generation GS-20 pyrosequencing platform (Roche), in 2006, Sogin and colleagues were able to obtain over 100,000 short rRNA gene reads per environmental sample and demonstrated that deep water marine microbial communities are one to two orders of magnitude more diverse than previously thought (Sogin *et al.*, 2006). Pyrosequencing of rRNA gene amplicons has now been used for in-depth diversity analysis of samples ranging from soil to human oral cavity (Keijser *et al.*, 2008; Roesch *et al.*, 2007). The first generation pyrosequencing platform was limited to read lengths of about 70 bases, while the new third-generation GS-20 FLX using Titanium chemistry produces reads of 400 bases with high quality (Roche product literature). At this longer length, the usefulness of these partial rRNA gene sequences for taxonomic characterization of a sample is not much less than that of full-length sequences (Wang *et al.*, 2007). In addition to the greatly increased sequencing 'depth' possible with pyrosequencing, individual samples can each be 'tagged' using primers incorporating different 'key' sequences and mixed before sequencing, with the key sequence used to sort the sequences by sample. This provides a greatly reduced cost per sample compared to Sanger sequencing, not only allowing sequencing to greater depth, but also allowing many more samples to be studied. Other new sequencing technologies such as Illumina, SOLiD (Applied Biosystems), and Helicos may offer even higher throughput and further increases in the rate of diversity discovery.

The interrelation between rRNA gene and whole genome sequences

Comparison of SSU rRNA to genome-based approaches

While SSU rRNA gene has (successfully) served as the backbone of bacterial and archaeal phylogeny and taxonomy to date (see previous section), there are several important pitfalls associated with the use of SSU rRNA gene for these purposes. First, SSU rRNA is just one of the genes in the genome and, actually, the most conserved one in terms of sequence conservation. Accordingly, it is possible that important levels of phenotypic differentiation may underlie almost identical SSU rRNA genes as this has been implicated to be the case, for example, in natural populations of marine *Vibrio* species (Hunt *et al.*, 2008; Thompson *et al.*, 2005). Thus, a comprehensive understanding of how well the relationships based on this single gene reflect the whole-cell/whole-genome relatedness is important. It is also important to precisely quantify the extent of horizontal transfer of rRNA genes as well as the sequence diversity and recombination[1] frequency among the (often) multiple rRNA operons within the same genome, as these processes can greatly affect the robustness of the phylogenetic signal of the rRNA gene. The advent of whole-genome sequencing a decade ago (Fig. 1.5) provided the opportunity to evaluate these issues as well as the robustness of the pre-existing SSU-rRNA-based phylogeny and taxonomy.

Several genomic approaches have been developed to evaluate the ribosomal phylogeny; all have at their core the construction of a phylogeny based on the whole or selected parts of the genome, followed by comparisons of this phylogeny to that based on the SSU rRNA gene. The methods for building the genomic phylogeny vary and include analysis of concatenated alignments of the all or a subset of the universally conserved protein-coding genes (Brown *et al.*, 2001; Ciccarelli *et al.*, 2006), Boolean matrices based on the distribution (e.g. presence or absence) of genes or pathways in the genome (Snel *et al.*, 1999), similarity matrices based on the sequence identity of orthologous genes (Wolf *et al.*, 2001), and variations of those (Gophna *et al.*, 2005). Each approach has its own advantages and disadvantages and there are typically differences, albeit usually relatively small, in the results of each approach (for reviews on this subject see

1 Recombination here refers to the hybridization, through a homologous mechanism, between two highly similar sequences, which gives rise to a new hybrid allele. Recombination is also the underlying process by which any horizontally transferred DNA fragment is incorporated into the genome of a recipient cell, through a homologous (related sequences) or non-homologous (unrelated sequences) mechanism.

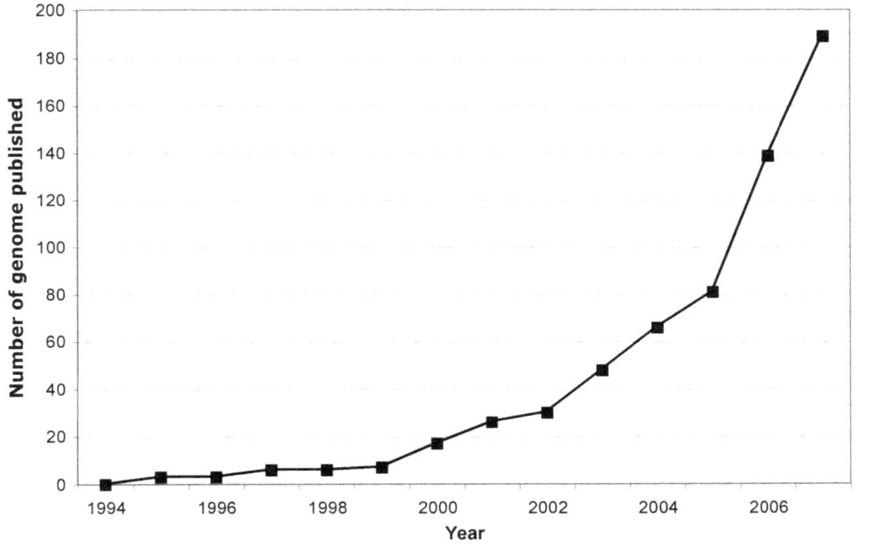

Figure 1.5 The growth of completed bacterial and archaeal genomes over time. Graph shows the number of genomes completely sequenced by the year of the public release of their genomic sequence to the National Center for Biotechnology Information (NCBI) database.

Delsuc *et al.*, 2005; Gogarten and Townsend, 2005). The major conclusion from all these studies is that the SSU rRNA-based methodology is sufficiently reliable and robust for evaluating relationships between organisms related at the genus level and higher (for the species level see above). Consequently, the ribosomal phylogeny remains a reasonable measure of relatedness and will continue to serve as the backbone of prokaryotic taxonomy; at least until a drop in sequencing cost makes genome sequencing affordable for everyday practice.

Departures from the topology of the ribosomal tree supported by several of the genome-based studies mentioned previously (Ciccarelli *et al.*, 2006) are probably minor and/or rather difficult to confirm (Doolittle, 1999; Doolittle and Bapteste, 2007). Several reasons account for the latter. Firstly, fossil records that can be used to calibrate and compare the phylogenies derived based on different genes are sparse for microorganisms. Therefore, the phylogenetic resolution, e.g. which scales of evolutionary history can be robustly resolved, and the accuracy of the derived phylogenies remain speculative. Secondly, the number of genes conserved in all microorganisms is very small, most likely smaller than fifty (Ciccarelli *et al.*, 2006; Koonin, 2003) Accordingly, the phylogenetic signal in the concatenated alignments of the sequences of all the universal genes is frequently inadequate, particularly for evaluating some of the ancestral nodes of the microbial tree of life. It is important to realize that the evolutionary history of Bacteria and Archaea is so long, estimated to ~3.5 billion years (reviewed in Hedges, 2002), that it probably surpasses the evolutionary time that is measurable by any of the universal genes. Last, recent work has elucidated that all genes in the genome, even the ribosomal ones (Miller *et al.*, 2005), are subject to horizontal gene transfer (HGT) and recombination events, and the number of such events for every gene is essentially unknown. Current algorithms are not proficient in detecting and quantifying HGT events; hence, it remains unknown how such processes may have obscured the derived phylogenies (Gogarten and Townsend, 2005). There is, however, a growing consensus that, due to reduced genetic barriers for homologous recombination (Eppley *et al.*, 2007; Fraser *et al.*, 2007; Majewski *et al.*, 2000) and the specificity of the vectors of HGT such as bacteriophages towards closely related hosts, HGT is more pronounced between closely related organisms. Therefore, although the real phylogeny of organisms is probably impossible to be accurately reconstructed, even with full genomic sequences available, an average 'consensus' phylogeny may be attainable (for a contrasting opinion see Doolittle and Bapteste, 2007).

Work in the labs of the authors has shown that the average identity of all shared genes between two organisms represents a simple, highly accurate and robust measurement of the whole-genome-level relatedness of the organisms, which offers several advantages over the SSU rRNA gene-based or other measurements (Goris *et al.*, 2007; Konstantinidis *et al.*, 2006b). This measurement is employed below to offer representative and illustrative examples of the strengths and weaknesses of the ribosomal gene approach as well as broad windows into the microbial biodiversity. Comparisons among the first 175 fully sequenced bacterial and archaeal genomes that sample most of the major phylogenetic lineages reveal that the SSU rRNA gene identity between any two organisms shows, on average, very good correspondence to the average amino acid identity (AAI) of the shared genes between the organisms (Konstantinidis and Tiedje, 2005b). For instance, the relationship between SSU rRNA and AAI is almost perfect (r^2 = ~0.99) and independent of the genome size or phylogenetic position of the organisms included in the analysis (Fig. 1.6, black curve), revealing that the SSU rRNA gene can reflect nicely the genome average relatedness. The SSU rRNA gene appears to have limited resolution between genomes showing higher than 80% AAI (discussed below), whereas the permissible substitutions in its sequence reach saturation around 60–65% identity, presumably due to functional constraints. The SSU rRNA gene appears to offer higher resolution than the genome-aggregate sequence identity between distantly related organisms, e.g. organisms related at the phylum or class level or showing < 45–50% AAI. It remains unclear, however, how robust and quantitative the signal of the SSU rRNA gene is at this level of relatedness for reasons presented above and because, at this

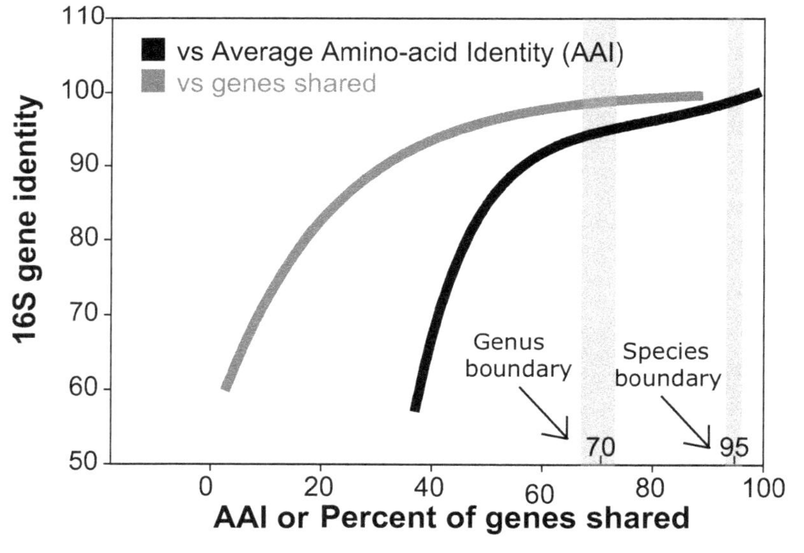

Figure 1.6 Relationship of the SSU rRNA gene identity to genome-aggregate average amino acid identity (AAI) and per cent of shared genes between any two bacterial or archaeal genomes. The equations describing the curves shown are: (I) SSU rRNA vs. AAI, $y = -5711.56 + 3993.87^*(\ln x) - 917.88^*(\ln x)^2 + 70.5(\ln x)^3$; and (II) SSU rRNA vs. per cent genes shared, $y = 54.2 + 46^*(1 - e^{-0.0483x})$. The fit of the curves to the mean values, i.e. the mean AAI or per cent genes shared value per unit of SSU rRNA gene identity, of the experimental data from 175 × 175 whole-genome comparisons (J. Bacteriol. 2005, 187(18): 6258–64) is perfect, with r^2~0.99 for both equations. The use of the mean values reduces the noise in the data (for all data see the original publication) and represents the 'average' relationship between SSU rRNA identity and AAI or per cent genes shared. Note that the relationship between SSU rRNA and AAI approximates linearity between closely related organisms, e.g. showing >90% AAI, whereas the relationship becomes logarithmic for more distantly related organisms.

level, the whole-genome-level relatedness is not very discriminative. Finally, there is a very strong correlation between SSU rRNA gene identity and the per cent of genes in the genome shared between any two genomes (Fig. 1.6, grey curve), suggesting that SSU rRNA identity is also predictive of the gene-content and, hence, biochemical similarity among prokaryotic organisms.

It is evident from the analysis, however, that the genomic comparisons show a much higher resolving power than the ribosomal RNA genes between closely related genomes since, for example, a 96–100% difference in the SSU rRNA gene identity is spread over 75–100% ANI units (Fig. 1.7). ANI represents the average nucleotide identity of shared genes and constitutes a more precise measurement of relatedness than AAI for evaluating short evolutionary time scales[2] (Konstantinidis and Tiedje, 2005b). This level of relatedness is nonetheless the most interesting for studying and evaluating the mechanisms of evolution and creation of genetic diversity.

Studying closely related organisms with genome sequencing

Perhaps the most important contribution of genome sequencing towards deepening our understanding of the microbial world relates to exposing the differences between closely related microorganisms. Comparative analyses of genomes of closely related organisms shows that microbial species encompass much greater

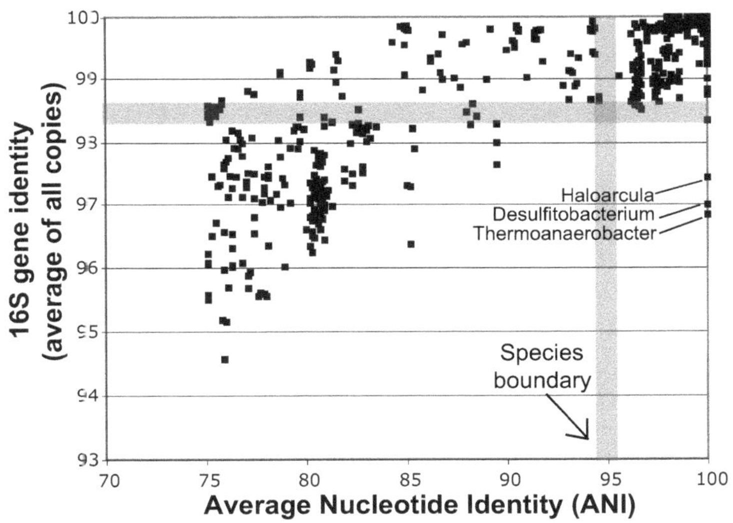

Figure 1.7 Relationship of the SSU rRNA gene identity to genome-aggregate average nucleotide identity (ANI) between two closely related genomes. Each data point represents a pair of genomes and shows the average SSU rRNA gene identity, calculated from all copies in the genome (y-axis), plotted against the genome-aggregate average nucleotide identity (ANI, x-axis) between the two genomes in the pair. The area that corresponds to the most frequently used standards for species demarcation (i.e. the 70% DNA–DNA hybridization, which translates to ~95% ANI) is denoted on the graph. Note that no pair of genomes with lower than 98.5% SSU rRNA gene identity are expected to show higher than 95% ANI, i.e. to belong to the same species according to the previous standards. The average sequence identity among all SSU rRNA gene copies within a genome is represented by the data points for which $x = 100$. On average, the intra-genomic identity of SSU rRNA genes is 99.9% (SD = 0.3, 410 genomes evaluated in total); the most pronounced outliers relative to the mean value are denoted on the graph.

2 It is important to realize that what is defined here as 'a short evolutionary scale' corresponds still to large evolutionary time. For instance, *Escherichia coli* shows ~80% ANI to *Salmonella enterica* and previous estimates suggest that these two lineages emerged from their last common ancestor about ~100 million years ago (Lawrence and Ochman, 1998). Although there is still a lot of uncertainty in the attempts to 'date' bacterial species, this data do reveal that species are often very old evolutionary entities.

genetic diversity than previously anticipated. For instance, fully sequenced representatives of the *E. coli* species, the best-studied microorganism, were found to differ in up to 30% of their genes (Welch *et al.*, 2002). Such extensive gene-content differences were very surprising for their time because, for example, the *E. coli* genomes are rarely differentiated based on phenotypic or other characteristics by traditional methods and have identical rRNA genes. Furthermore, the functional annotation of the genome-specific genes provided novel insights into the genetic determinants that may underlie the differentiated pathogenic potential of some of these genomes (Brzuszkiewicz *et al.*, 2006; Welch *et al.*, 2002). Similar studies in other species not only confirm the technical limitations and limited resolving power of traditional methods such as the SSU rRNA gene sequencing for studying intraspecies diversity but also provide means for evaluating the individual level (genotype), a level that has been very elusive in microbiological research, yet so critical for advancing our knowledge.

In addition to identifying genes that differ, genome sequencing has also contributed significantly towards identifying the genes shared by a group of organisms and, hence, provides a system-level understanding of the group considered. For instance, the genomes of lactic acid-producing bacteria, a group of major importance in the commercial production of many fermented foods and beverages, are characterized by a broad repertoire of transporters. These transport genes may facilitate the efficient carbon and nitrogen acquisition from the nutritionally rich environments the lactobacilli inhabit (Makarova *et al.*, 2006). Further, sequencing provided a new approach for identifying the important mutations in the enzymes or proteins between closely related organisms and elucidating how selection is acting on the evolution of new functions and variants. Novel insights into the evolution and biology of important pathogenic bacteria and viruses such as the SARS virus (Marra *et al.*, 2003) have emerged from such applications of genome sequencing. For example, sequencing of chickenpox viruses from vaccine samples discovered two mutations, causing amino acid substitutions in the major transactivating protein IE62, which have been naturally selected and are presumably responsible for vaccine-associated rash (Quinlivan *et al.*, 2007). Such findings have major implications for public health and for promoting public acceptance of vaccines or medicine.

Genomic sequences have also helped in evaluating and standardizing the traditional techniques for studying species-level genetic diversity. Although the relative resolving power of the most popular typing techniques such as pulsed field gel electrophoresis (PFGE), restriction fragment length polymorphism (RFLP), REP and BOX PCR, and DNA–DNA hybridization (DDH) are more or less known, what is actually measured by these methods remains less clear (Nemoy *et al.*, 2005; Rademaker *et al.*, 2000). Using the genomic sequence as reference standards, these issues can now be assessed. In perhaps the most important recent example, DNA–DNA hybridizations among fully sequenced strains showed that the 'gold standard' for defining bacterial species, the 70% DDH, corresponds well to 95% average nucleotide identity (ANI), the best genome-derived parameter found to correlate with the DDH values (Goris *et al.*, 2007; Konstantinidis and Tiedje, 2005a). In other words, bacterial species frequently encompass strains that may show up to 5% difference in their aggregate genomic nucleotide relatedness.

Notably, pairs of organisms with higher than 95% ANI also show higher than ~98.5% SSU rRNA gene identity as well (Fig. 1.7), revealing a new more stringent standard, compared to the 97% previously proposed (Stackebrandt and Goebel, 1994), on the rRNA gene identity that can be employed for describing novel species when genomic sequences or DDH values are not available. It is important to note, however, that the rRNA gene standard is not very rigorous since, although there are no organisms showing > 95% ANI and having < 98.5% identity in their rRNA gene sequences, the reverse is not always true (see Fig. 1.7). For the latter cases, an approach of higher resolution is required to confirm how may distinct species may exist among organisms with > 98.5% SSU rRNA identity. Stackebrandt and Ebers recently arrived at the same conclusions and standards when they carefully compared the SSU rRNA gene identities among many organisms to their corresponding DDH values, which were collected from the literature (Stackebrandt and Ebers, 2006). Furthermore, the sequence divergence among the (frequently) multiple SSU rRNA gene copies within the same genome is very small, typically < 0.2% (Acinas *et al.*, 2004b; Klappenbach *et al.*, 2001), but a few exceptions to this rule have also been noted. For example, (only) three genomes (from the 175 analysed, in total) showed more than 1.5% nucleotide difference among the SSU rRNA genes they harbour (denoted in Fig. 1.7). Therefore, the intra-genomic SSU rRNA gene diversity only infrequently represents a major source of diversity when applying the 98.5% identity standard for studying cultivated organisms in the laboratory and/or in environmental SSU rRNA gene surveys.

MLS(T/A), a cost-effective intermediate solution

The limited resolution of traditional typing methods, including the sequencing of a single genetic marker such as the ribosomal RNA gene, led microbiologists to seek alternative approaches. One such alternative approach that has emerged as the method of choice for performing species-level studies, as it represents a reasonable compromise between SSU rRNA gene and whole genome sequencing, is the multilocus sequence typing (MLST) method (Maiden *et al.*, 1998). The method involves the PCR amplification of several, typically six to ten, loci from the genome followed by phylogenetic analysis of genomes based on the amplified loci, and can be optimized to process several organisms in a relatively short period of time. The primary criteria for selecting the loci are: (1) being universally present in all strains targeted, (2) offering highly conserved sites for PCR-primer design, and (3) being unlinked in the genome to avoid hitchhiking effects of horizontal transfer or recombination events. Initially, the analysis was based on the uniqueness of the restriction fragments of the amplified loci, whereas this typing analysis has been recently replaced by phylogenetic analysis of the concatenated alignments of the sequenced amplicons. The latter development of the method, facilitated by the drop in sequencing cost, is frequently called multilocus sequence analysis (MLSA) and the MLST terminology is arguably reserved for the original application of the method that involved analysis of fragment types.

The robustness of the MLSA approach relative to whole genomes for studying genetic diversity among closely related bacteria has been recently evaluated (Konstantinidis *et al.*, 2006a,b). The analyses of several important bacterial groups show that even a random combination of six to eight genes would offer a robust reconstruction among the organisms considered provided that the organisms show 80–95% or less ANI, as is the case for organisms of closely or distantly related species of the same genus. The selection of genes for the MLSA application becomes more critical when evaluating the intraspecies diversity. At this level, the criteria used for gene selection are not adequate and an understanding of the phylogenetic informativeness and resolution of the genes employed for MLSA is required. The previous study also provided a methodology to achieve the latter when a few genomic sequences of the species are available. In addition to examining diversity, MLSA data have also been interrogated to provide insights into the role of homologous recombination in shaping microbial diversity patterns. In this case, the individual gene-based phylogeny is compared to the average phylogeny of all genes in the MLSA for incongruities that may be attributable to recombination events specific to the gene in question (Feil, 2004; Feil *et al.*, 2001; Hanage *et al.*, 2005; Sheppard *et al.*, 2008). The emerging findings from such studies suggest that different microbial groups show different rates of recombination reflecting, at least in part, the variation in the selective pressure(s) of the environment in which they live (Feil *et al.*, 2001; Wirth *et al.*, 2006). Owing to the yet-limited coverage of the genome achieved by the MLSA approach, however, the robustness of these findings as well as their general applicability to the whole genome level remain to be seen.

Novel insights emerging from genome sequencing

The fluidity of microbial genomes

Genome sequencing has also revealed that microbial genomes are more fluid and dynamic than previously anticipated. For instance, comparisons among 70 genomes that represented 30 bacterial species showed that a substantial fraction of the variable gene content (~40% on average) within species involves mobile elements such as prophages, plasmids and transposases and the hypothetical[3] genes associated with these elements (Konstantinidis and Tiedje, 2005a). This and other data (e.g. Lawrence and Ochman, 1998; Nelson *et al.*, 1999; Nesbo *et al.*, 2006; Welch *et al.*, 2002) suggest a major role for horizontal gene transfer for genome evolution. Nonetheless, the actual importance of all these mobile and hypothetical genes for the cell remains speculative. A growing body of literature suggests that the majority of them in any given genomic sequence constitute junk and ephemeral parts of the genome (Lerat and Ochman, 2005; Ochman and Davalos, 2006). Given enough time for evolution, however, the continuous recycling of this large gene pool may bring a new ecologically important gene into the genome. Several cases where the latter has been documented are already known, although identifying phenotypically important genes over a background of presumably junk or transient DNA remains particularly challenging. Among the best-documented examples is the toxin genes in the genomes of bacteriophages of pathogenic species of *Staphylococcus* (Boyd and Brussow, 2002), which may help the bacterium to overcome a host's defence mechanisms during infection. In a similar example from a different ecological context, viruses of the cyanobacterium *Prochlorococcus*, probably the most important microbial photosynthetic organism in open oceans, carry photosynthetic genes that are believed to supplement the photosynthetic machinery of the cyanobacterium host and, hence, confer a selective advantage to *Prochlorococcus* (Lindell *et al.*, 2005). Without doubt the interplay between the vectors of horizontal transfer, the genome, and the environment is a particularly interesting one for study and one frontier with which we can now begin to grapple.

Another way that fluidity has been realized in microbial genomes is through comparisons of gene synteny, i.e. conservation of the position of two or more genetic loci on the chromosome between different genomes. Bacterial genomes loaded with mobile elements, which can excise and move to another location through a variety of mechanisms, or repetitive elements, which can recombine through homologous recombination and cause genomic deletions and rearrangements, show a dramatic decrease in synteny with increased evolutionary distance between the compared genomes (for a review see Bentley and Parkhill, 2004). Mobile elements are usually found in multiple copies in the genome; thus, they can also serve as homologous recombination hot spots. The decrease may be so dramatic that almost no synteny is observed at the genus level and beyond. On the contrary, genomes that are lacking such elements may show remarkable synteny even when separated by large evolutionary time. The latter case typically involves species with reduced genome sizes that live inside eukaryotic host cells (endosymbionts) or in very stable environments. In fact, it has been hypothesized that the evolution of these reduced genomes has proceeded through recombination events and deletions triggered by the repetitive or mobile elements that existed in their ancestor genomes (Moran, 2002). The presence of such elements in the genomes of their larger genome-sized relatives is in support of this reductive evolution scenario. Further, due to the large population sizes and strong selection pressures in the environment, there is a preference for deletions, as opposed to insertions, for bacterial genomes (Mira *et al.*, 2001). Therefore, it appears as if there is a positive correlation between genome size and mobile/repetitive element/content, and hence, genomic fluidity and adaptability, although exceptions have been recently discovered. For example, sequencing of *Rickettsia felis* identified the first conjugative plasmid in an obligate intracellular parasite and a genome highly enriched in mobile elements relative to related *Rickettsia* or other symbiotic genomes (Ogata *et al.*, 2006). On the other extreme, genomes of *Burkholderia*, a genus renowned for its metabolic versatility and for harbouring three chromosomes (most bacteria have one circular chromosome), show different degrees of synteny for each of their three chromosomes. In particular, the largest chromosome is significantly more syntenic and conserved than the smaller ones, revealing varied evolutionary

3 Hypothetical denotes open reading frames that lack evidence of being protein-coded, e.g. they are not phylogenetically conserved.

pressures within the same genome (Chain *et al.*, 2006). These examples simply confirm the fact that microbes are so diverse that rules may not be universally applicable and we need to understand better the ecology of each species in order to advance our understanding of how genomes are evolving in different environmental contexts.

Gene and genetic diversity

As has been indicated in the preceding sections, microbial species show extensive gene content diversity. A more systematic study suggests that genomes of same bacterial species may differ in up to 35% of their gene content (Konstantinidis and Tiedje, 2005a). Accordingly, the total gene repertoire within a single bacterial species, the pan-genome, may not be easily exhaustible by sequencing (Tettelin *et al.*, 2005) and the total number of unique genes within a bacterial species can easily surpass the genes found, for example, in the human genome. Furthermore, almost any bacterial genome sequenced to date has 10 to 30% of its genes not related to any gene deposited in the public databases (Nelson *et al.*, 2000), revealing that the sequencing efforts to date are far from covering the total natural gene diversity on Earth. The same conclusions are reproduced when evaluating the protein family level as well. Several recent studies have shown that the accumulation curve of novel protein families is far from saturation or even decreasing in rate, revealing an enormous protein universe (Koonin *et al.*, 2002; Venter *et al.*, 2004; Yooseph *et al.*, 2007), the great majority of which is, needless to say, microbial. Nonetheless, a few preliminary observations have indicated that the recovery of new protein structures (the structure is generally more evolutionarily conserved than the sequence upon which families are based) has started to decrease in rate (Falkowski *et al.*, 2008; Levitt, 2007), indicating that the protein universe may, in fact, be exhaustible.

Microbes not only show enormous gene-content diversity but also (genomic) sequence diversity. For instance, comparisons among 175 genomes showed that organisms from different domains, i.e. Archaea vs. Bacteria, or from different phyla typically show <45% average amino acid identity (AAI) in the proteins they share (Konstantinidis and Tiedje, 2005b). It is important to realize that the sequence diversity within a single of these deep lineages is comparable, if not exceeding, the sequence diversity within the whole eukaryotic domain. Currently, there are about 30 genomes representing an equal amount of deep lineages that show <45% AAI to any other genome available. In fact, such deep lineages continue to grow in number with more genomes sequenced, and this trend will likely persist according to the number of new lineages found per year. For instance, there were 15 (of 42 in total), 24 (of 146), and 29 (of 432) fully sequenced genomes representing deep lineages at the end of 2000, 2003, and 2006, respectively. [A conventional cut-off of 45% AAI was used here to define comparable 'deep lineages' instead of using the designated higher levels of prokaryotic taxonomy, i.e. phylum, class, etc. (Brenner *et al.*, 2001), since the latter levels are not directly comparable in terms of the sequence diversity they encompass (Konstantinidis and Tiedje, 2005b)]. These results are in good agreement with the results from the analysis of the ribosomal RNA genes described in an earlier section. The results suggest that although the most abundant deep lineages in nature have probably been sampled already, there is still a large number of deep lineages that remain to be discovered either in extreme or under-sampled environments.

Despite the daunting gene/genetic diversity of the microbial world several universal trends are becoming apparent and are summarized below. Bacterial and archaeal genomes vary by more than an order of magnitude, ranging from ~0.5 Mbp to more than 10 Mbp. Yet the amount of coding DNA as a fraction of the total genome remains constant at ~90%, revealing that bacterial genomes are very compact and gene-rich (Mira *et al.*, 2001; Nelson *et al.*, 2000). Further, the distribution of genome sizes in nature is far from random and correlates with the breadth of the ecological niche of the organism. Bacteria with the smallest genome sizes known tend to have narrow ecological niches, e.g. being symbiotic organisms, or obligatory pathogens of a restricted variety of hosts. On the contrary, species with large genome sizes are typically more ecologically and metabolically diverse and dwell in complex environments such as the soil. Accordingly, several important functional gene-content shifts accrue with genome size, which reflect the difference in the breadth of the ecological niche. Large genome-sized organisms tend to have a higher representation of metabolic, regulatory and transport genes in their genomes at the expense of genes related to informational and cellular processes (Konstantinidis and Tiedje, 2004; Ranea *et al.*, 2004; van Nimwegen, 2003).

Summary and perspectives for the future

The information emerging from whole genome sequences has substantially broadened our understanding of the microbial world; for example, it has validated and improved microbial taxonomy and has revealed that microbial species are much more diverse than previously anticipated. However, lots of questions still remain unanswered: what is the value of all this genome and gene diversity, how redundant or useful is this diversity, and what is the relative importance of the mechanisms creating diversity? Moreover, although shotgun genomic sequencing has been recently applied successfully to environmental samples to provide access to the 'uncultivated majority' of microorganisms, the great majority of microbial groups and environments remain undersampled. The continuous reduction in sequencing cost and the improvements in the sequencing technologies (e.g. read length) will soon provide the necessary means to crack the 'genome of the biosphere'. The genomic information alone, however, is not going to be very useful for addressing the previous imperative questions. For the latter, the genomes must be linked to populations and their environment, and interrogated along with gene expression (Frias-Lopez *et al.*, 2008; Lo *et al.*, 2007) and time-series data to assess not only the presence of genomic diversity but also the activity derived from the genomes and their evolution over time. Only then will a robust understanding and modelling of microbial diversity emerge. Clearly, a new exciting era of investigation for the microbial sciences lay ahead, fuelled largely by the explosion in genomic and related technologies.

Acknowledgements

The authors' work has been supported by the following United States agencies: DOE Office of Biological and Environmental Research, NSF, NIEHS and USDA. We thank Qiong Wang and Donna McGarrell for their assistance with this manuscript.

References

Achtman, M., and Wagner, M. (2008). Microbial diversity and the genetic nature of microbial species. Nat. Rev. Microbiol. *6*, 431–440.

Acinas, S.G., Klepac-Ceraj, V., Hunt, D.E., Pharino, C., Ceraj, I., Distel, D.L., and Polz, M.F. (2004a). Fine-scale phylogenetic architecture of a complex bacterial community. Nature *430*, 551–554.

Acinas, S.G., Marcelino, L.A., Klepac-Ceraj, V., and Polz, M.F. (2004b). Divergence and redundancy of 16S rRNA sequences in genomes with multiple rrn operons. J. Bacteriol. *186*, 2629–2635.

Ashelford, K.E., Chuzhanova, N.A., Fry, J.C., Jones, A.L., and Weightman, A.J. (2005). At least 1 in 20 16S rRNA sequence records currently held in public repositories is estimated to contain substantial anomalies. Appl. Environ. Microbiol. *71*, 7724–7736.

Barns, S.M., Cain, E.C., Sommerville, L., and Kuske, C.R. (2007). Acidobacteria phylum sequences in uranium-contaminated subsurface sediments greatly expand the known diversity within the phylum. Appl. Environ. Microbiol. 73, 3113–3116.

Bentley, S.D., and Parkhill, J. (2004). Comparative genomic structure of prokaryotes. Annu. Rev. Genet. *38*, 771–792.

Boyd, E.F., and Brussow, H. (2002). Common themes among bacteriophage-encoded virulence factors and diversity among the bacteriophages involved. Trends Microbiol. *10*, 521–529.

Brenner, D., Staley, J., and Krieg, N. (2001). Bergey's Manual of Systematic Bacteriology, Vol. 1, 2nd edn (New York, Springer-Verlag).

Brown, J.R., Douady, C.J., Italia, M.J., Marshall, W.E., and Stanhope, M.J. (2001). Universal trees based on large combined protein sequence datasets. Nat. Genet. *28*, 281–285.

Brzuszkiewicz, E., Bruggemann, H., Liesegang, H., Emmerth, M., Olschlager, T., Nagy, G., Albermann, K., Wagner, C., Buchrieser, C., and Emody, L., *et al.* (2006). How to become a uropathogen: comparative genomic analysis of extraintestinal pathogenic *Escherichia coli* strains. Proc. Natl. Acad. Sci. U.S.A. *103*, 12879–12884.

Chain, P.S., Denef, V.J., Konstantinidis, K.T., Vergez, L.M., Agullo, L., Reyes, V.L., Hauser, L., Cordova, M., Gomez, L., and Gonzalez, M., *et al.* (2006). *Burkholderia xenovorans* LB400 harbors a multi-replicon, 9.73-Mbp genome shaped for versatility. Proc. Natl. Acad. Sci. U.S.A. *103*, 15280–15287.

Ciccarelli, F.D., Doerks, T., von Mering, C., Creevey, C.J., Snel, B., and Bork, P. (2006). Toward automatic reconstruction of a highly resolved tree of life. Science *311*, 1283–1287.

Cohan, F. (2002). What are bacterial species? Annu. Rev. Microbiol. *56*, 457–487.

Cole, J.R., Chai, B., Farris, R.J., Wang, Q., Kulam-Syed-Mohideen, A.S., McGarrell, D.M., Bandela, A.M.,

Cardenas, E., Garrity, G.M., and Tiedje, J.M. (2007). The ribosomal database project (RDP-II): introducing myRDP space and quality controlled public data. Nucleic Acids Res. 35, D169-D172.

Cole, J.R., Wang, Q., Cardenas, E., Fish, J., Chai, B., Farris, R.J., Kulam-Syed-Mohideen, A.S., McGarrell, D.M., Marsh, T., Garrity, G.M., *et al.* (2009). The Ribosomal Database Project: improved alignments and new tools for rRNA analysis. Nucleic Acids Res. 37, D141–D145; doi:10.1093/nar/gkn1879.

Delsuc, F., Brinkmann, H., and Philippe, H. (2005). Phylogenomics and the reconstruction of the tree of life. Nat. Rev. Genet. *6*, 361–375.

Doolittle, W.F. (1999). Phylogenetic classification and the universal tree. Science *284*, 2124–2129.

Doolittle, W.F., and Bapteste, E. (2007). Inaugural Article: Pattern pluralism and the Tree of Life hypothesis. Proc. Natl. Acad. Sci. U.S.A. *104*, 2043–2049.

Eckburg, P.B., Bik, E.M., Bernstein, C.N., Purdom, E., Dethlefsen, L., Sargent, M., Gill, S.R., Nelson, K.E., and Relman, D.A. (2005). Diversity of the human intestinal microbial flora. Science *308*, 1635–1638.

Eppley, J.M., Tyson, G.W., Getz, W.M., and Banfield, J.F. (2007). Genetic exchange across a species boundary in the archaeal genus ferroplasma. Genetics *177*, 407–416.

Falkowski, P.G., Fenchel, T., and Delong, E.F. (2008). The microbial engines that drive Earth's biogeochemical cycles. Science *320*, 1034–1039.

Feil, E.J. (2004). Small change: keeping pace with microevolution. Nat. Rev. Microbiol. *2*, 483–495.

Feil, E.J., Holmes, E.C., Bessen, D.E., Chan, M.S., Day, N.P., Enright, M.C., Goldstein, R., Hood, D.W., Kalia, A., and Moore, C.E., *et al.* (2001). Recombination within natural populations of pathogenic bacteria: short-term empirical estimates and long-term phylogenetic consequences. Proc. Natl. Acad. Sci. U.S.A. *98*, 182–187.

Fraser, C., Hanage, W.P., and Spratt, B.G. (2007). Recombination and the nature of bacterial speciation. Science *315*, 476–480.

Frias-Lopez, J., Shi, Y., Tyson, G.W., Coleman, M.L., Schuster, S.C., and Chisholm, S.W., *et al.* (2008). Microbial community gene expression in ocean surface waters. Proc. Natl. Acad. Sci. U.S.A. *105*, 3805–3810.

Garrity, G.M., Winters, M., Kuo, A.W., and Searles, D.B. (2002). Taxonomic Outline of the Prokaryotes, Release 2.0 2nd edn (New York, Springer-Verlag).

Garrity, G.M., Lilburn, T.G., Cole, J.R., Harrison, S.H., Euzeby, J., and Tindall, B.J. (2007). The Taxonomic Outline of Bacteria and Archaea. Michigan State University Board of Trustees. *TOBA Release 7.7, March 2007*, http://www.taxonomicoutline.org/.

Gogarten, J.P., and Townsend, J.P. (2005). Horizontal gene transfer, genome innovation and evolution. Nat. Rev. Microbiol. *3*, 679–687.

Gophna, U., Doolittle, W.F., and Charlebois, R.L. (2005). Weighted genome trees: refinements and applications. J. Bacteriol. *187*, 1305–1316.

Goris, J., Konstantinidis, K.T., Klappenbach, J.A., Coenye, T., Vandamme, P., and Tiedje, J.M. (2007). DNA–DNA hybridization values and their relation to whole genome sequence similarities. Int. J. Syst. Evol. Microbiol. *57*, 81–91.

Hanage, W.P., Fraser, C., and Spratt, B.G. (2005). Fuzzy species among recombinogenic bacteria. BMC Biol. *3*, 6.

Harris, J.K., Kelley, S.T., and Pace, N.R. (2004). New perspective on uncultured bacterial phylogenetic division OP11. Appl. Environ. Microbiol. *70*, 845–849.

Hedges, S.B. (2002). The origin and evolution of model organisms. Nat. Rev. Genet. *3*, 838–849.

Horton, T.R., and Bruns, T.D. (2001). The molecular revolution in ectomycorrhizal ecology: peeking into the black-box. Molecular Ecology *10*, 1855–1871.

Hunt, D.E., David, L.A., Gevers, D., Preheim, S.P., Alm, E.J., and Polz, M.F. (2008). Resource partitioning and sympatric differentiation among closely related bacterioplankton. Science *320*, 1081–1085.

Keijser, B.J.F., Zaura, E., Huse, S.M., van der Vossen, J.M.B.M., Schuren, F.H.J., Montijn, R.C., tenCate, J.M., and Crielaard, W. (2008). Pyrosequencing analysis of the oral microflora of healthy adults. J. Dent. Res. *87*, 1016–1020.

Klappenbach, J.A., Saxman, P.R., Cole, J.R., and Schmidt, T.M. (2001). rrndb: the Ribosomal RNA Operon Copy Number Database. Nucleic Acids Res. *29*, 181–184.

Konstantinidis, K.T., and Delong, E.F. (2008). Genomic patterns of recombination, clonal divergence and environment in marine microbial populations. ISME J. *2*, 1052–1065.

Konstantinidis, K.T., and Tiedje, J.M. (2004). Trends between gene content and genome size in prokaryotic species with larger genomes. Proc. Natl. Acad. Sci. U.S.A. *101*, 3160–3165.

Konstantinidis, K.T., and Tiedje, J.M. (2005a). Genomic insights into the species definition for prokaryotes. Proc. Natl. Acad. Sci. U.S.A. *102*, 2567–2572.

Konstantinidis, K.T., and Tiedje, J.M. (2005b). Towards a genome-based taxonomy for prokaryotes. J. Bacteriol. *187*, 6258–6264.

Konstantinidis, K.T., Ramette, A., and Tiedje, J.M. (2006a). The bacterial species definition in the genomic era. Phil. Trans. R. Soc. Lond. B. Biol. Sci. *361*, 1929–1940.

Konstantinidis, K.T., Ramette, A., and Tiedje, J.M. (2006b). Toward a more robust assessment of intraspecies diversity, using fewer genetic markers. Appl. Environ. Microbiol. *72*, 7286–7293.

Koonin, E.V. (2003). Comparative genomics, minimal gene-sets and the last universal common ancestor. Nat. Rev. Microbiol. *1*, 127–136.

Koonin, E.V., Wolf, Y.I., and Karev, G.P. (2002). The structure of the protein universe and genome evolution. Nature *420*, 218–223.

Lapage, S.P. (1992). International Code of Nomenclature of Bacteria (1990 Revision) (Washington, ASM Press).

Lawrence, J.G., and Ochman, H. (1998). Molecular archaeology of the *Escherichia coli* genome. Proc. Natl. Acad. Sci. U.S.A. *95*, 9413–9417.

Lerat, E., and Ochman, H. (2005). Recognizing the pseudogenes in bacterial genomes. Nucleic Acids Res. *33*, 3125–3132.

Levitt, M. (2007). Growth of novel protein structural data. Proc. Natl. Acad. Sci. U.S.A. *104*, 3183–3188.

Lindell, D., Jaffe, J.D., Johnson, Z.I., Church, G.M., and Chisholm, S.W. (2005). Photosynthesis genes in marine viruses yield proteins during host infection. Nature *438*, 86–89.

Lo, I., Denef, V.J., Verberkmoes, N.C., Shah, M.B., Goltsman, D., and DiBartolo, G., *et al.* (2007). Strain-resolved community proteomics reveals recombining genomes of acidophilic bacteria. Nature *446*, 537–541.

Ludwig, W., Schleifer, K.-H., and Whitman, W.B. (2008). Revised road map to the phylum firmicutes. Bergey's manual of systematic bacteriology. In Bergey's Manual of Systematic Bacteriology (New York, Springer-Verlag).

Maiden, M.C., Bygraves, J.A., Feil, E., Morelli, G., Russell, J.E., Urwin, R., Zhang, Q., Zhou, J., Zurth, K., and Caugant, D.A. (1998). Multilocus sequence typing: a portable approach to the identification of clones within populations of pathogenic microorganisms. Proc. Natl. Acad. Sci. U.S.A. *95*, 3140–3145.

Majewski, J., Zawadzki, P., Pickerill, P., Cohan, F.M., and Dowson, C.G. (2000). Barriers to genetic exchange between bacterial species: Streptococcus pneumoniae transformation. J. Bacteriol. *182*, 1016–1023.

Makarova, K., Slesarev, A., Wolf, Y., Sorokin, A., Mirkin, B., Koonin, E., Pavlov, A., Pavlova, N., Karamychev, V., and Polouchine, N., *et al.* (2006). Comparative genomics of the lactic acid bacteria. Proc. Natl. Acad. Sci. U.S.A. *103*, 15611–15616.

Marra, M.A., Jones, S.J., Astell, C.R., Holt, R.A., Brooks-Wilson, A., Butterfield, Y.S., Khattra, J., Asano, J.K., Barber, S.A., and Chan, S.Y., *et al.* (2003). The Genome sequence of the SARS-associated coronavirus. Science *300*, 1399–1404.

Miller, S.R., Augustine, S., Olson, T.L., Blankenship, R.E., Selker, J., and Wood, A.M. (2005). Discovery of a free-living chlorophyll d-producing cyanobacterium with a hybrid proteobacterial/cyanobacterial small-subunit rRNA gene. Proc. Natl. Acad. Sci. U.S.A. *102*, 850–855.

Mira, A., Ochman, H., and Moran, N.A. (2001). Deletional bias and the evolution of bacterial genomes. Trends Genet. *17*, 589–596.

Moran, N.A. (2002). Microbial minimalism: genome reduction in bacterial pathogens. Cell *108*, 583–586.

National Research Council (2007). The New Science of Metagenomics: Revealing the Secrets of Our Microbial Planet (Washington, DC, National Academies Press), pp. 170.

Nelson, K.E., Clayton, R.A., Gill, S.R., Gwinn, M.L., Dodson, R.J., Haft, D.H., Hickey, E.K., Peterson, J.D., Nelson, W.C., and Ketchum, K.A., *et al.* (1999). Evidence for lateral gene transfer between Archaea and bacteria from genome sequence of Thermotoga maritima. Nature *399*, 323–329.

Nelson, K.E., Paulsen, I.T., Heidelberg, J.F., and Fraser, C.M. (2000). Status of genome projects for non-pathogenic bacteria and archaea. Nat. Biotechnol. *18*, 1049–1054.

Nemoy, L.L., Kotetishvili, M., Tigno, J., Keefer-Norris, A., Harris, A.D., and Perencevich, E.N., *et al.* (2005). Multilocus sequence typing versus pulsed-field gel electrophoresis for characterization of extended-spectrum beta-lactamase-producing *Escherichia coli* isolates. J. Clin. Microbiol. *43*, 1776–1781.

Nesbo, C.L., Dlutek, M., and Doolittle, W.F. (2006). Recombination in Thermotoga: implications for species concepts and biogeography. Genetics *172*, 759–769.

Ochman, H., and Davalos, L.M. (2006). The nature and dynamics of bacterial genomes. Science *311*, 1730–1733.

Ogata, H., La Scola, B., Audic, S., Renesto, P., Blanc, G., Robert, C., Fournier, P.E., Claverie, J.M., and Raoult, D. (2006). Genome sequence of *Rickettsia bellii* illuminates the role of amoebae in gene exchanges between intracellular pathogens. PLoS Genet. *2*, e76.

Olsen, G.J., Larsen, N., and Woese, C.R. (1991). The ribosomal RNA database project. Nucleic Acids Res. *19*, 2917–2021.

Quinlivan, M.L., Gershon, A.A., Al Bassam, M.M., Steinberg, S.P., Larussa, P., Nichols, R.A., and Breuer, J. (2007). From the Cover: Natural selection for rash-forming genotypes of the varicella-zoster vaccine virus detected within immunized human hosts. Proc. Natl. Acad. Sci. U.S.A. *104*, 208–212.

Rademaker, J.L., Hoste, B., Louws, F.J., Kersters, K., Swings, J., and Vauterin, L., *et al.* (2000). Comparison of AFLP and rep-PCR genomic fingerprinting with DNA–DNA homology studies: Xanthomonas as a model system. Int. J. Syst. Evol. Microbiol. *50 Pt 2*, 665–677.

Ranea, J.A., Buchan, D.W., Thornton, J.M., and Orengo, C.A. (2004). Evolution of protein superfamilies and bacterial genome size. J. Mol. Biol. *336*, 871–887.

Rappe, M.S., and Giovannoni, S.J. (2003). The uncultured microbial majority. Annu. Rev. Microbiol. *57*, 369–394.

Roesch, L.F.W., Fulthorpe, R.R., Riva, A., Casella, G., Hadwin, A.K.M., Kent, A.D., Daroub, S.H., Camargo, F.A.O., Farmerie, W.G., and Triplett, E.W. (2007). Pyrosequencing enumerates and contrasts soil microbial diversity. The ISME Journal *1*, 283–290.

Rossello-Mora, R., and Amann, R. (2001). The species concept for prokaryotes. FEMS Microbiol. Rev. *25*, 39–67.

Rusch, D.B., Halpern, A.L., Sutton, G., Heidelberg, K.B., Williamson, S., and Yooseph, S., *et al.* (2007). The Sorcerer II Global Ocean Sampling Expedition: Northwest Atlantic through Eastern Tropical Pacific. PLoS Biol. *5*, e77.

Sangwan, P., Chen, X., Hugenholtz, P., and Janssen, P.H. (2004). Chthoniobacter flavus gen. nov., sp. nov., the first pure-culture representative of subdivision two, Spartobacteria classis nov., of the phylum Verrucomicrobia. Appl. Environ. Microbiol. *70*, 5875–5881.

Sheppard, S.K., McCarthy, N.D., Falush, D., and Maiden, M.C. (2008). Convergence of Campylobacter species:

implications for bacterial evolution. Science *320*, 237–239.

Sibley, C.G., Comstock, J.A., and Ahlquist, J.E. (1990). DNA hybridization evidence of hominoid phylogeny: a reanalysis of the data. J. Mol. Evol. *30*, 202–236.

Snel, B., Bork, P., and Huynen, M.A. (1999). Genome phylogeny based on gene content. Nat. Genet. *21*, 108–110.

Sogin, M.L., Morrison, H.G., Huber, J.A., Welch, D.M., Huse, S.M., Neal, P.R., Arrieta, J.M., and Herndl, G.J. (2006). Microbial diversity in the deep sea and the underexplored 'rare biosphere'. Proc. Natl. Acad. Sci. U.S.A. *103*, 12115–12120.

Stackebrandt, E., and Ebers, J. (2006). Taxonomic parameters revisited: tarnished gold standards. Microbiology Today *Nov*, 152–155.

Stackebrandt, E., and Goebel, B.M. (1994). Taxonomic note: a place for DNA-DNA reassociation and 16S rRNA sequence analysis in the present species definition in bacteriology. Int. J. Syst. Bacteriol. *44*, 846–849.

Stackebrandt, E., Frederiksen, W., Garrity, G.M., Grimont, P.A.D., Kampfer, P., Maiden, M.C.J., Nesme, X., Rossello-Mora, R., Swings, J., and Truper, H.G., *et al.* (2002). Report of the ad hoc committee for the re-evaluation of the species definition in bacteriology. Int. J. Syst. Evol. Microbiol. *52*, 1043–1047.

Staley, J.T. (1997). Biodiversity: are microbial species threatened? Curr. Opin. Biotechnol. *8*, 340–345.

Staley, J.T. (2006). The bacterial species dilemma and the genomic-phylogenetic species concept. Phil. Trans. R. Soc. Lond. B. Biol. Sci. *361*, 1899–1909.

Tettelin, H., Masignani, V., Cieslewicz, M.J., Donati, C., Medini, D., Ward, N.L., Angiuoli, S.V., Crabtree, J., Jones, A.L., and Durkin, A.S., *et al.* (2005). Genome analysis of multiple pathogenic isolates of *Streptococcus agalactiae*: implications for the microbial 'pan-genome'. Proc. Natl. Acad. Sci. U.S.A. *102*, 13950–13955.

Thompson, J.R., Pacocha, S., Pharino, C., Klepac-Ceraj, V., Hunt, D.E., and Benoit, J., *et al.* (2005). Genotypic diversity within a natural coastal bacterioplankton population. Science *307*, 1311–1313.

Tringe, S.G., von Mering, C., Kobayashi, A., Salamov, A.A., Chen, K., Chang, H.W., Podar, M., Short, J.M., Mathur, E.J., and Detter, J.C., *et al.* (2005). Comparative metagenomics of microbial communities. Science *308*, 554–557.

Tyson, G.W., Chapman, J., Hugenholtz, P., Allen, E.E., Ram, R.J., Richardson, P.M., Solovyev, V.V., Rubin, E.M., Rokhsar, D.S., and Banfield, J.F. (2004). Community structure and metabolism through reconstruction of microbial genomes from the environment. Nature *428*, 37–43.

van Nimwegen, E. (2003). Scaling laws in the functional content of genomes. Trends Genet. *19*, 479–484.

Venter, J.C., Remington, K., Heidelberg, J.F., Halpern, A.L., Rusch, D., Eisen, J.A., Wu, D., Paulsen, I., Nelson, K.E., and Nelson, W., *et al.* (2004). Environmental genome shotgun sequencing of the Sargasso Sea. Science *304*, 66–74.

Wang, Q., Garrity, G.M., Tiedje, J.M., and Cole, J.R. (2007). Naïve Bayesian classifier for rapid assignment of rRNA sequences into the new bacterial taxonomy. Appl. Environ. Microbiol. *73*, 5261–5267.

Ward, D.A. (2006). A Macrobiological Perspective on Microbial Species. Microbe Magazine *1*, 269–278.

Wayne, L.G., Brenner, D.J., Colwell, R.R., Grimont, P.A.D., Kandler, O., Krichevsky, M.I., Moore, L.H., Moore, W.E.C., Murray, R.G.E., and Stackebrandt, E., *et al.* (1987). Report of the Ad Hoc Committee on reconciliation of approaches to Bacterial Systematics. Int. J. Syst. Bacteriol. *37*, 463–464.

Welch, R.A., Burland, V., Plunkett, G., III, Redford, P., Roesch, P., Rasko, D., Buckles, E.L., Liou, S.R., Boutin, A., and Hackett, J., *et al.* (2002). Extensive mosaic structure revealed by the complete genome sequence of uropathogenic *Escherichia coli*. Proc. Natl. Acad. Sci. U.S.A. *99*, 17020–17024.

Wheeler, D.L., Chappey, C., Lash, A.E., Leipe, D.D., Madden, T.L., Schuler, G.D., Tatusova, T.A., and Rapp, B.A. (2000). Database resources of the National Center for Biotechnology Information. Nucleic Acids Res. *28*, 10–14.

Whitaker, R.J., Grogan, D.W., and Taylor, J.W. (2003). Geographic barriers isolate endemic populations of hyperthermophilic archaea. Science *301*, 976–978.

Wilmotte, A., and Herdman, M. (2001). Phylogenetic relationships among the Cyanobacteria based on 16S rRNA sequences, Vol 1, 2nd edn (New York, Springer-Verlag).

Wirth, T., Falush, D., Lan, R., Colles, F., Mensa, P., Wieler, L.H., Karch, H., Reeves, P.R., Maiden, M.C., Ochman, H., *et al.* (2006). Sex and virulence in *Escherichia coli*: an evolutionary perspective. Mol. Microbiol. *60*, 1136–1151.

Woese, C.R., Kandler, O., and Wheelis, M.L. (1990). Towards a natural system of organisms: proposal for the domains Archaea, Bacteria, and Eucarya. Proc. Natl. Acad. Sci. U.S.A. *87*, 4576–4579.

Wolf, Y.I., Rogozin, I.B., Grishin, N.V., Tatusov, R.L., and Koonin, E.V. (2001). Genome trees constructed using five different approaches suggest new major bacterial clades. BMC Evol. Biol. *1*, 8.

Yooseph, S., Sutton, G., Rusch, D.B., Halpern, A.L., Williamson, S.J., and Remington, K., *et al.* (2007). The Sorcerer II Global Ocean Sampling expedition: expanding the universe of protein families. PLoS Biol. *5*, e16.

Genomics and Metagenomics: History and Progress

2

Karen E. Nelson, Peter A. Bryan and Bryan A. White

Abstract

The extensive suite of molecular-based approaches developed over the past decade has enabled the field of metagenomics, the study of uncultured microorganisms. Paramount to metagenomic analysis is the use of high-throughput DNA sequencing technologies, which with the advent of low cost next-generation methods is transforming metagenomics. The application of metagenomics, to both global environments and microbes associated with a living host, has facilitated study of the functional ecology of environmental microorganisms. Novel functional genes and environmental functional signatures can be retrieved using metagenomics, and these can form the basis of hypothesis driven analyses of uncultured microorganisms. As with any technology, the daunting task is to understand and apply the growing number of metagenomic sequences in the context of microbial ecology and evolution.

Introduction

The early genomics studies that began appearing in the 1980s progressed exponentially to the current state where we have completed the genome sequencing projects for several hundred microbial and numerous eukaryotic species (http://www.genomesonline.org/) in addition to thousands of viral genomes. Current estimates are that there are hundreds of partial genomes available and many other genome sequencing projects are also in progress http://www.jgi.doe.gov/sequencing/allinoneseqplans.php). The real launch of the genomic era, however, began in the early 1990s with the availability of the complete genome sequence of *Haemophilus influenzae*. This study which was mind shattering at the time has now become routine protocol in many labs in the USA and worldwide.

Genomics really came of age when we began to witness a greater level of microbial diversity within species than previously anticipated, lateral gene transfer, and the significance of phage and viral genomics. The field of genomics which gave us the first genome of a free living organism also laid the foundation for generating genomic sequence data from whole environments without first using a culturing step, a field of research now known as 'metagenomics'. The term metagenomics was first used in the late 1990s, and was defined as the genomic analysis of microorganisms by direct extraction and cloning of DNA from an assemblage of microorganisms. The availability of 'next generation' sequencing technologies such as 454 pyrosequencing have made it such that a cloning step is no longer essential for metagenomic projects (see http://www.454.com/).

History of sequencing

Two fundamental methods for sequencing of DNA were developed in the late 1970s. These were the Maxam–Gilbert chemical sequencing method (Maxam and Gilbert, 1977) and the Sanger dideoxy chain termination method (Sanger *et al.*, 1977). Both methods depended on the same principle: the generation of a nested set of single stranded DNA molecules having one invariant end with the other ends differing by increments of single bases although they vary in approach (for a review see Ballou and Ninfa, 1998).

The Maxam–Gilbert method employed enzymes and is degradative, while the Sanger method depends on incorporation of dideoxynucleotides and the synthesis of complementary strand DNA (Karp *et al.*, 2005). In both methods, four reactions, each specific for one base, generates a set of fragments, which, when fractionated in parallel on the basis of length, allow for the identification of the bases. Pyrosequencing (also called 454 sequencing; see Chapter 4) was originally developed in the mid 1990s (Ronaghi *et al.*, 1996, 1998) and has been continuously developed since then (www.pyrosequencing.com; http://www.roche.com/media/). This method uses light signals to generate DNA sequence data. A primer is hybridized to a single stranded (ss) DNA template and then incubated with enzymes and the substrates, adenosine 5′-phosphosulphate (APS) and luciferin. The first of the deoxynucleotide triphosphates (dNTP) is added to the reaction, and DNA polymerase catalyses incorporation of dNTP into the DNA strand. Each incorporation event is accompanied by the release of pyrophosphate (PPi) (Ronaghi *et al.*, 1996; Ronaghi *et al.*, 1998). ATP sulphurylase converts PPi to ATP in the presence of adenosine 5′-phosphosulphate, and apyrase continuously degrades unincorporated dNTPs and excess ATP. The complementary DNA strand is built and the nucleotide sequence is determined from the signal peak in the program (Ronaghi *et al.*, 1996; Ronaghi *et al.*, 1998).

The newer or 'next-gen' sequencing technologies include that available from 454 Life Sciences (http://www.454.com/), which has become widely used in metagenomic sequencing projects. The elimination of cloning vectors and their associated biases in terms of the clonability of certain DNA fragments is a major advantage in using this system (Hyman, 1988; Ronaghi *et al.*, 1996, 1998; Margulies *et al.*, 2005). This sequencing technology also readily reads through secondary structures, and has capacity to produce very large amounts of sequence – current estimates are greater than 100 Mb per run (for more information please see http://www.454.com/enabling-technology/index.asp), and the estimates for the capacity for generating sequencing reads continue to increase with improvements in the technology. One current disadvantage of this approach is shorter read lengths when compared to traditional Sanger sequencing. However, the lengths of the generated sequences have been increasing from the initial GS20 platform (~100 bases) to the GSFLX (~250 bases) and preliminary results from the latest version called 'Titanium' (Roche; http://www.roche.com/media/media_releases/med_dia_2008-10-01.htm) suggest that read lengths will be significantly longer. The GS *Titanium* series reagents and software have sequencing reads with an improved average length of 400 bp and a fivefold throughput increase to 400–600 million bp per run (http://www.454.com/about-454/news/index.asp?display=detail&id=104).

Initially, these 'next-generation' technologies were met with a bit of scepticism, primarily because of the shorter read lengths of these sequences, compared to the traditional Sanger-based approach. Very early, however, the additional value that came from including these methods with ongoing sequencing approaches became evident. In one of the early studies that investigated the abilities for these approaches to be successfully used in sequencing, Goldberg and colleagues investigated the use and cost effectiveness of merging the traditional Sanger sequencing approaches with next-gen sequencing ('A Sanger/pyrosequencing hybrid approach for the generation of high-quality draft assemblies of marine microbial genomes'; Goldberg *et al.*, 2006). By combining both 3730*xl* Sanger sequencing (www.appliedbiosystems.com), where longer reads are traditionally obtained, with the shorter reads from 454 sequencing, they were able to compare merged assemblies to the same genomes that had been generated in the traditional way with Sanger sequencing strategies alone (Goldberg *et al.*, 2006). In those early days, they were able to show that 454 sequencing technology presented a great improvement over Sanger methodology when used alone, by virtue of the absence of a cloning bias and an ability to read through highly variable regions including hard stops. However, they found that the shorter reads that were available at the time with the first generation of GS20 454 sequencing technology, did not reliably span repetitive areas, and an absence of paired ends limited scaffolding information (see below for a definition of scaffolding). The GS20 version of 454 technology was however capable of sequencing regions of the genome where Sanger sequencing was ineffective. The merging of the two techniques also resulted in an 86% gap reduction in most of the genomes included in the study (Goldberg *et al.*, 2006). The results from this study demonstrated that the combination of the two sequencing technologies resulted in a higher quality genome assembly, and that 454 technology significantly reduced work and manual labour necessary for traditional closing/finishing of genomes (Goldberg *et al.*, 2006).

Although the next-generation sequencing technologies, which also include the Solexa/Illumina 1G Genome Analysis System (www.illumina.com) and Applied Biosystems SOLiD Sequencing (www.appliedbiosystems.com), currently present average read lengths that are much shorter than the traditional methods, a far higher number of sequence reads can be produced in a single day or on a single run by these technologies (Table 2.1). If history serves us well, it is anticipated that these next-generation technologies will eventually generate as good, or even longer, read lengths as some of the traditional methods. This entire topic is nicely reviewed (Schuster, 2008). Additional 'next-gen' technologies that will be released in the near term include that from Helicos (www.helicosbio.com) and Complete Genomics (www.completegenomics.com) (see Metzker, 2005, for an overview of emerging short read technologies).

In order to get meaningful information from raw sequence data, the information needs to be 'assembled' using available bioinformatics tools whereby overlapping sequences are put back together. Over the years, assembly software has been developed and refined, especially to address challenges that are presented by having repetitive elements in these genomes. Earlier assembly software includes the Celera Assembler developed by Gene Myers, Granger Sutton, Art Delcher, and others at Celera Genomics in 1998 and the TIGR Assembler that was also developed by Granger Sutton.

New assembly algorithms have had to be developed to address the challenges presented by the data types that are being generated by the newer 'next-generation' sequencing technologies, especially as the next-gen technologies which are not clone based, and which generate on average shorter sequencing reads do not allow (until relatively recently with 454) for the ends of a read (paired-ends) to be accessed. Assembly of sequences using mating of the paired-ends of these reads leads to generally larger/longer assemblies that in turn assist with the scaffolding and orientation of the genomic information.

The Newbler assembler is available with the software for 454 sequencing machines (www.454.com/downloads/protocols/1_paired_end.pdf). Also a new set of algorithms, such as 'Velvet,' developed by Zerbino and Birney (2008), have been developed to work with high coverage, very short read length (25–50 bp) datasets such as those generated on Solexa. Velvet can assemble very short reads and paired-ends up to 50 kb on prokaryotic data.

Granger Sutton's group at the J. Craig Venter Institute (JCVI) (Miller *et al.*, 2008) have modified the Celera Assembler to use combinations of ABI 3730 and 454 FLX reads: CABOG (for Celera Assembler with the Best Overlap Graph) appears to be very good at resolving homopolymers, high read coverage, and heterogeneous read lengths, and on preliminary test studies it resulted

Table 2.1 Projected sequencing capability comparison

	Throughput (million bp/day)	Throughput (million bp/run)	Throughput (reads/day)	Read length (bp)
3730*xl*	1–2	0.8	8000	> 800
454 Titanium*	1000	500	2,500,000	>350
SOLiD	>300	3000–10,000	>10,000,000	25–35
Solexa	≥200	1500–3,000	>10,000,000	25–35

Courtesy of Yu-hui Rogers at the J. Craig Venter Institute (JCVI) in Rockville, Maryland.

in the longest assemblies when compared to other available assembly software (Miller *et al.*, 2008). CABOG is available as open-source from http://wgs-assembler.sf.net under the GNU Public License (Miller *et al.*, 2008).

From single species to complex metagenomic libraries for sequencing

As mentioned above, genome sequencing projects were traditionally based on the Sanger sequencing method. Hundreds of prokaryotic and eukaryotic genomes including that of humans were generated through the use of Sanger sequencing technologies. The basic approach to these sequencing projects was to sequence clones that were created in different cloning vectors, resulting in multiple clone libraries with different fragment insert lengths (Frangeul *et al.*, 1999). Initially, this tedious process included the sequencing of ordered-clone libraries for overlapping clones (Frangeul *et al.*, 1999). More than one library insert size was recommended to improve the ability to completely cover the organisms' genome. Whereas the sequencing of smaller sized insert libraries (1.5–3 kb) gave a greater depth of coverage, the larger-insert libraries provided sufficient sequence lengths for significant scaffolding (see above) (Frangeul *et al.*, 1999). Some of the early microbial sequencing projects including that of *Escherichia coli* (Blattner *et al.*, 1997), as well as many of the larger eukaryotic projects took advantage of previously generated genetic and physical maps as well as ordered clone libraries in cosmids, lambda clones, YACs (yeast artificial chromosome) or BAC (bacterial artificial chromosome) clones (Frangeul *et al.*, 1999) for the sequencing projects.

When investigating communities where multiple species are present, the situation becomes much far more complicated and a range of tools and approaches has been developed to study these environments (see Chapters 4 and 5 for more information about these fingerprinting approaches). Microbial diversity in a range of environments has been estimated by sequencing or fingerprinting of small subunit ribosomal RNA (SSU rRNA) targets (Pace, 1997; Theron and Cloete, 2000; Torsvik and Øvreås, 2002; see Chapter 1). SSU rRNA analysis undoubtedly continues to provide us with a wealth of information about microbial community structure in a range of ecosystems. The latest release of the Ribosomal Database Project (RDP 10.5; http://rdp.cme.msu.edu/) for example contains 706,103 aligned and annotated 16S rRNA sequences. It is evident, however, that SSU data in isolation, although a very useful tool to study microbial diversity, is limited to the resolution and information provided by this sequence. For example, the physiological role that is being played by individual species that have been identified by sequencing of 16S rRNA genes cannot be defined, although some level of analogy can be drawn from available genetic information on closely related species; see for example (Gill *et al.*, 2006). It is however apparent from whole genome sequencing and microarray analysis that species that appear to be closely related from 16S rRNA gene sequences may have tremendous differences in genome composition (Mongodin *et al.*, 2005; DeBoy *et al.*, 2006). Please refer to Chapter 1 for more information about these approaches.

Fingerprinting approaches based on SSU rRNA genes (Chapter 4) revealed that the majority of the microorganisms inhabiting environmental samples were unknown and had never been cultivated. Therefore, other methods were necessary to determine the functional roles of these microbes in their respective communities. This dilemma led to the attractive option of sequencing and analysing large DNA fragments retrieved directly from the environment and helped launch the metagenomics era (Beja *et al.*, 2000, 2002; Quaiser *et al.*, 2002; Liles *et al.*, 2003). Initial attempts were made primarily by the group of Ed DeLong where they focused on the creation of large insert libraries from small quantities of DNA that were derived from environmental samples to characterize genes from uncultivated species (DeLong, 1992; Beja *et al.*, 2000). In the first of a series of reports, they analysed a 40 kb genomic DNA fragment from a planktonic marine archaeon that had been cloned into a fosmid library. By using large fragments that have a ribosomal or other phylogenetically relevant sequence present as a marker, it was possible to relate the sequenced fragment to previously characterized species. DeLong and workers subsequently constructed BAC libraries with an average insert size of 80 kb, and a maximum insert size of 155 kb (Beja *et al.*, 2000). Béjà and colleagues used large genomic fragments (≈80 kb) from a diverse population of largely uncultivated microorganisms to assign functional properties to archaeal clones (Béjà *et al.*, 2000). This group demonstrated the value and utility of BAC libraries for providing genetic information on uncultivated species, and laid the foundation for subsequent metagenomic studies in other environments. This analysis of microbial communities, which is based on the metabolic potential or gene content, has recently been termed a 'gene-centric' approach (Hugenholtz and Tyson, 2008). In this approach, genes that are over-represented, or found frequently in a community, provide a metabolic context to the microbiome that is often not evident from the phylogenetic content.

Communities that have been subject to metagenomic sequencing

As with 16S rRNA based approaches for studying mixed environments, all the metagenomic studies to date have demonstrated an unanticipated level of microbial diversity (Rondon *et al.*, 2000; Beja *et al.*, 2002a,b; Voget *et al.*, 2003) (Fig. 2.1). Among the communities that have been the subject of comprehensive metagenomic studies are soils (Voget *et al.*, 2003; Tringe *et al.*, 2005), the human gastrointestinal tract (Gill *et al.*, 2006; Kurokawa *et al.*, 2007), the oceans (Venter *et al.*, 2004), whale fall (Tringe *et al.*, 2005), the rumen (Brulc *et al.*, 2009), the chicken caecum (Qu *et al.*, 2008), the termite hindgut (Warnecke *et al.*, 2007) and acid mine drainage systems (Tyson *et al.*, 2004). Some of these will be further discussed in separate sections below.

Soil metagenomics

Among some of the earliest metagenomic studies are those that investigated a limited number of Sanger-based sequenced clones. For example, Voget and workers (Voget *et al.*, 2003) examined a soil metagenomic library to identify more than 15 different genes encoding novel biocatalysts.

Goodman's group constructed two metagenomic/BAC libraries that totalled more than 1 Gbp of DNA. 16S rRNA gene analysis showed a wide variety of phyla present including the low-G+C, Gram-positive Acidobacterium, Cytophagales, and Proteobacteria. The BAC vector was also used to express genes from the inserts, with a range of phenotypes be expressed including antibacterial, lipase, amylase, nuclease, and haemolytic activities.

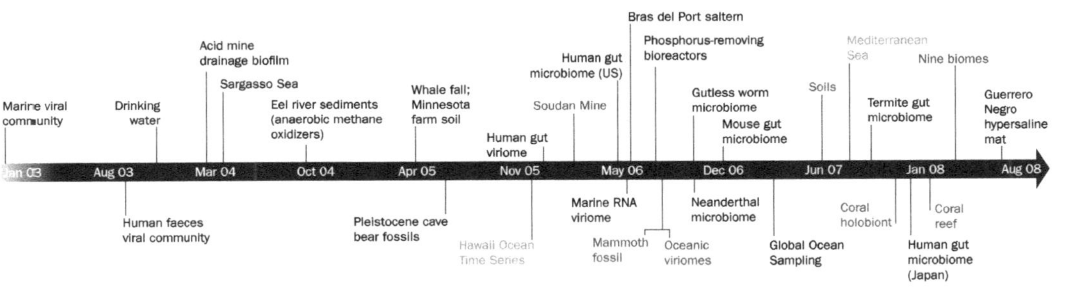

Figure 2.1 Timeline of sequence-based metagenomic projects showing the variety of environments sampled since 2002. The oceanic viriomes (all viruses in a habitat) (August 2006) were from the Sargasso Sea, Gulf of Mexico, coastal British Columbia and the Arctic Ocean. The nine biomes (March 2008) were stromatolites, fish gut, fish ponds, mosquito viriome, human-lung viriome, chicken gut, bovine gut and marine viriome. The different technologies used are dye-terminator shotgun sequencing (black), fosmid library sequencing (light grey) and pyrosequencing (dark grey). (Graphic based on data sets represented at http://www.genomesonline.org.) (Reprinted with permission from Hulgenholtz and Tyson 2008, copyright Nature 2008.)

This same dataset was later studied for the characterization of clones from which structural analysis revealed these were triaryl cations that exhibited broad-spectrum antibiotic activity against Gram-negative and Gram-positive organisms. This was one of the earlier studies based directly on metagenomic sampling that revealed the successful expression of DNA extracted directly from soil (Gillespie, 2003). At around the same time, Knietsch and colleagues used metagenomic soil libraries that had been derived from meadow, sugar beet field, cropland (1,267,000 clones; ~ 4.05 Gbp of environmental DNA) to screen for the production of carbonyls from short-chain (C_2 to C_4) polyols. Three gene products had limited similarity to known alcohol oxidoreductases. Additional soil metagenomic studies have resulted in the identification of novel molecules that are associated with various catabolic and anabolic steps.

Metagenomic analysis of soil communities was extended by Tringe *et al.* (2005), who used quantitative gene content analysis (gene-centric metagenomics) to compare 100 Mbp of sequences from soil with 75 Mbp of sequences from three whale fall carcasses. Analysis of these nutrient rich environments was compared to Sargasso Sea sequences (see below; Venter *et al.*, 2004) and Acid Mine drainage community (see below; Tyson *et al.*, 2004). These data showed that this kind of analysis could reveal habitat-specific gene finger prints. For example, soil was enriched for genes involved in plant cell wall degradation (cellobiose phosphorylase), but lacked bacteriorhodopsin orthologues which were found in the Sargasso Sea samples. Analysis of gene operons for different functional systems also differed between these samples.

Global ocean metagenomics

The largest metagenomic datasets to date are those that have been presented by Venter and colleagues (JCVI, Rockville, Maryland) on global ocean samples that began with a seminal paper released in 2003 (Venter *et al.*, 2004). In that study, whole-genome shotgun sequencing of microbial populations collected on tangential flow and impact filters from seawater samples from the Sargasso Sea generated a total of 1.045 billion base pairs (bp) of non-redundant sequence (Venter *et al.*, 2004). This metagenomic sequence data was annotated and analysed with a variety of bioinformatics approaches to elucidate the gene content, diversity, and relative abundance of the organisms within the samples. Resulting estimates were that the sequences were derived from at least 1800 species, based on sequence relatedness, including 148 previously unknown bacterial phylotypes. The 1.6 Gbp dataset also included more than 1.2 million new genes. Variation in species present and stoichiometry suggested substantial oceanic microbial diversity (Venter *et al.*, 2004).

Venter and colleagues followed this study with a massive expedition, data collection and analysis study that was published in 2007 (http://collections.plos.org/plosbiology/gos-2007.php). For this study, 'gigabasepairs' of data represented genes from microbes collected in waters off Canada, the Panama Canal, the Galapagos Islands, The Pacific and the South Pacific Gyre (Parthasarathy *et al.*, 2007). The assemblies and associated annotated peptides are available (http://www.ncbi.nlm.nih.gov/Genbank) and the data are also freely available through CAMERA – Cyberinfrastructure for Advanced Marine Microbial Ecology Research and Analysis (http://camera.calit2.net) (Parthasarathy *et al.*, 2007). The metagenomic dataset that was generated from this Global expedition included the above-mentioned Sargasso Sea dataset and represents the largest dataset collected to date comprising close to 7.7 million sequencing reads.

Acid mine drainage community

Another ecosystem that has been extensively studied is an Acid Mine drainage biofilm. Banfield and colleagues used metagenomic approaches to study the microbial community structure of this community that has resulted in a number of publications (Tyson *et al.*, 2004). In one of their first publications, they revealed almost complete genomes of *Leptospirillum* group II and *Ferroplasma* type II. Partial genomes from other species were also recovered, and single-nucleotide polymorphisms (SNP) were found to be the predominant form of variation among related strains. The *Ferroplasma* type II genome was proposed to be the result of three genomes that had been through a series of homologous recombination events that had resulted in some mosaic genomes (Tyson *et* al., 2004). The relative simplicity of this system, with few dominant community members, allowed for the recovery of near-complete genomes of *Leptospirillum* and *Ferroplasma* species, as well as the partial recovery of three other species from 75 Mbp of sequence data.

Deep subsurface

Edwards and colleagues (Edwards *et al.*, 2006) used the pyrosequencing technique to generate environmental genome sequences from two adjacent sites in the Soudan Mine, Minnesota, USA. The authors were able to highlight significant differences in metabolic potential in each environment, particularly in terms of the distinct biochemistries. They concluded that pyrosequencing would be widely used to sequence environmental samples because of the speed, cost, and technical advantages (Edwards *et al.*, 2006). Indeed, Dinsdale *et al.* used functional metagenome profiling of 1,040,665 454 pyrosequences from 45 microbial metagenomes and 853,470 pyrosequences from 45 viral metagenomes to determine the metabolic processes that impart microbial function and that are important for growth and survival in any given environment or biome. This meta-genecentric approach allowed them to define the key processes important both for understanding and ecosystems., and proposed that the frequency of a gene encoding a particular metabolic function reflects its relative importance in an environment. Moreover, results from this meta-analysis suggest that instead of changing the organisms present, variation in gene content via horizontal gene transfer between similar microbes controls gene distribution within an environment.

Animal studies

Many animals of a wide range of orders have a portion of the digestive system, adapted to accommodate a fermentation, which aids in the digestion of plant material and provides them with a variety of nutritional benefits. The microbiome inhabiting the rumen are characterized by Bacteria (up to 10^{11} viable cells per ml), Archaea (0.5–3% of the total microbial population), a variety of protozoa (Clarke, 1977; Clarke and Bauchop, 1977), anaerobic fungi, and bacteriophage (Lin *et al.*, 1997; Ziemer *et al.*, 2000). (Klieve and Bauchop, 1988; Klieve and Swain, 1993). Suppressive subtractive hybridization (SSH- see further details below) has demonstrated an unexpectedly large difference in archaeal community structure between the rumen microbial populations of two steers fed identical diets and housed together (Galbraith *et al.*, 2004). However, no functional differences were detected that were related to plant cell wall hydrolysis. In contrast, Ferrer *et al.* (2005) isolated 22 clones from a bovine rumen microbiome library and detected four clones that potentially represent previously undescribed families of glycoside hydrolases, as well as a novel polyphenol oxidase (laccase) (Beloqui *et al.*, 2006).

To further address the complexity of the rumen microbiome, we have used the gene-centric comparative metagenomic strategy with direct 454 pyrosequencing for the comparison of the fibre-adherent rumen microbiome from three bovines (Brulc *et al.*, 2009). 454 technology and the subsystems-based annotations available in the SEED database (see below) were used to gain a better understanding of the metabolic potential of these metagenomes. For the three animals included in this study that were on identical diets, the community structures were markedly different with respect to nutrient utilization and glycoside hydrolase content. From 178,713, 264,849 and 345,317 reads from the different animals, (35,115,534 bases) from the fibre-associated rumen metagenomes, the authors were able to detail the different metabolic profiles that are associated with the different microbial populations in the animals (Brulc *et al.*, 2009) and conclude that, in contrast to the termite hindgut (see below) the diet drives a microbiome that is suited to the degradation of plant cell walls from forages.

In a similar study, the same group of researchers studied the complex microbiome of the caeca of chickens, which is known to play an important role in feed conversion and well-being of these animals (Qu *et al.*, 2008). Here, the caecal microbiome of a chicken that was pathogen free was compared to that derived from a chicken that had been challenged with *Campylobacter jejuni*. From 55,364,266 bases of random sampled pyrosequence data gene tags were identified using the SEED subsystems-based annotations and were primarily Firmicutes, Bacteroidetes and Proteobacteria. The study suggested that mobile

DNA elements are a major functional component of caecal microbiomes contributing to the evolution of the metavirulomes for the species that are present and that this is a contributing factor in disease resistance to zoonotic pathogens (Qu *et al.*, 2008).

In a study focused on termite gastrointestinal communities, Warnecke and colleagues (Warnecke *et al.*, 2007) conducted a metagenomic analysis of the microbes present in the hindgut paunch of a *Nasutitermes* species. Here they were able to demonstrate the presence of diverse bacterial genes for the metabolism of cellulose and xylan, primarily those from wood. Symbiotic processes characterized included metabolic processes associated with hydrogen and carbon dioxide processing as well as nitrogen-fixation (Warnecke, *et al.*, 2007).

Alvinella pompejana is an endemic inhabitant of deep-sea hydrothermal vents, and is characterized by a filamentous microflora that lines the dorsal spaces. A recent metagenomic study was conducted on a single worm (with sequences from 2 other genomic libraries included in the analysis), and revealed 27 highly related strains of bacteria (Grzymski *et al.*, 2008). These were dominated by two clusters of sequences (35% and 30%, respectively, based on the V1–V2 aligned sequences). The average contig size from this metagenomic dataset was 1462 bp in length and 103,371 proteins could be predicted from the complete dataset (Grzymski *et al.*, 2008). Clusters of orthologous groups (COGs) analysis of the proteins indicated that the episymbionts are metabolically adapted for growth in hydrothermal vent ecosystems. The episymbiont metagenome was enriched for heterotrophic and aerobic metabolisms and in genes encoding the complete rTCA cycle, sulphur oxidation, and denitrification.

Jeff Gordon's group (Washington University, St. Louis, MO, USA) has conducted a series of studies investigating the microbial flora of mice using both 16S rRNA sequencing only and metagenomic approaches. They initially focused on genetically obese ob/ob mice, lean ob/+ and wild-type siblings, and their ob/+ mothers that had all been maintained on a polysaccharide-rich diet. They initially generated 5088 bacterial 16S rRNA gene sequences from the distal intestine, and found that most of the mouse gut species were unique. At the division level, there were similarities with the human gastrointestinal flora primarily through the high level of Firmicutes and Bacteroidetes in these mouse samples (Ley *et al.*, 2005). From these results, the authors suggested that obesity influenced the diversity of the microbial species that are present in the gastrointestinal tract, and they also suggested that manipulation of this population of microbial species may aid in the regulation of energy balance in overweight people (Ley *et al.*, 2005).

Follow-up studies (also detailed in Chapter 7) have investigated the interplay between the gastrointestinal ecology and energy balance of animals on a Western diet. Here they found that obesity that was induced by the diet resulted in an increased proportion of one class of the Firmicutes, and that this same population could be reduced by manipulation of the diet. Transplantation of the microbial populations from the obese mice to lean germ-free mice resulted in a higher level of the deposition of fat than when these microbial populations were taken from lean donors (Turnbaugh *et al.*, 2008). A significant component of the work from Gordon's group (and others) has laid the foundation for understanding the human microbiome and the interactions between humans and their microbial inhabitants.

The human microbiome

It is by now clearly evident that the human body hosts a range of microbial species and communities (see Chapter 7 for a more detailed discussion on this topic). In the initial metagenomic analysis of the human gastrointestinal tract (Gill *et al.*, 2006), a total of 139,521 sequences were generated (representing 65,059 and 74,462 sequences respectively with faecal samples from two healthy humans). This resulted in approximately 78 Mbp of unique DNA sequence, and a total of 6000 16S rRNA gene clones were also sequenced. The two subjects were free of any medications. The combined assemblies resulted in 17,668 contigs or 14,572 scaffolds that ranged from 1000 to 57,894 bp, with the largest contig 44,747 bp. 56,292 reads (40% of the total) which accounted for 45,078,063 bp of DNA sequence did not assemble into contigs. 50,164 ORFs were predicted from this dataset, and of these we obtained 19,866 unique database matches (only 5700 of these unique predictions could be identified in both subjects).

The datasets were compared to the genomes of two completely available genomes that are known to be associated with the human gastrointestinal tract (*Bifidobacterium longum* and *Methanobrevibacter smithii*; Fig. 2.2). A high level of nucleotide heterogeneity could be observed, suggesting that these sequence reads were derived from multiple closely related strains including other *Bifidobacterium* phylotypes in the gastrointestinal tract.

From the comparative genome analyses, we were able to identify novel virulence factors, and recreate a metabolic reconstruction of some of the ongoing processes in the gastrointestinal tract (Gill *et al.*, 2006). We also explored the metabolic potential of the microbiota using the Kyoto Encyclopedia of Genes and Genomes (KEGG) pathways and COGs, both tools allow for an investigation of metabolic capabilities. Relative abundance of pathways could be deduced with odds ratios in COG and KEGG categories, where a ratio of 1 suggests that the DNA sample has the same proportion of given category as a dataset to which it is being compared where as a ratio greater than one indicates enrichment in the test sample. We obtained 62,036 hits to the COG database with COG diversity greater than that described for an Acid Mine drainage but less than that described for whale fall, soil, and the Sargasso Sea samples. This preliminary study when combined with many other ongoing studies of the human body (see Chapter 7) hint at tremendous diversity that we will encounter in the near future.

The National Institutes of Health launched a human microbiome initiative (http://nihroadmap.nih.gov/hmp/) with primary goals to determine if there is a core human microbiome, to understand the changes in the human microbiome that can be correlated with human health, and to develop new technological and bioinformatics tools to support these goals. Initial sequencing initiatives for this program are in place, and include the development of a microbial genome reference set of at least 600 genomes with a goal of attaining a reference set of 1000 genomes. Metagenomic sequencing is

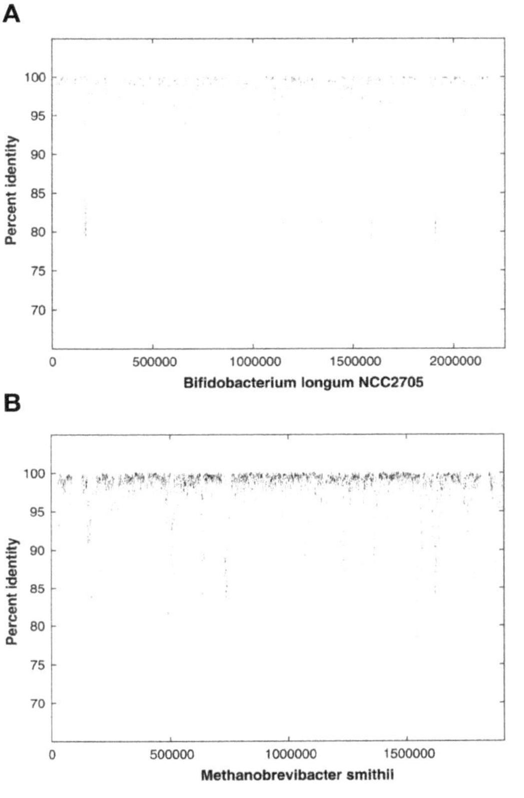

Figure 2.2 Comparison of random metagenome reads with completed genome of Bifidobacterium longum and Methanobrevibacter smithii. **A** Per cent identity plot (PIP) of alignments of shotgun reads along the genome of B. longum strain NCC2705. The x axis represents the coordinate along the genome, and the y axis represents the per cent identity of the match. **B** Per cent identity plot (PIP) of the alignment of shotgun reads along the draft genome of M. smithii. The x axis represents the coordinate along a pseudomolecule formed by concatenating all contigs in the M. smithii draft assembly. The y axis represents the per cent identity of the match. The variation in the per cent identity of the matches between the shotgun reads from subjects 7 and 8 as compared with the genome sequences of B. longum NCC2705 suggests considerable diversity among Bifidobacterium-like organisms within our samples. Alignments of the reads to the draft genome of M. smithii exhibit a much narrower range of per cent identity (89% of alignments were at 95% or better identity as compared with 48% for B. longum), consistent with lower levels of diversity among archaeal members of the gastrointestinal tract (reprinted with permission from Gill *et al.*, 2006, copyright 2006 Science).

also being used to characterize the microbial communities from 15–18 body sites from 250 individuals. The NIH has put in place additional resources for a Data Analysis and Coordination Center for the Human Microbiome Project (http://hmp.nih.gov/), the Development of New Tools for the Computational Analysis of Human Microbiome Project Data and the Development of New Technologies Needed for Studying the Human Microbiome. Ethical, Legal, and Social Implications of Human Microbiome Research are also being funded. Finally, several Human Microbiome Demonstration Projects will be funded to investigate the relationships between various diseases and the microbial inhabitants in the body and at a particular body site. According to NHGRI policy for data release, all of these reference genomes and the metagenomic data will be readily released to the broader scientific community. The NIH effort has expanded to be a world-wide effort including Asia and Europe (http://www.nih.gov/news/health/oct2008/nhgri-16.htm) (Blow, 2008).

Bioinformatics tools for metagenomic comparisons

The sequencing of individual genomes has allowed for the development of various tools and databases for addressing complex biological questions. These tools have to a certain extent allowed for accurate predictions on genome organization, gene content, novel genes, gene families, biochemical pathways, transport and metabolism, gene transfer events etc., and functional genomic studies. Many of the tools that have been developed for the computational analysis of single genomes have, out of necessity, been applied directly to the analysis of environmental metagenomic data. In many cases, however, the analysis has proved certain challenges related to the assembly of the data, phylogenetic analysis, sequence polymorphisms, comparisons between multiple communities (Chen and Pachter, 2005), and overall interpretation of individual species from mixed incomplete genomes. In addition, challenges include the gene identification and prediction processes.

In traditional single genome sequencing projects, the genome data from closely related species serves as 'scaffolds' for the newly assembled data. In metagenomics projects, where novel species will be encountered, there is no scaffold backbone for the data to be aligned to, and very often the sequences become unassembled fragments, with no real matches, and to which conventional approaches for the identification of the sequences cannot be applied (see for example (Gill *et al.*, 2006; Noguchi *et al.*, 2006)). For example, unassembled reads make up approximately 50% of the assembled sequences in the Sargasso Sea dataset (Venter *et al.*, 2004; Chen and Pachter, 2005). Less that 2% of the Sargasso Sea sequence data matched at 90% identity or higher with sequences in other databases (Tress *et al.*, 2006). Tress and co-workers proposed that the high proportion of sequence fragments encountered in the Sargasso Sea study limits the ability to make conclusions with respect to distribution of sequences, and that these sequence fragments may bias any subsequent studies (Tress *et al.*, 2006).

Binning has been used in some metagenomic projects to sort the data based on GC composition (Venter *et al.*, 2004). Genomic features that include phylogenetic markers and or codon biases can also be used to separate out 'similar' sequences. However, gene transfer events limit the interpretation of these sequences, as very distant species may harbour similar pieces of genetic information (Chen and Pachter, 2005).

Predicted genes can be Blast searched against a number of databases including COGs and KEGG databases. In the metagenomic study of the human gastrointestinal tract by Gill and colleagues (Gill *et al.*, 2006) COG- and KEGG-based analyses revealed an enrichment for glycan, amino acid, and xenobiotic metabolism, methanogenesis, as well as 2-methyl-D-erythritol 4-phosphate pathway-mediated biosynthesis of vitamins and isoprenoids.

Schuster and workers have described a new program, MEGAN, for the analysis of metagenomic data (Huson *et al.*, 2007). The data to be analysed is imported as raw DNA sequences and subsequently compared to other datasets with BLAST, and MEGAN then analyses the taxonomic content of the data that is assigned with a simple algorithm (Huson *et al.*, 2007). The result is that the data can be put through a preliminary analysis without an initial assembly, and a statistical output can be obtained (Huson *et al.*, 2007). The DOE Joint Genome Institute also maintains a metagenome tool in the Integrated Microbial Genomes (IGM) system, which is termed IMG/M (Markowitz *et al.*, 2008). This tool is used for analysing the functional capability of microbial communities using reference isolate genomes to annotate metagenome sequences.

Noguchi and colleagues have recently described 'MetaGene: prokaryotic gene finding from environmental genome shotgun sequences' (Noguchi *et al.*, 2006). This program is designed to predict genes from fragmented genomic sequences (Noguchi *et al.*, 2006), and primarily uses di-codon frequencies in its analyses. The program extracts all open reading frames (ORFs) from a dataset and given a score from the base composition and length of the ORF. A high score for ORFs is derived from orientations and distances of neighbouring ORFs in addition to the scores for the ORFs themselves (Noguchi *et al.*, 2006). The program is estimated to have a sensitivity of 95% and a specificity of 90% for artificial shotgun sequences (Noguchi *et al.*, 2006). MetaGene runs in a very short time, and the authors state that analysis of the sequences derived from the Sargasso Sea dataset took only 15 minutes. The software is currently available at http://metagene.cb.k.u-tokyo.ac.jp/ (Venter *et al.*, 2004; Noguchi *et al.*, 2006).

Overbeek and colleagues developed the SEED database that is a curated database of microbial genomes with an interface of subsystems (Overbeek *et al.*, 2005) (http://theseed.uchicago.edu/FIG/index.cgi). RAST (Aziz *et al.*, 2008; Meyer *et al.*, 2008) is a freely available automated open source system that is based off of the SEED foundation, and can be reliably used with assembled and unassembled metagenomic sequence data for analysis. RAST provides functional assignments by comparisons to protein and nucleotide databases. Summaries of the phylogenies and functional potential of these metagenomes can also be accessed (see http://metagenomics.nmpdr.org). MG-RAST is a SEED-based environment that allows for analysis of 454 and Sanger-based metagenome datasets over the web. The server provides the annotation of sequence fragments, their phylogenetic classification (phylogroup), functional classification of samples (functional gene attributes or environmental gene tags), and comparison between multiple metagenomes. The server also computes an initial metabolic reconstruction for the metagenome and allows comparison of metabolic reconstructions of metagenomes and genomes. The subsystems-based annotations available in the SEED database, enable a rapid gene-centric annotation based on biochemical pathways, fragments of pathways, and clusters of genes that function together, or any group of genes that considered to be related. The SEED database arranges metabolic pathways into a hierarchical structure, where all of the genes required for a specific metabolic task are arranged into subsystems. At the highest level of organization, the subsystems include both catabolic and anabolic functions. Moving down the hierarchy reveals the individual pathways for each metabolic process. Therefore, the per cent of sequences assigned to each major SEED metabolic category in each sample can be used provide a reliable and consistent annotation within and between metagenomes. Comparing metagenome samples using the SEED database can lead to the identification of signature functions associated with each metagenome sample through statistical techniques that are not only robust but are rapid enough to perform with hundreds of thousands or millions of data points per sample. Currently, the user simply has to upload sequences and or assemblies in fasta format. Beneficial aspects of RAST and MG-RAST are that they will handle pyrosequence datasets, and that the system are designed such that new analysis tools can be easily incorporated into the system (Aziz *et al.*, 2008; Meyer *et al.*, 2008).

Wet lab tools for metagenomics

Suppressive subtractive hybridization

The technique of suppressive subtractive hybridization (SSH) allows for the identification of unique DNA sequences present in one sample relative to another. This process is accomplished by subtracting the sequences that are common to both and amplifying those that are unique. SSH exploits the suppression PCR effect seen when long inverted terminal repeats attached to DNA fragments selectively suppress amplification of undesirable sequences in the PCR procedure

(Diatchenko *et al.*, 1996; Gurskaya *et al.*, 1996; Jin *et al.*, 1997; Akopyants *et al.*, 1998; Diatchenko *et al.*, 1999). In the SSH method, DNAs from the two species in question are fully digested with a frequent cutter to yield fragments of average size between 250 and 2000 bp. Digested tester DNA is divided into two aliquots, each of which is ligated to a separate adaptor. Driver DNA is added to each aliquot of the tester DNA, both are then denatured and allowed to anneal to complementary sequences. The two subtracted tester DNA aliquots are then pooled, allowed to anneal and additional driver DNA is added for a second subtraction. DNA's unique to the tester from either aliquot anneal to each other while non-unique DNA's have already been bound. A PCR is run on a small amount of the subtracted DNA. Different DNA complexes are formed which do not allow PCR amplification or if they do, allow only linear amplification. These include complexes with the same adaptor at either end (these form loops because of self-complementary stretches in the adaptors). Complexes of unique DNA's containing different adaptors at either end, allow for logarithmic amplification. A nested PCR is run using a small amount of the primary PCR product as template. The resultant clone libraries are enriched for unique sequences from the two genomes that are being compared. The advantages of SSH over more conventional subtractive methods are that the intermediate steps for physical separation of single and double-strand DNAs are eliminated, only one subtractive hybridization round is required, and more than 1000-fold enrichment for unique sequences can be achieved (Nesbo *et al.*, 2002).

White and colleagues (Galbraith *et al.*, 2004) were the first to apply the SSH technique to a mixed microbial community, when they compared the metagenomes of microbial consortia from rumen samples. The SSH study was conducted with samples of total genomic DNA obtained from rumen fluid samples of two hay-fed steers. Unique fragments were found in almost all (95 of 96) of the randomly clones. In this study, the authors demonstrated that SSH allows for a clear discerning and identification of DNA fragments that are unique to one complex community relative to another. Here, the approach is to hybridize genomic samples from two related sources followed by sequencing and data analysis of the genomic DNA that is highly enriched in sequences unique to either source. In addition, the resulting enriched clones can be used as probes to screen other genomic libraries, or in expression studies.

Single cell approaches

Fluorescent *in situ* hybridization (FISH), (see Chapter 10 for more details), can be used to aid in the isolation and sequencing of previously uncultured species by being used in conjunction with fluorescence-activated flow cytometry followed by whole-genome amplification of the desired cells. Cells are sorted to allow generation of sufficient genomic DNA (femtogram quantities are sufficient) (Lasken, 2007) from the desired target to use the high fidelity proofreading *phi*29 DNA polymerase in a multiple displacement amplification (MDA) reaction for whole genome amplification (Spits, 2006). MDA has been used effectively to amplify genomic DNA from single cells for sequencing individual microbial genomes. The genomes would become the reference genomes needed to accurately assign random reads from metagenome projects to their respective metabolic partner, thus building metagenomes virtually one genome at a time. The utility of reference genomes has recently been demonstrated in studies derived from the human microbiome initiative. Reference genomes were used to analyse the repertoire of glycoside hydrolases in reference genomes (Lozupone *et al.*, 2008) and then for analysis of the core gut microbiome from twins (Turnbaugh *et al.*, 2008).

Conclusions

The field of genomics has rapidly expanded to the point where we are now able to sequence complex communities without the need for first culturing the individual species that are present in these environments. Venter and colleagues have described an unprecedented level of microbial diversity in samples collected from a global ocean sampling expedition (Rusch *et al.*, 2007). Metagenomics provides the capacity to characterize genetic diversity present in samples irrespective and regardless of the availability of laboratory culturing techniques. Markers that are identified from these mixed populations can be used to monitor individual species, or provide insight into metabolic capabilities within these environments [see for example the study on the human gastrointestinal tract (Gill *et al.*, 2006)].

Future trends

In the future, the magnitude of sequencing projects will necessitate the development of high-throughput screening approaches based on mRNA expression, proteomics and metabolomic profiling. Only in this manner can metagenomics couple with other 'omics' approaches to screen and elucidate pathways and metabolite systems hitherto not known or understood.

In addition to sequence-based approaches for data analysis of metagenomic samples, functional approaches are going to be increasingly important to analyse these datasets. Here, the metagenomic samples can be screened for particular expressed traits, and to date, new antibiotics and enzymes have been among the early findings (Schloss and Handelsman, 2003; Handelsman, 2004). Clone libraries have been screened for specific activities including antibacterial, lipase, amylase, nuclease, and haemolytic activities from soils (Rondon *et al.*, 2000), hydrolases and polyphenol oxidases, and cyclodextrinases from rumen communities (Ferrer *et al.*, 2005, 2007; Beloqui *et al.*, 2006), and natural products in the form of small molecules (MacNeil *et al.*, 2001). Zhou *et al.* (2007) investigated the diversity of cellulases in the termite *Reticulitermes flavipes* and characterized four cellulases that code for proteins that are highly similar to previously known endoglucanases, exoglucanases and xylanases. One of the four contains the first intron sequence to be identified from a termite symbiont cellulase (Zhou *et al.*, 2007). With quantitative real-time PCR (qRT-PCR) they were able to show where the individual enzymes were being expressed (Zhou *et al.*, 2007).

Handelman's group (University of Wisconsin, Madison, WI) has recently reported on the functional profiling of remote Alaskan soils (Allen *et al.*, 2008). They demonstrated that there is a great reservoir of regular, divergent and bifunctional beta-lactamases that function in *Escherichia coli*. The results from this study also that suggest that resistance genes can be readily harboured in unpolluted soils when no selective pressure from anthropogenic activity is present.

In a recent publication from Schuster's group (Urich *et al.*, 2008), the fact is highlighted that metagenomic studies limit the conclusions that can be made on gene expression, and as such the functional roles of these genes in their natural environment. They propose that mass analysis of the transcriptome of these environments presents the possibility of deciphering community activity. They have generated cDNA from a soil community, and directly sequenced it by 454-pyrosequencing. These reads were analysed with the MEGAN program (Huson *et al.*, 2007). From one soil sample, they obtained 258,411 RNA-tags of which 193,219 tags had valid taxonomic information. This highlights the need for new numerical techniques, as well as traditional phylogenetic and multivariate approaches. This, combined with metatranscriptomics and metaproteomics, will help unlock the ecological potential of metagenomics in microbial ecology, and change the way in which we understand microbial communities in their natural habitat.

Web resources

A report by the National Research Council (NRC) for a Global Metagenomics Initiative (http://www.genomeweb.com/issues/news/139188-1.html; http://www8.nationalacademies.org/onpinews/newsitem.aspx?RecordID=11902) states that metagenomics 'presents the greatest opportunity to revolutionize understanding of the microbial world'. The report highlights the importance of organisms in performing functions including balancing the atmosphere's composition, fighting disease, and supporting plant growth. The report proposes detailed analysis of carefully chosen microbial communities worldwide. Novel techniques and community databases would be obvious benefits of such a system as well as groups of scientists working together to disseminate advances, agree on standards, and develop guidelines for best practices, the committee said. This report was sponsored by the National Science Foundation, National Institutes of Health, and the U.S. Department of Energy.

References

Andersson, A.F., Lindberg, M., Jakobsson, H., Backhed, F., Nyren, P., and Engstrand, L. (2008). Comparative analysis of human gut microbiota by barcoded pyrosequencing. PLoS ONE 3, e2836.

Aziz, R.K., Bartels, D., Best, A.A., DeJongh, M., Disz, T., Edwards, R.A., Formsma, K., Gerdes, S., Glass, E.M., Kubal, M., *et al.* (2008). The RAST Server: rapid annotations using subsystems technology. BMC Genomics *9*, 75.

Bik, E.M., Eckburg, P.B., Gill, S.R., Nelson, K.E., Purdom, E.A., Francois, F., Perez-Perez, G., Blaser, M.J., and Relman, D.A. (2006). Molecular analysis of the bacterial microbiota in the human stomach. Proc. Natl. Acad. Sci. U.S.A. *103*, 732-737.

Brulc, J.M., Antonopoulos, D.A., Berg, M.E., Wilson, M., K., Edwards, R.E., Frank, E.D., Emerson, J.B., Nelson, K.E., and White, B.A. (2007). Comparative metagenomics of the rumen microbiome by random sampling pyrosequencing.

DeBoy, R.T., Mongodin, E.F., Emerson, J.B., and Nelson, K.E. (2006). Chromosome evolution in the Thermotogales: large-scale inversions and strain diversification of CRISPR sequences. J. Bacteriol. *188*, 2364-2374.

Dekio, I., Hayashi, H., Sakamoto, M., Kitahara, M., Nishikawa, T., Suematsu, M., and Benno, Y. (2005). Detection of potentially novel bacterial components of the human skin microbiota using culture-independent molecular profiling. J Med Microbiol *54*, 1231-1238.

Dicksved, J., Halfvarson, J., Rosenquist, M., Jarnerot, G., Tysk, C., Apajalahti, J., Engstrand, L., and Jansson, J.K. (2008). Molecular analysis of the gut microbiota of identical twins with Crohn's disease. ISME J *2*, 716-727.

Eckburg, P.B., Bik, E.M., Bernstein, C.N., Purdom, E., Dethlefsen, L., Sargent, M., Gill, S.R., Nelson, K.E., and Relman, D.A. (2005). Diversity of the human intestinal microbial flora. Science *308*, 1635-1638.

Edwards, R.A., Rodriguez-Brito, B., Wegley, L., Haynes, M., Breitbart, M., Peterson, D.M., Saar, M.O., Alexander, S., Alexander, E.C., Jr., and Rohwer, F. (2006). Using pyrosequencing to shed light on deep mine microbial ecology under extreme hydrogeologic conditions. BMC Genomics 7, 57.

Fierer, N., Hamady, M., Lauber, C.L., and Knight, R. (2008). The influence of sex, handedness, and washing on the diversity of hand surface bacteria. Proc. Natl. Acad. Sci. U.S.A.

Fleischmann, R.D., Adams, M.D., White, O., Clayton, R.A., Kirkness, E.F., Kerlavage, A.R., Bult, C.J., Tomb, J.F., Dougherty, B.A., Merrick, J.M., *et al.* (1995). Whole-genome random sequencing and assembly of Haemophilus influenzae Rd. Science *269*, 496-512.

Galbraith, E.A., Antonopoulos, D.A., and White, B.A. (2004). Suppressive subtractive hybridization as a tool for identifying genetic diversity in an environmental metagenome: the rumen as a model. Environ Microbiol *6*, 928-937.

Gao, Z., Tseng, C.H., Pei, Z., and Blaser, M.J. (2007). Molecular analysis of human forearm superficial skin bacterial biota. Proc. Natl. Acad. Sci. U.S.A. *104*, 2927-2932.

Gill, S., Pop, M., DeBoy, R., Eckburg, P., Hance, I., Relman, D., Fraser, C., and Nelson, K. (2006a). Metagenomic analysis of the Human Gastrointestinal Microbiome.

Gill, S.R., Pop, M., Deboy, R.T., Eckburg, P.B., Turnbaugh, P.J., Samuel, B.S., Gordon, J.I., Relman, D.A., Fraser-Liggett, C.M., and Nelson, K.E. (2006b). Metagenomic analysis of the human distal gut microbiome. Science *312*, 1355-1359.

Goldberg, S.M., Johnson, J., Busam, D., Feldblyum, T., Ferriera, S., Friedman, R., Halpern, A., Khouri, H., Kravitz, S.A., Lauro, F.M., *et al.* (2006). A Sanger/pyrosequencing hybrid approach for the generation of high-quality draft assemblies of marine microbial genomes. Proc. Natl. Acad. Sci. U.S.A. *103*, 11240-11245.

Handelsman, J. (2004). Metagenomics: application of genomics to uncultured microorganisms. Microbiol Mol Biol Rev *68*, 669-685.

Hugenholtz, P., Goebel, B.M., and Pace, N.R. (1998). Impact of culture-independent studies on the emerging phylogenetic view of bacterial diversity. J Bacteriol *180*, 4765-4774.

Huson, D.H., Auch, A.F., Qi, J., and Schuster, S.C. (2007). MEGAN analysis of metagenomic data. Genome Res.

Ley, R.E., Backhed, F., Turnbaugh, P., Lozupone, C.A., Knight, R.D., and Gordon, J.I. (2005). Obesity alters gut microbial ecology. Proc. Natl. Acad. Sci. U.S.A. *102*, 11070-11075.

Li, M., Wang, B., Zhang, M., Rantalainen, M., Wang, S., Zhou, H., Zhang, Y., Shen, J., Pang, X., Wei, H., *et al.* (2008). Symbiotic gut microbes modulate human metabolic phenotypes. Proc. Natl. Acad. Sci. U.S.A. *105*, 2117-2122.

Manichanh, C., Rigottier-Gois, L., Bonnaud, E., Gloux, K., Pelletier, E., Frangeul, L., Nalin, R., Jarrin, C., Chardon, P., Marteau, P., *et al.* (2006). Reduced diversity of faecal microbiota in Crohn's disease revealed by a metagenomic approach. Gut *55*, 205-211.

Meyer, F., Paarmann, D., D'Souza, M., Olson, R., Glass, E.M., Kubal, M., Paczian, T., Rodriguez, A., Stevens, R., Wilke, A., *et al.* (2008). The metagenomics RAST server – a public resource for the automatic phylogenetic and functional analysis of metagenomes. BMC Bioinformatics 9, 386.

Miller, J.R., Delcher, A.L., Koren, S., Venter, E., Walenz, B.P., Brownley, A., Johnson, J., Li, K., Mobarry, C., and Sutton, G. (2008). Aggressive Assembly of Pyrosequencing Reads with Mates. Bioinformatics.

Mongodin, E.F., Hance, I.R., Deboy, R.T., Gill, S.R., Daugherty, S., Huber, R., Fraser, C.M., Stetter, K., and Nelson, K.E. (2005). Gene transfer and genome plasticity in Thermotoga maritima, a model hyperthermophilic species. J Bacteriol *187*, 4935-4944.

Nelson, K.E., Clayton, R.A., Gill, S.R., Gwinn, M.L., Dodson, R.J., Haft, D.H., Hickey, E.K., Peterson, J.D., Nelson, W.C., Ketchum, K.A., *et al.* (1999). Evidence for lateral gene transfer between Archaea and bacteria from genome sequence of Thermotoga maritima. Nature *399*, 323-329.

Nelson, K.E., Eisen, J.A., and Fraser, C.M. (2001). Genome of Thermotoga maritima MSB8. Methods Enzymol *330*, 169-180.

Nelson, K.E., Zinder, S.H., Hance, I., Burr, P., Odongo, D., Wasawo, D., Odenyo, A., and Bishop, R. (2003). Phylogenetic analysis of the microbial populations in the wild herbivore gastrointestinal tract: insights into an unexplored niche. Environ Microbiol *5*, 1212-1220.

Noguchi, H., Park, J., and Takagi, T. (2006). MetaGene: prokaryotic gene finding from environmental genome shotgun sequences. Nucleic Acids Res *34*, 5623-5630.

Overbeek, R., Begley, T., Butler, R.M., Choudhuri, J.V., Chuang, H.Y., Cohoon, M., de Crecy-Lagard, V., Diaz, N., Disz, T., Edwards, R., *et al.* (2005). The subsystems approach to genome annotation and its use in the project to annotate 1000 genomes. Nucleic Acids Res. *33*, 5691-5702.

Parthasarathy, H., Hill, E., and Maccallum, C. (2007). Global Ocean Sampling Collection. PLoS Biol *5*, e83.

Qu, A., Brulc, J.M., Wilson, M.K., Law, B.F., Theoret, J.R., Joens, L.A., Konkel, M.E., Angly, F., Dinsdale, E.A., Edwards, R.A., *et al.* (2008). Comparative metagenomics reveals host specific metavirulomes and horizontal gene transfer elements in the chicken cecum microbiome. PLoS ONE 3, e2945.

Relman, D.A., and Falkow, S. (2001). The meaning and impact of the human genome sequence for microbiology. Trends Microbiol *9*, 206-208.

Rusch, D.B., Halpern, A.L., Sutton, G., Heidelberg, K.B., Williamson, S., Yooseph, S., Wu, D., Eisen, J.A., Hoffman, J.M., Remington, K., *et al.* (2007). The Sorcerer II Global Ocean Sampling Expedition: Northwest Atlantic through Eastern Tropical Pacific. PLoS Biol *5*, e77.

Sanders, W.E., and Sanders, C.C. (1984). Modification of normal flora by antibiotics: effects on individuals and the environment. In Koot, R.K., and Sande, M.A., eds. New Dimensions in Antimicrobial Chemotherapy, (New York: Churchill Livingstone), pp. 217–241.

Schmeisser, C., Stockigt, C., Raasch, C., Wingender, J., Timmis, K.N., Wenderoth, D.F., Flemming, H.C., Liesegang, H., Schmitz, R.A., Jaeger, K.E., *et al.* (2003). Metagenome survey of biofilms in drinking-water networks. Appl Environ Microbiol *69*, 7298-7309.

Sebat, J.L., Colwell, F.S., and Crawford, R.L. (2003). Metagenomic profiling: microarray analysis of an environmental genomic library. Appl Environ Microbiol *69*, 4927-4934.

Seksik, P., Sokol, H., Lepage, P., Vasquez, N., Manichanh, C., Mangin, I., Pochart, P., Dore, J., and Marteau, P. (2006). Review article: the role of bacteria in onset and perpetuation of inflammatory bowel disease. Aliment Pharmacol Ther *24 Suppl 3*, 11-18.

Sokol, H., Lepage, P., Seksik, P., Dore, J., and Marteau, P. (2006a). Temperature gradient gel electrophoresis of fecal 16S rRNA reveals active Escherichia coli in the microbiota of patients with ulcerative colitis. J Clin Microbiol *44*, 3172-3177.

Sokol, H., Seksik, P., Rigottier-Gois, L., Lay, C., Lepage, P., Podglajen, I., Marteau, P., and Dore, J. (2006b). Specificities of the fecal microbiota in inflammatory bowel disease. Inflamm Bowel Dis *12*, 106-111.

Sokol, H., Pigneur, B., Watterlot, L., Lakhdari, O., Bermudez-Humaran, L.G., Gratadoux, J.J., Blugeon, S., Bridonneau, C., Furet, J.P., Corthier, G., *et al.* (2008). Faecalibacterium prausnitzii is an anti-inflammatory commensal bacterium identified by gut microbiota analysis of Crohn disease patients. Proc. Natl. Acad. Sci. U.S.A. *105*, 16731-16736.

Sokol, H., Seksik, P., Furet, J.P., Firmesse, O., Nion-Larmurier, I., Beaugerie, L., Cosnes, J., Corthier, G., Marteau, P., and Dore, J. (2009). Low counts of Faecalibacterium prausnitzii in colitis microbiota. Inflamm Bowel Dis.

Spiegelman, D., Whissell, G., and Greer, C.W. (2005). A survey of the methods for the characterization of microbial consortia and communities. Can J Microbiol *51*, 355-386.

Sundset, M.A., Edwards, J.E., Cheng, Y.F., Senosiain, R.S., Fraile, M.N., Northwood, K.S., Praesteng, K.E., Glad, T., Mathiesen, S.D., and Wright, A.D. (2009). Molecular diversity of the rumen microbiome of norwegian reindeer on natural summer pasture. Microb. Ecol. *57*, 335-348.

Sundset, M.A., Praesteng, K.E., Cann, I.K., Mathiesen, S.D., and Mackie, R.I. (2007). Novel rumen bacterial diversity in two geographically separated sub-species of reindeer. Microb. Ecol. *54*, 424-438.

Tettelin, H., Masignani, V., Cieslewicz, M.J., Donati, C., Medini, D., Ward, N.L., Angiuoli, S.V., Crabtree, J., Jones, A.L., Durkin, A.S., *et al.* (2005). Genome analysis of multiple pathogenic isolates of *Streptococcus agalactiae*: implications for the microbial "pan-genome". Proc. Natl. Acad. Sci. U.S.A. *102*, 13950-13955.

Tettelin, H., Masignani, V., Cieslewicz, M.J., Eisen, J.A., Peterson, S., Wessels, M.R., Paulsen, I.T., Nelson, K.E., Margarit, I., Read, T.D., *et al.* (2002). Complete genome sequence and comparative genomic analysis of an emerging human pathogen, serotype V *Streptococcus agalactiae*. Proc. Natl. Acad. Sci. U.S.A. *99*, 12391-12396.

Tringe, S.G., von Mering, C., Kobayashi, A., Salamov, A.A., Chen, K., Chang, H.W., Podar, M., Short, J.M., Mathur, E.J., Detter, J.C., *et al.* (2005). Comparative metagenomics of microbial communities. Science *308*, 554-557.

Turnbaugh, P.J., Backhed, F., Fulton, L., and Gordon, J.I. (2008). Diet-induced obesity is linked to marked but reversible alterations in the mouse distal gut microbiome. Cell Host Microbe *3*, 213-223.

Turnbaugh, P.J., Hamady, M., Yatsunenko, T., Cantarel, B.L., Duncan, A., Ley, R.E., Sogin, M.L., Jones, W.J., Roe, B.A., Affourtit, J.P., *et al.* (2009). A core gut microbiome in obese and lean twins. Nature *457*, 480-484.

Tyson, G.W., Chapman, J., Hugenholtz, P., Allen, E.E., Ram, R.J., Richardson, P.M., Solovyev, V.V., Rubin, E.M., Rokhsar, D.S., and Banfield, J.F. (2004). Community structure and metabolism through reconstruction of microbial genomes from the environment. Nature *428*, 37-43.

Venter, J.C., Remington, K., Heidelberg, J.F., Halpern, A.L., Rusch, D., Eisen, J.A., Wu, D., Paulsen, I., Nelson, K.E., Nelson, W., *et al.* (2004). Environmental genome shotgun sequencing of the Sargasso Sea. Science *304*, 66-74.

Voget, S., Leggewie, C., Uesbeck, A., Raasch, C., Jaeger, K.E., and Streit, W.R. (2003). Prospecting for novel biocatalysts in a soil metagenome. Appl Environ Microbiol *69*, 6235-6242.

Warnecke, F., Luginbuhl, P., Ivanova, N., Ghassemian, M., Richardson, T.H., Stege, J.T., Cayouette, M., McHardy, A.C., Djordjevic, G., Aboushadi, N., *et al.* (2007). Metagenomic and functional analysis of hindgut microbiota of a wood-feeding higher termite. Nature *450*, 560-565.

Willing, B., Halfvarson, J., Dicksved, J., Rosenquist, M., Jarnerot, G., Engstrand, L., Tysk, C., and Jansson, J.K. (2008). Twin studies reveal specific imbalances in the mucosa-associated microbiota of patients with ileal Crohn's disease. Inflamm Bowel Dis.

Zerbino, D.R., and Birney, E. (2008). Velvet: algorithms for de novo short read assembly using de Bruijn graphs. Genome Res *18*, 821-829.

Metaproteomics: Techniques and Applications

3

Brian D. Dill, Jacque C. Young, Patricia A. Carey and Nathan C. VerBerkmoes

Abstract

Microbial ecology is currently experiencing a renaissance spurred by the rapid development of molecular techniques and 'omics' technologies in particular. As never before, these tools have allowed researchers in the field to produce a massive amount of information through *in situ* measurements and analysis of natural microbial communities, both vital approaches to the goal of unravelling the interactions of microbes with their environment and with one another. While genomics can provide information regarding the genetic potential of microbes, proteomics characterizes the primary end-stage product, proteins, thus conveying functional information concerning microbial activity. Advances in mass spectrometry instrumentation and methodologies, along with bioinformatics approaches, have brought this analytic chemistry technique to relevance in the biological realm due to its powerful applications in proteomics. Mass spectrometry-enabled proteomics, including 'bottom-up' and 'top-down' approaches, is capable of supplying a wealth of biologically relevant information, from simple protein cataloguing of the proteome of a microbial community to identifying post-translational modifications of individual proteins.

Introduction

Microbial communities are ubiquitous on our planet, inhabiting a variety of ecosystems including those that are not habitable by higher organisms. These ecosystems range from hostile acid mine drainage streams to the frozen arctic and also include more hospitable habitats such as those found in soils and oceans. Microbes play a crucial role in every aspect of their environment by recycling essential nutrients that make up living systems. These diverse communities comprise various functional guilds, each playing specific roles in the environment. Certain microbes, such as cyanobacteria, function in the primary production of the carbon cycle by taking up carbon dioxide and converting it to organic molecules. Other microbes are essential for biodegradation, the decomposition of organic material to carbon dioxide and water. In addition, some microorganisms have important functions in the soil including suppression of plant pathogens and provision of plants with nutrients such as nitrogen and phosphorus (see Chapter 7). In addition, all open surfaces of the human body are associated with microbes, and our entire intestinal track is populated by an array of microbes that are essential to our general well being (see Chapter 9). Mammals and all higher organisms survive through continual synergistic relationships with the microbial communities that naturally populate them and war with those pathogenic microbes that would try and invade. It has been estimated that we have approximately 10 times more microbial cells than human cells in our bodies and orders of magnitude more unique microbial genes (Turnbaugh *et al.*, 2007). This realization has spurred a worldwide scientific effort to classify and understand, at the molecular level, the microbial communities that inhabit our bodies (Turnbaugh *et al.*, 2007; Chapters 2 and 9). Because of their ubiquity and their diverse metabolic capabilities, microbial communities are thus the keystones of life.

Microbial ecologists have been dedicated to the study of microbes in their natural environment for decades, with the rationale that an estimated 95–99% of microbes cannot be cultured. Furthermore, the study of isolates outside of their natural habitat and in isolation is insufficient to capture a true biological picture. Thus, the field strives for *in situ* measurements, taken directly from natural environments. In the infancy of this field, geochemical, phenotypic, and spatial/temporal measurements were made, but direct biomolecular measurements at the system level were not possible. In the 1980s and 1990s, molecular fingerprinting techniques based on unique ribosomal RNA patterns were developed and widely used to characterize the community memberships of environments (see Chapters 1 and 4). These methods rapidly advanced microbial ecology, as direct temporal and spatial monitoring of community membership and dynamics could now be easily monitored; however, these methods primarily reveal the composition of members inhabiting the community at a gross level, but are less valuable for elucidating the functional roles of the community.

Microbial ecology and physiology are currently being revolutionized by the application of high-throughput molecular biology techniques applied directly to natural environmental samples inhabited by communities of microbes. The advent of metagenomic DNA sequencing (Chapter 2) and the ability to sequence whole genomes from microbes to plants and animals has forever changed biology. The field of microbial physiology is rapidly changing due to the information obtained from sequencing and annotation of microbial isolate genomes. This process has, to a large extent, been fully developed, with hundreds of microbial genome sequences completed and hundreds more being characterized at a breathtaking pace (http://img.jgi.doe.gov/cgi-bin/m/main.cgi). The development of technologies to use this information and to further probe the functional components of microbes at a global level (functional genomics) is also rapidly evolving.

In 2004, two landmark papers were published, one analysing an acid mine drainage microbial community (Tyson *et al.,* 2004) and the other an ocean water microbial community (Venter *et al.,* 2004; see Chapter 8). These studies spawned a new subfield of microbial ecology by demonstrating the viability of direct shotgun DNA sequencing of microbes from environmental samples, termed cultivation independent genomics (Banfield *et al.,* 2005) or metagenomics (Tringe *et al.,* 2005; see Chapter 2). Since these original studies, a myriad of natural communities have been sequenced, creating a massive collection of unique genes dwarfing original sequencing projects of isolates. The question is quickly becoming what to do with all the data? Comparative genomics of communities both spatially and temporally separated (Tringe *et al.,* 2005) can reveal insights but does not fully address the fundamental biochemical questions: What genes are being expressed by what species or strains and at what levels? How does gene expression change over time and space as the environment changes? Clearly high-throughput molecular tools are needed to complement genomics to fully understand microbial communities at the functional level.

A natural extension of genomics (the study of the complete set of genes for an organism) is the characterization of the end-stage products, most of which are proteins. Proteomics is defined as the study of the entire suite of proteins expressed by a cell type, an organism, or a community (the latter being categorized as metaproteomics). Proteome analyses, whether in simple microbes, yeast, or higher organisms, presents greater technical challenges than genomics sequencing efforts but promises to provide higher levels of information on the physiology of the system than either genomics or transcriptomics (the study of the transcriptional profiles expressed by an organism or a community). Proteomics involves a variety of approaches and methodologies tasked with objectives beyond simple protein identification. The field can be divided into six major applications (Fig. 3.1) (note: this list is not all-inclusive of the field):

- *Protein cataloguing*: simply identifying proteins present in a cell under a given growth condition.
- *Expression analyses*: determining differences between two or more growth states, time points, environmental settings, etc.

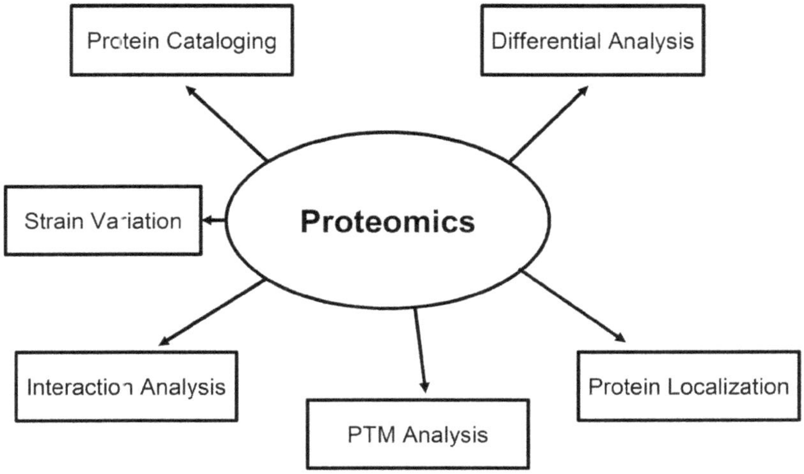

Figure 3.1 Major applications of MS-based proteomics. Proteomics consists of a number of sub-disciplines centered on the common theme of analysing proteins from biological systems.

- *Protein localization*: locating proteins spatially in a cell, biofilm, etc.
- *Post-translational modification analyses:* discovering small molecule adducts on proteins and their effects on function.
- *Protein–protein interaction analyses:* characterizing the components of protein machines or interaction networks.
- *Strain variation analyses:* identifying changes in proteins at the amino acid level over time and space.

Methods used for protein analysis

The field of proteomics, as it currently exists, is diversified and complex with no single measurement platform, experimental approach, or desired result; therefore, it's a constantly evolving field. A list of general proteomic methods is given in Box 3.1. The start of proteomics can probably be most accurately attributed to the invention of 2D-PAGE (two-dimensional polyacrylamide gel electrophoresis) (O'Farrell, 1974; Klose, 1975). Although this approach was initially limited by lack of quality protein identification methods, it provided the best separations of proteins and the possibility to visualize 1000s of proteins on a single gel. Simultaneously, developments during the 1980s and the 1990s in the seemingly unrelated field of mass spectrometry would revolutionize the field of proteomics, creating a completely new approach termed *MS-based proteomics.*

Multiple developments in the field of mass spectrometry positioned the techniques to be very well suited to the emerging field of proteomics. The development of soft ionization techniques, which liberated large biomolecules such as peptides and proteins into the gas phase, was the first key. Electrospray ionization (ESI or ES) allowed for the direct desorption and ionization of peptides and proteins into the gas phase from a liquid matrix (Fenn *et al.*, 1989). Matrix-assisted laser desorption/ionization (MALDI) allowed for the direct desorption and ionization of peptides and proteins into the gas phase from a solid matrix (Hillenkamp *et al.*, 1991; Nakanishi *et al.*, 1994). MALDI is typically connected to time of flight instruments (TOF), which are well suited for the pulsed form of ions produced via MALDI.

The development of rapid scanning tandem mass spectrometers and their application to peptide and protein sequencing was the second major key (Hunt *et al.*, 1981; Hunt *et al.*, 1986, Biemann *et al.*, 1986; Biemann *et al.*, 1988). This was accomplished by a process called tandem mass spectrometry or mass spectrometry/mass spectrometry (MS/MS), utilizing collisional-induced dissociation (CID). In MS/MS, peptides or proteins within a narrow mass/charge (m/z) window are isolated in the gas phase inside the mass spectrometer; collisional energy is added to the peptide via injection of a stable gas, causing uniform and predictable breaks along the peptide

Box 3.1 General proteomic methods

- **2D-PAGE (two-dimensional polyacrylamide gel electrophoresis):** Variation on SDS-PAGE in which proteins generate a 2D array of spots. Proteins are first separated by pI on a strip, and this is followed by applying this strip to a slab gel and separating in a second dimension by molecular weight.
- **'Shotgun' or 'bottom-up' proteomics:** Proteomic approach involving analysis of peptides following proteolytic digestion of proteins (as opposed to 'top down' analysis of intact proteins).
- **Ionization:** Process of inducing peptides (or other analytes of interest) to enter a charged gas phase to enter the mass spectrometer.
    - **MALDI (matrix-assisted laser desorption/ionization):** Ionization of proteins or peptides from a solid matrix using a laser pulse.
    - **ESI (electrospray ionization):** Ionization of proteins or peptides in solution by spraying through an electrical charged nozzle.
- **Mass/charge (m/z) measurement**
    - **TOF (Time of flight):** Measures m/z by detecting the time required for an ion travelling through a tube to strike a detector.
    - **Ion trap**: Involves capture and manipulation of ions by electrical charge/magnetic field within an ion trap to determine m/z.
- **Peptide mass fingerprinting:** Method of identifying proteins by inferring the presence of peptides directly by their respective m/z.
- **MS/MS (tandem MS):** The isolation, activation, and fragmentation of peptides in a mass spectrometer to provide additional information via the uniform and predictable breaks induced along the peptide backbone.

backbone. Fragment ions are created, which correspond to the partial breaking of the peptides from both the N- and C-terminus. This information can be directly related to genomic sequence information or directly interpreted through *de novo* methods as discussed below.

The third major key advancement was the development of low flow (200–300 nl/min) single- and multi-dimensional high-performance liquid chromatography (HPLC) techniques (McCormack *et al.*, 1997; Washburn *et al.*, 2001). Low flow LC was necessary for direct coupling to nano-electrospray and rapid scanning mass spectrometers and had the added benefit of greater sensitivity and resolution. These three steps, along with computational advances, laid the groundwork for the rapidly developing field of proteomics.

The initial focus of microbial proteomics was to couple the high quality separation technique of 1D- or 2D-PAGE with mass spectrometry (Shevchenko *et al.*, 1996). The basic process involved digestion of proteins within excised gel slices with trypsin followed by peptide extraction and analysis by mass spectrometry. This utilized MALDI-TOF and an informatics technique termed peptide mass fingerprinting or, alternatively, analysis of the peptides directly by nano-spray-MS/MS or nano-LC-ES-MS/MS. As these methods matured, analyses of whole microbial proteomes by 2D-PAGE followed by MS analyses became very common. As the technology of 2D-PAGE-MS became more widely accepted, it also became clear that fundamental limitations of the technology, such as limited dynamic range and difficulties in resolving basic and membrane proteins, would limit deep and definitive characterization of even simple microbial proteomes. The concurrent development of pure liquid based separations of proteolytic digests sometimes termed 'shotgun' proteomics or 'bottom-up' proteomics would prove to circumvent many of the shortcomings of the 2D-PAGE-MS technique (Washburn *et al.*, 2001). A direct comparison of gel-based vs. pure liquid-based approaches and the transition from gel-based to mainly liquid-based separation techniques has been well reviewed (Peng and Gygi, 2001; Mann *et al.*, 2001; Liu *et al.*, 2002; VerBerkmoes *et al.*, 2004).

The rapid advancements of multidimensional liquid-based separations coupled to rapid scanning tandem mass spectrometers have lead to many new studies of whole microbial proteomes with greater depth, accuracy, and levels of quantitation than ever observed before (Washburn *et al.*, 2001; VerBerkmoes, 2002; Lipton *et al.*, 2002; Peng, 2003; Corbin *et al.*, 2003; VerBerkmoes *et al.*, 2006). With development of even faster scanning mass spectrometers (Schwartz *et al.*, 2002; Hu *et al.*, 2005), it is possible to identify 1500–2500 proteins from a single growth state of a microbial isolate in 1–2 days with minimal false discoveries (less than 1%), as well as make quantitative comparisons on hundreds to thousands of these proteins (Brown *et al.*, 2006, Chourey *et al.*, 2006; Callister *et al.*, 2006; Thompson *et al.*, 2007; Luo *et al.*, 2007). By analysing multiple growth states, it is possible to identify 50–90% of the predicted proteome from microbial isolates.

The increase in mass spectrometry performance over the past five years has been breathtaking, and the field has seen jumps in the complexity of samples that can be characterized to great depth as well as an increase in the quality of the data (Fig. 3.2). The original line of commercial ion traps (three-dimensional traps) connected to multi-dimensional LC systems could provide good dynamic range across a large protein complex, such as the ribosome, where almost every subunit could be identified with high sequence coverage and numerous post-translational modifications (PTMs) could be determined (Strader, 2004). The next generation ion traps (linear ion traps) provided a greater jump in dynamic range; two to four times the number of proteins could be identified from an identical sample in the same amount of time (Schwartz *et al.*, 2002). The development of the hybrid linear ion trap Orbitrap was a landmark for all fields of proteomics; it is currently possible to keep the same dynamic range, durability, and speed of a linear ion trap but now with high mass accuracy and high mass

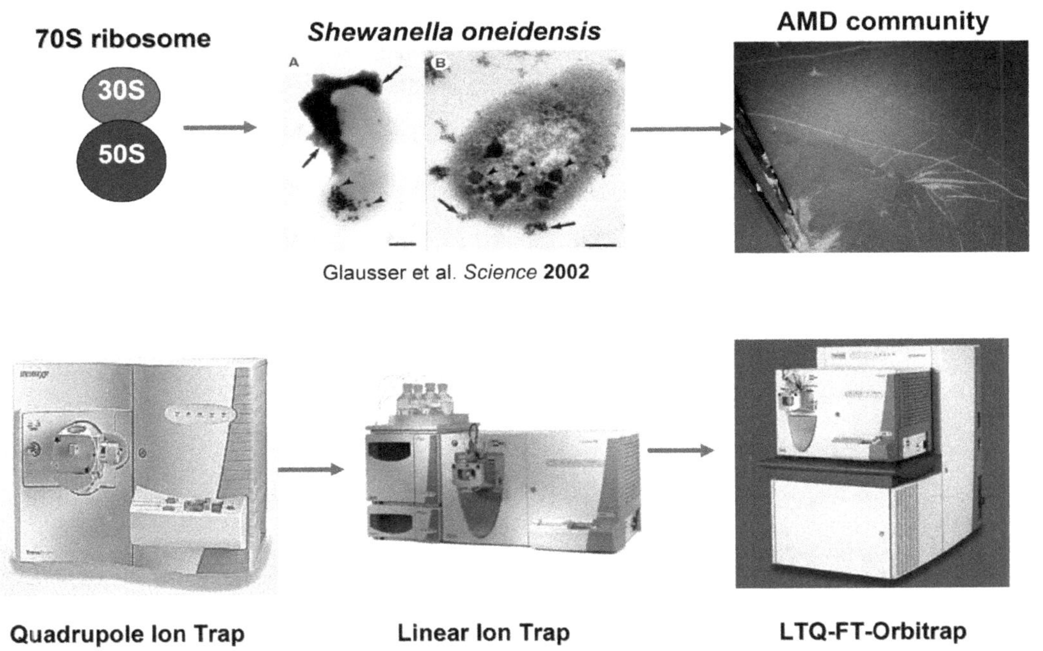

Figure 3.2 The evolution of microbial proteomics alongside the ion trap mass spectrometer. Powered by developments in mass spectrometry instrumentation allowing high-throughput, accurate, high resolution measurements, analysis of microbial proteomics has advanced from analysis of intracellular complexes such as the ribosome, through clonal cultures of microbial isolates, to sampling environmental microbial communities *in situ*. Mass spectrometer photographs courtesy of Thermo Fisher Scientific.

resolution (Hu *et al.*, 2005). This has proved essential for the analysis of simple to medium complexity microbial communities as will be discussed below, but this instrument still lacks the speed and overall dynamic range necessary for the great complexity of microbial communities such as those inhabiting the GI tract, soil (reviewed in Kent and Triplett, 2002; Chapter 6), and bodies of water (reviewed in DeLong, 1997 and 2005; Chapter 8). What will next generations of mass spectrometers be capable of, and how quickly will they become available? These are unanswered questions at this point, but the challenges and potential technical solutions are discussed in this chapter.

Bottom-up or shotgun proteomics

One of the great attractions of the shotgun proteomics approach is its overall simplicity. Once the basic techniques are mastered, they can be adapted to various systems and analysis schemes. A general protocol for the process is given in Box 3.2. The general methodology starts with cellular lysis in the presence of protease inhibitors or under denaturing conditions. Typical cell lysis procedures can be used, such as probe tip sonication, bead-beating, French press, freeze-thaw cycles, or mortar and pestle; however, addition of enzymes such as lysozyme should be avoided due to downstream complications. At this point, an optional cellular fractionation can be done, which enables deeper proteome coverage (assuming sufficient biomass is available for an effective fractionation). Next, proteins are denatured and digested into peptides of (generally) 7–30 amino acids by trypsin or other proteases, and, finally, the peptide solutions are desalted by solid phase extraction and concentrated.

Quality digestions are key to obtaining good mass spectrometric data; under or overdigested peptides will not be detected or will confound computational identification. Specific enzymes such as trypsin are most commonly used, but multi-enzyme digests or non-specific digests can be used to handle proteins that are poorly digested by traditional enzymes. For example, proteins poor in lysine and arginine, such as

Box 3.2 Typical sample preparation for bacterial shotgun proteomics

1. Culture bacterial cells at some level of growth (mid-log phase for example) or harvest microbes *in situ* from environment.
2. Harvest cells by centrifugation and wash twice with 50 mM Tris pH 7.6 (many alternative techniques needed for environmental samples).
3. Lyse cells (sonication or French press for example) in 50 mM Tris pH 7.6 with 10 mM EDTA. Keep samples ice cold; if doing top-down analysis, add additional protease inhibitors. Single tube methods for very small samples are now available but require technical modifications (Thompson *et al*, 2008).
4. Pellet unbroken cells by centrifugation (5000 g x 15 min). Whole cell lysates can be separated into soluble and membrane fractions, this greatly increase dynamic range, as there is generally only 20–30% overlap between well-separated fractions.
5. Aliquot and freeze lysed samples at −80°C until further processing.
6. Quantify total protein amount by BCA assay.
7. Denature and reduce proteins with 6M guanidine and 5 mM DTT at 60°C for 1 hour. Then, dilute in 50 mM Tris (pH 7.6) with 5 mM $CaCl_2$ to obtain a final guanidine concentration of less than 1 M (to prevent denaturing trypsin in the next step).
8. Sequencing grade trypsin is added at 1:50–1:100 (estimated per total amount of protein) and digestions are run for 16 hours with gentle rocking at 37°C. Trypsin can be added a second time with an additional 6–12 hour incubation.
9. Add 10 mM DTT to the solution and incubate for 1 hour at 37°C as a final reduction step.
10. Centrifuge at low speed (5 min at 3000 g) to remove unbroken cells, cellular debris, etc.
11. Supernatant should then be immediately desalted with a C18 solid phase extraction and concentrated by centrifugal evaporator to ~10 μg/μl and filtered with a 45 μM spin filter to remove insoluble materials.
12. Run immediately or store aliquots at −80°C.

many membrane proteins, are not effectively digested by trypsin (which specifically cleaves following lysine and arginine residues), so non-specific proteases such as pepsin or Proteinase K can be effective alternatives. Another important consideration, particularly in processing samples containing environmental contaminants or organisms collected from an extreme environment, is whether the protease is active under the conditions of the sample (for example, is the solution within an acceptable pH range).

When it comes to the optimal amount of protein to use in this technique, the old adage 'more is better' is generally true, although any liquid chromatography method will 'crash' if a large excess is used. Owing to restrictions on the amount of available sample, which is particularly an issue with many environmental samples (as opposed to the ability to grow a large culture when working with an isolate), environmental microbiologists continually modify their methods to use smaller and smaller protein quantities; however, the overall dynamic range of the method is directly linked to sample quantity. The more sample that can be loaded or injected into a system, the more levels of fractionation can be used, and the lower abundance peptides will become more accessible. Generally, if 1–10 mg of wet biomass can be obtained for a microbial community, then chances are good that a successful proteome can be measured. For any microbial community, trial and error method development is generally needed to determine the best methodology to prepare the samples. For all proteomics applications, quality sample preparation is critical to quality results, and the optimization of sample preparation methods should not be overlooked.

Following optional off-line fractionation (such as cellular fractionation or separations based on protein or peptide chemistry), peptides are injected into an HPLC system, and peptides are on-line separated in one, two, or three dimensions before ionization into the mass spectrometer. The choice of the level of separation is based on the types of samples, overall complexity, depth of coverage desired, and time available to process the sample. These are complex decisions that must be made with logical analytical guidelines to obtain the best possible information from a sample. Obviously, the more dimensions of separation, the longer the experiment will take, and a trade off must be made between total number of samples that can be run for each experiment and the time required.

The choices for which types of separation method to use are also extremely important; multi-dimensional separations should be as orthogonal as possible, meaning each dimension of separation should use independent chemical properties. With peptides, there are only a limited number of options, most commonly size, charge, and hydrophobicity. Generally, reverse phase separation (based on hydrophobicity) is used as the final dimension before electrospray, because its associated solvents (water, organics such as methanol and acetonitrile, and formic or acetic acid) are very compatible with the electrospray process. Electrospray is adversely affected by non-volatile salts and detergents.

The choice of the first dimension of separation is much more diverse, with many different options in the literature. The most common method seen in shotgun proteomics is the use of strong cation exchange (SCX) for the first dimension of separation of the peptides. The reasons are the large binding capacity of SCX materials, and its separation is based on charge, which is orthogonal to hydrophobicity. The first dimension of SCX separation can be off-line, where fractions are collected, or on-line, where the SCX column is directly connected to the reverse phase (RP) column and the nanospray tip. This eloquent method, as first described by Washburn *et al.* (2001) and updated by McDonald *et al.* (2002), used a combined SCX-RP packed into a fused silica column with a nanospray tip; this is still one of the simplest, most straightforward, and robust approaches to two-dimensional chromatography to date. In this method, complex peptide mixtures are separated physically by their net charge and hydrophobicity by multi-dimensional HPLC and are sequentially ionized into the mass spectrometer by nanospray. The mass spectrometer records the m/z of the parent and fragment ions. During the entire chromatographic run, the mass spectrometer oscillates between full scan (where it is acquiring m/z values of peptides entering the mass spectrometer) and MS/MS mode (where selected peptides are fragmented by CID as they elute from the chromatographic column) (Fig.

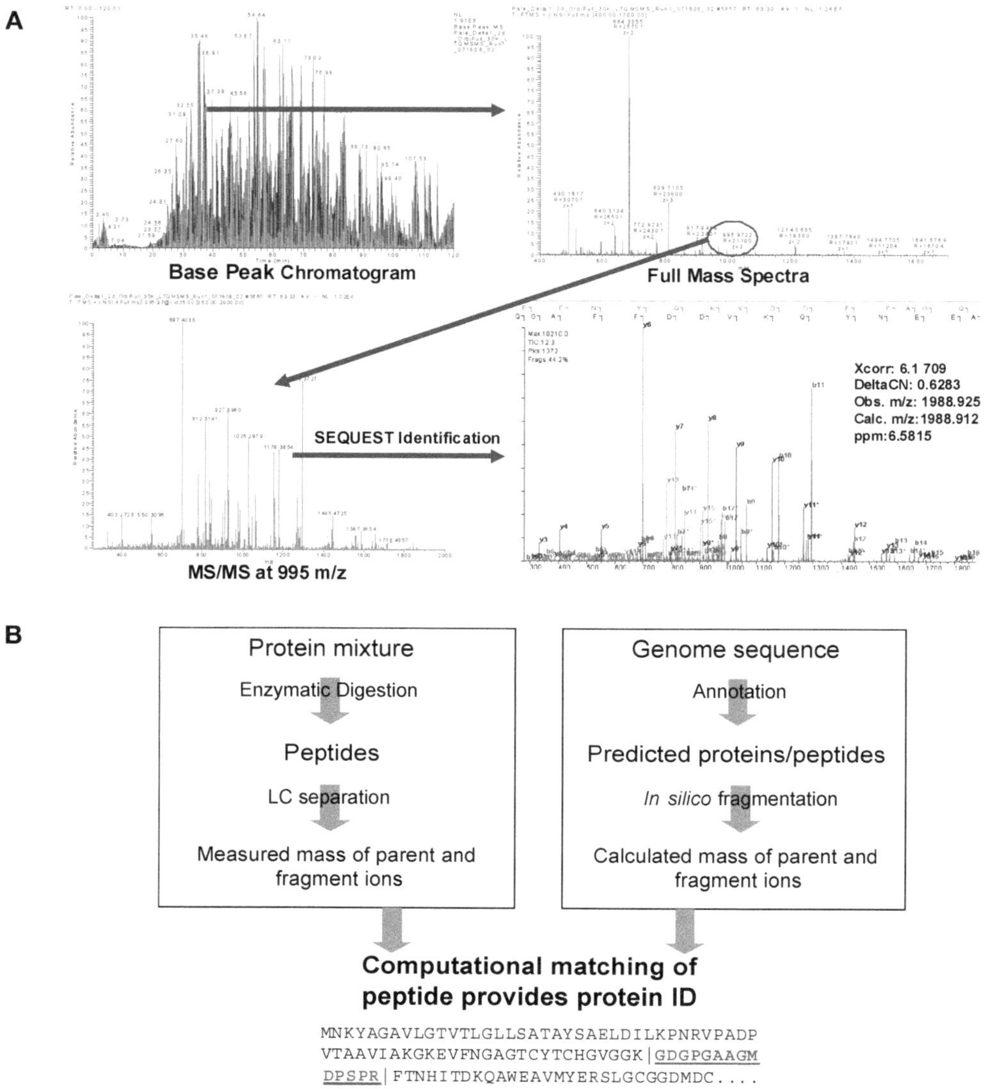

Figure 3.3 Overview of the bottom-up or shotgun proteomics approach. **A** The multi-dimensional separation of a complex peptide mixture. Nano-electrospray-MS/MS is conducted on peptides as they elute from a one-dimensional (reverse phase) or two-dimensional (strong cation exchange in combination with reverse phase) separation via on-line HPLC (panel 1). Full mass spectra are collected from ionized peptides, indicating the mass/charge ratio (m/z) of species and their relative intensities (panel 2). Following data dependent selection of ions measured in the full scan (typically the most intense 3–5 peaks per full scan), ions in the gas phase are serially isolated and fragmented, producing a tandem mass spectrum of daughter ions (panel 3). Computational model spectra are compared to the measured spectra, and a scoring algorithm indicates the fitness of the match (panel 4). **B** Computational identification of peptides. Following the collection of MS/MS data (left panel), computational methods are used to automate peptide/protein identification (right panel). Protein sequences are predicted from genome sequences, the predicted proteins are *in silico* digested to predict peptides, and the peptide fragmentation patterns of ions isolated in the gas phase of the mass spectrometer are predicted. The pattern of this ladder of fragment ions is the key to determining a computational match between the measured peptide's fragmentation and the computationally predicted fragmentation. Search engines focus on comparing theoretical MS/MS spectra with the observed spectra to determine matches between predicted and observed spectra. Cross-correlation scores or expectation values indicate the quality of the match and are thus used for filtering the falsely discovered peptides, and the difference between the observed and predicted full scan m/z of the parent peptide can be used for further confirmation of a match.

3.3A). The two-dimensional LC-MS/MS process is completely automated, and spectra are collected in a data-dependent manner. High-abundance peptides are typically targeted first, followed by lower abundance peptides.

Another approach to add an additional dimension of separation (and thus improve dynamic range) is the separation of proteins *prior* to digestion, followed by downstream separations of peptides. Since the chemical properties of proteins and their peptides will vary, significant improvements in separations can be made while using parallel separations (i.e. off-line fractionation of proteins by charge prior to digestion, followed by on-line separation of peptides by charge).

To avoid targeting the same peptides repeatedly during an LC-MS/MS run, a method called dynamic exclusion is employed. In this method, peptides that have been targeted once for MS/MS are not targeted again within a defined time window (generally 1–2 minutes). These peptides' masses are added to what is called the dynamic exclusion list after being targeted for MS/MS, which excludes them from being targeted again within the time window. Without dynamic exclusion, abundant peptides would be over-sampled at the expense of ignoring less abundant peptides. In a general microbial whole cell proteome, it would be difficult to see beyond peptides arising from very abundant proteins such as ribosomal proteins, chaperones including GroEL and GroES, ATP synthase proteins, and flagellin proteins.

The key to the shotgun proteomics methodology is the direct comparison of the observed mass and fragmentation patterns of the peptides obtained from the mass spectrometer with predicted masses and predicted fragmentation patterns from computational predicted proteomes (Fig. 3.3B). This process is done via database search programs such as SEQUEST (Eng *et al.*, 1994) and MASCOT (Perkins *et al.*, 1999). These computationally predicted proteomes are generated via DNA sequencing and annotation of individual microbes (reviewed in Binnewies *et al.*, 2006) or whole communities (reviewed in Tringe *et al.*, 2005). Thus, the analysis of the peptides via LC-MS/MS is intimately tied to the informatics. An entire field (proteome informatics) has been spawned to tackle the mountains of data created in typical proteomics studies and is discussed in further detail below.

The analysis of whole proteomes and protein complexes by mass spectrometry can provide very useful qualitative information, but what is needed in this field is accurate quantitative data between temporally or spatially resolved samples. Due to inherent matrix effects and sample losses, quantitation is a difficult challenge in shotgun proteomics and one of the heaviest areas of research (reviewed in Ong *et al.*, 2005). Recent developments in stable isotope labelling have allowed for accurate, relative quantitation of proteins in two different samples, such as *E. coli* grown under high salt and low salt conditions. In these experiments, certain proteins can be compared with counterparts from a different growth condition to obtain a relative expression level of up- or down-regulation. An absolute level of protein expression remains difficult to determine in a high-throughput method. Four main methodologies for relative quantitation employing stable isotope labels have been developed over the last few years: ICAT, iTRAQ, ^{18}O water labelling, and metabolic labelling (sometimes referred to as SILAC) (reviewed in Ong *et al.*, 2005). Each has inherent advantages and disadvantages that will be even more challenging in moving from isolates to natural communities. Another popular method for quantitation involves label-free methods, which rely on intrinsic values obtained in the course of the experiment such as intensities or areas of peptide peaks (Old *et al.*, 2005), spectral counts (Liu *et al.*, 2004), and normalized spectral abundance factors (NSAFs) (Florens *et al.*, 2006) to quantify peptides and thus proteins. They have grown in popularity due to simplicity, cost considerations, and the fact they can be used on any sample assuming proper experimental design is implemented. There is a strong effort in the proteome informatics community, both academic and commercial, to develop better tools and statistics for label-free methods (Ishihama *et al.*, 2005; Zhang *et al.*, 2006; Nesvizhskii *et al.*, 2007; Lu *et al.*, 2007).

The rapid development of MS-based proteomics for microbial isolates set the stage for the integration of this technology into the evolving field of metagenomics. As will be discussed in more detail below, genomic datasets are essential for proteomics techniques. As metagenomic datasets became available, the potential for large-scale deep characterization of microbial proteomes directly from natural microbial communities became clearly possible. Prior to this transition, proteomics efforts in microbial communities were primarily limited to 2D-PAGE followed by MS analyses of a few spots, revealing a handful of proteins but limited information on the system as a whole.

Proteome bioinformatics

Proteome bioinformatics is a field of science that has recently come about due to the need to handle the large datasets created by MS-based proteomics experiments. A typical proteome bioinformatics pipeline consists of the following steps: (1) extraction of MS and MS/MS spectra from instrument-generated capture files into text files, (2) assignment of precursor ion charge state(s) to each tandem mass spectrum, (3) identification of a peptide for each tandem mass spectrum with a search engine such as SEQUEST, MASCOT, and X!Tandem (Eng *et al.*, 1994; Perkins *et al.*, 1999; Craig *et al.*, 2004), (4) assembly and filtering of the output files; DTASelect (Tabb *et al.*, 2002) was one of the first and most widely used programs for this application, (5) comparisons of run-to-run variations and sample-to-sample changes with statistical tools and visualization software, (6) archival of the datasets into database systems such as Oracle or MySQL, and (7) creation of web-based compilations to give easy access to collaborators. This web-based data access allows direct interface with datasets for collaborators anywhere in the world. Upon request, datasets can be easily re-searched, re-filtered, sorted, and compared as necessary. One of the main advantages of proteomics data is that if stored correctly, it can be reanalysed many times with different parameters, search engines, and databases. Below are some examples of some web-based datasets from Ram *et al.* (2005), Lo *et al.* (2007), and VerBerkmoes *et al.* (2008).

http://compbio.ornl.gov/biofilm_amd/
http://compbio.ornl.gov/biofilm_amd_recombination/
http://compbio.ornl.gov/human_gut_microbial_metaproteome/

The most common method for searching proteomics data is the direct comparison of observed MS/MS spectra with theoretical spectra generated from a genomic database (Fig. 3.3B). Most common search engines such as SEQUEST, MASCOT, and X!Tandem use some sort of cross-correlation or probabilistic function to do the comparison between the observed MS/MS spectra and the theoretical spectra. The theoretical spectra are generally produced from specific predicted peptides' cleavages and then sorted by mass. The search engine can use the mass of the peptide's parent ion to minimize the list of candidate peptides to a narrow mass region. While the mass of the peptide is generally important in finding the candidate peptides, the real power of search engines come from comparing the theoretical MS/MS spectra with the observed spectra. Each observed spectrum contains a ladder of the fragment ions produced after isolation and fragmentation (generally by CID) of the peptide inside the mass spectrometer. This ladder of fragment ions is the key to the technology and contains specific sequence information from the peptide that was isolated and fragmented. The caveat is that directly reading the sequence from the spectrum is more difficult, especially for lower quality MS/MS spectra (roughly ½ to ¾ of MS/MS spectra of real peptides are not of ultra high quality). The search engines allow rapid and accurate identification of the MS/MS spectra so long as the peptide that generated the MS/MS spectra is in the database being searched. Herein lies the problem for metaproteomics; very often in isolate microbial proteomics studies there is an exact match or nearly an exact match between the genome and proteome, but, as discussed below, in proteomics of microbial communities this is very often not the case. Clearly some proteins and thus peptides will have changed between a reference metagenome and the metaproteomes being analysed; therefore, there is a need for other methods to analyse MS/MS data from metaproteomic studies (discussed below).

Typical search engines assign theoretical peptide sequences to almost every tandem mass spectrum. Filtering programs, such as DTASelect, determine probable peptides based on user criteria, while assembly programs associate each peptide sequence with one or more proteins. Typical filtering and assembly programs also report several important measures for each peptide identified: the peptide's sequence, the associated protein(s) identified, if the peptide is unique to a single protein, the peptide's score (probability or cross correlation), and the number of spectra matching the peptide. Designation as a unique peptide indicates that a peptide's amino acid sequence is contained only in the sequence of a given protein, among those in the database. This measure is especially important to metaproteomics since their protein databases are highly likely to contain redundant peptides due to genomic overlap of species and strains. See the following for an example of a DTASelect file containing links to all the identified spectra from a metaproteome (one sample from the gut metaproteome) (VerBerkmoes *et al.*, 2008):

http://compbio.ornl.gov/human_gut_microbial_metaproteome/ms_analysis/

Database searching programs, such as SEQUEST, identify peptides by 'looking up the answer in the back of the book' (Sadygov *et al.*, 2004), where the genome sequence is the answer section of the book. This approach has worked extremely well for genomically characterized microbial isolates. It also works adequately for organisms whose sequences are close to the genomically characterized type, because proteins can be identified using the subset of peptides that do not contain amino acid substitutions. The largest challenge for proteomics of natural microbial consortia is likely to arise when candidate peptide sequences are not available (i.e. the microbes' proteins differ in sequence from any genomically characterized strain). Little is known about population level heterogeneity of microbes in the environment. Given their fast evolutionary rates, it is likely that 'the answer in the back of the book' will be inadequate. As genomic re-sequencing of every sample is not practical, it will be necessary to augment database searching with *de novo* sequence tagging methods. Strictly speaking, these are pseudo-*de novo* methods because they seek to identify peptides that are related, by one or more amino acid substitution(s), to an existing sequence (rather than being completely unknown).

De novo sequencing methods attempt to determine the sequence of a peptide directly from the MS/MS spectra without a database, while sequence tagging approaches attempt to produce a short tag that can be matched against a database; in either case the use of high mass accuracy data will be necessary. When a peptide is dissociated in a mass spectrometer, the m/z ratios of its fragments are measured and recorded. As the amide bonds are the most likely peptide fragmentation sites, the fragments often form a ladder in the MS/MS scan with mass differences that correspond to amino acid masses. A series of abundant peaks representing fragmentation sites in a spectrum enables inference of amino acid sequences. Although the *de novo* sequencing and sequence tagging approaches seem straightforward, these methods have not been widely applied in proteomics due to data quality and software limitations.

Fig. 3.4 illustrates how peptide and amino acid sequence results from database searching and *de novo* sequence tagging methods are integrated to reveal proteins with amino acid polymorphisms due to microbial strain variation. The top of the figure displays that several strains of proteins, A, E, and C, are digested into peptides. The left-hand side depicts a typical database searching program: (1) the peptides are fragmented by MS/MS, (2) on the far left, the protein sequence database is *in silico* digested into peptides from which theoretical MS/MS spectra are generated, (3) the observed spectra are searched against the theoretically generated spectra, (4) the observed spectra are then assigned to candidate peptides from the database, and (5) finally, the database results are coupled with the *de novo* results. On the right hand side of Fig. 3.4, a *de novo* sequence tagging program is illustrated: (1) the peptides are fragmented by MS/MS and (2) the *de novo* sequence tagging program assembles short confident amino acid sequences. The bottom of Fig. 3.4 demonstrates how the database peptide results and *de novo* tags are merged and correlated with candidate proteins using a sequence database as a reference. Thus, protein sequences containing amino acid polymorphisms can be reconstructed with the help of *de novo* tags. Therefore, database search programs provide the majority of a

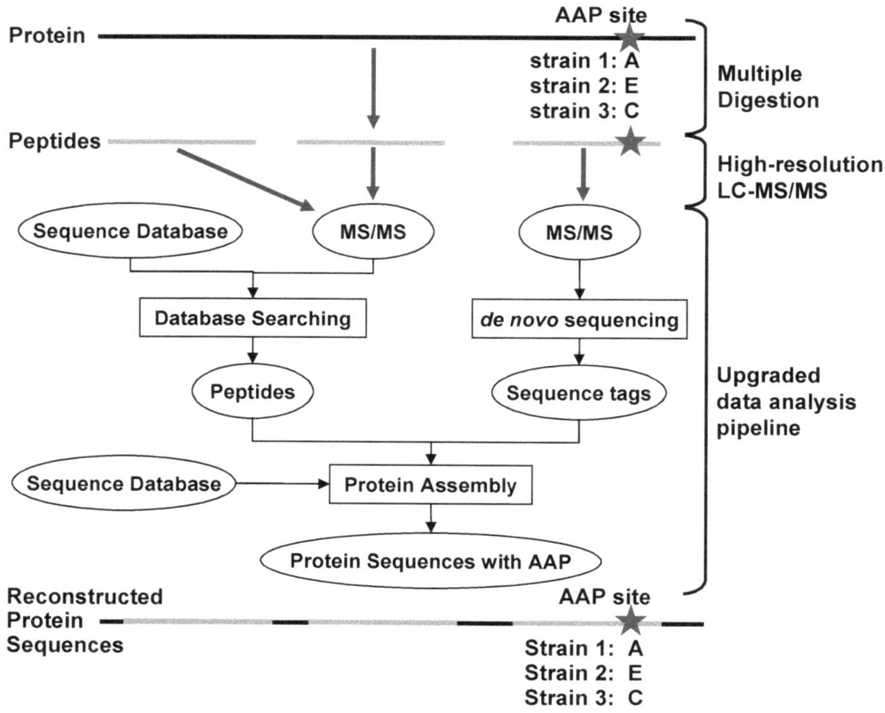

Figure 3.4 Overview of the *de novo* method for identifying amino acid polymorphisms (AAPs). The traditional approach to protein identification relies solely on the presence of the target sequence in a search database. This prevents identification of peptides with substitutions of even a single amino acid from that in the database sequence due to the high resolution of mass measurements of modern mass spectrometers. Computational generation of sequence tags from tandem spectra allow identification of sequences similar to, but not identical to, database sequences. Indications of AAPs include 'missing' or underrepresented peptides on a high peptide coverage protein or high abundance unmatched spectra can be targeted.

protein's sequence, while *de novo* sequence tags provide the short amino acid sequences required to reconstruct amino acid polymorphisms for microbial communities.

PEAKS is probably the most widely used *de novo* sequencing program today (Ma *et al.*, 2003). It examines each tandem mass spectra and provides a full peptide sequence while ensuring the mass of the sequence reported matches the parent mass within a user set error. Full sequencing *de novo* methods have not been widely applied to pipelines since the methods frequently report a low score for most spectra, even spectra for which a database searching program assigns a confident peptide. On the other hand, PepNovo (another commonly used *de novo* method) reports short amino acid sequences, tags, for spectra (Frank *et al.*, 2005). Tagging programs are frequently used to reduce the search space for database searching programs. Tagging *de novo* programs are also highly important, since frequently the b_1 and y_1 ions (the ions from the C and N termini of a peptide) are missing from tandem spectra, making full sequencing extremely difficult to impossible. However, by producing short sequences of amino acids as opposed to full peptides, results are far more likely to be matched to multiple proteins and even multiple peptides. Thus metaproteomics benefits greatly from both programs that produce full sequences and tags, and future studies employing these methods will most likely use both approaches in tandem.

The acid mine drainage model system

Ram *et al.* (2005) detailed the first large-scale proteome study of a natural microbial community inhabiting acid mine drainage (AMD) streams. This study targeted a location within the Richmond Mine near Redding in northern California, which contains extremely acidic waters (pH < 1.0) with a high heavy metal content (submolar concentrations of $FeSO_4$ and millimolar levels of Zn, Cu, As, and other metals (Baker *et al.*, 2003). The processes occurring within the Richmond system are broadly typical of the AMD generating processes that occur at other sites around the world. The AMD environment provides an ideal environment to study a microbial community *in situ*, due to the low number of species able to survive in this extreme niche, the relatively controlled environmental factors, which are well characterized, and the large amount of biomass that grows on the water surface. Using genomic sequences from a distinct but similar AMD biofilm (Tyson *et al.*, 2004), it was possible to match shotgun MS/MS spectra to peptides and thus proteins derived from the community genomic database. In the study over 2000 proteins were identified with high confidence (false positive or discovery rate of less than 1%), not only verifying that proteome measurements can be conducted on complex natural microbial communities, but specifically yielding protein information for the five most dominant organisms. Proteome coverage was greatest for the most abundant organism, *Leptosprillium* group II, with 50% of its proteome measured. For this species, it was possible to directly map proteins back into genomic context (proteogenomics) providing a starting point for functional prediction and testing.

The goal for most first stage proteomics efforts is large-scale identification and discovery – a hypothesis generating approach. In the study by Ram *et al.* (2005), over 570 unknown proteins were identified. Many of these were abundant and localized in the membrane or extracellular fraction, and thus in close contact with the low pH. Two of those proteins, both directly purified from numerous natural AMD biofilms (one in the membrane and one in the extracellular fraction), were extremely abundant as estimated from spectral counts and normalized spectral abundance factors. These proteins were targeted for further in-depth biochemical characterization leading to detailed biochemical data and proposed functions for each as cytochromes important in iron reduction (the primary energy source in the AMD system). The first protein, Cyt_{572}, was shown to be an unusual membrane-bound cytochrome, which provides a critical function within the ecological niche of acidophilic microbial communities (Jeans *et al.*, 2008). The second protein, Cyt_{579}, was shown to be an extracellular soluble cytochrome, which acts as an electron transfer protein shuttling electrons derived from Fe(II) oxidation (Singer *et al.*, 2008). These two studies were groundbreaking, because data were taken directly from community genomic and proteomic datasets to target interesting abundant hypothetical proteins for purification and functional characterization directly from the natural environment. This type of directed biochemical study is key to functional genomics but limited in throughput. Other types of techniques such as informatics, high-throughput structure analysis, and protein function arrays are needed to help characterize the diversity of novel enzymes likely to be found in nature.

Sequence variation is characteristic of natural populations and is, thus, a ubiquitous problem in proteomic analyses. Genetic variation within microbial communities separated by space and/or time presents a major challenge for proteomic studies because peptide identification depends on an exact match between the predicted and observed masses of peptides and their fragmentation products. A single amino acid substitution will prevent peptide identification unless more advanced approaches are utilized (i.e. *de novo* sequence tagging, discussed above). While genetic variation can confound peptide identification, this potential handicap can be harnessed to explore strain variation and yield information regarding evidence of genetic exchange. While PCR-based methods for genetic typing are widely used in microbial ecology, proteomics-inferred genome typing (PIGT) is similarly capable of strain-typing, through proteomic data, which conveys information for both genome type and activity. Furthermore, high mass accuracy measurements allow sensitive and confident typing of single amino acid polymorphisms (SAAPs), except isoleucine and leucine, which are isobaric; single nucleotide polymorphisms can be problematic in PCR-based approaches due to the lower stringency of primer hybridization.

The PIGT methodology was first demonstrated in Lo *et al.* (2007) and then expanded in Wilmes *et al.* (2008b) and Denef *et al.* (2008). In the first study, Lo *et al.* utilized a proteomic approach for typing of a genetically uncharacterized biofilm from the ABend location in the Richmond mine for the purpose of tracking recombination between two closely related strains. This biofilm, like that previously characterized from the 'UBA' and '5-way' locations, is dominated by a *Leptospirillum* group II bacterium. Genomic data from sequencing of UBA and 5-way biofilms allowed the generation of databases for both the UBA and 5-way *Leptospirillum* II strains. Following shotgun proteomic analysis, ABend biofilm-derived *Leptospirillum* II peptides were attributed to either the UBA or 5-way *Leptospirillum* II genome type by identifying unique and non-unique peptides between each strain (Fig. 3.5). Unique peptides were then used to differentiate proteins as derived from the UBA or 5-way genome type; proteins were assigned to either UBA or 5-way strains, determined from a comparison of the per cent protein coverage by peptides matching either the UBA or 5-way databases. These protein assignments were mapped back to their genomic context, providing a picture of relative genetic contributions of each strain type. This analysis revealed insights into recombination events between the UBA and 5-way *Leptospirillum* II types to produce the hybrid ABend. For example, genomic blocks of high conservation were often found to contain housekeeping genes, while blocks of high variation were found in lower expressed genes. Thus, PIGT is capable not only of strain typing, but also mapping areas of recombination between two strains.

Proteomics in complex microbial communities and environments

Recent findings from acid mine drainage microbial communities continue to highlight that this model system is far less simple than originally believed, but it still represents a system with far less complexity than that found in ground waters, oceans, or soil samples. This concept was thoroughly reviewed by Wilmes and Bond (2006a), and they highlighted the massive increase of potential unique protein species going from simpler communities such as AMD biofilms and activated sludge to more complex systems such as ocean water and soil. The large number of unique taxa producing their own unique protein products across a wide dynamic range creates a technical challenge for modern proteomics. The tissues of mammalian systems have long been the holy grail of advanced proteomics methods due to their heterogeneity, large dynamic range of expressed proteins, intron–exon splicing, and a myriad of potential post-translational modifications. Microbial proteomics had been considered much simpler, but this concept is starting to change with the emergence of metaproteomics and its application to complex microbial communities. A central goal of proteomics has always been the complete detection of all unique protein products in a sample, but that goal seems almost impossible in even moderately complex systems.

So is it pointless to even try? Can measurements of thousands of proteins, the limit of what can be achieved with current technologies, be of any use, if the sample contains hundreds of thousands to millions of unique proteins? Will it ever be possible to accurately determine the original proteins from measured peptides and, more importantly, the organisms? How can we quantify proteins between two complex environmental samples, when not only protein abundances are changing but organism abundances are fluctuating? Owing to the inherent specificity of the proteomics technique, will it be necessary to have a matching metagenome for every metaproteome sample? Will it be possible to build metabolic maps and networks to see how these are perturbed by environmental events to truly begin to understand how these communities function at the molecular level? The answers to these questions are challenging, and the necessary technological improvements will not occur overnight; however, as discussed below, rapid advancements are occurring, the specific technical challenges can be identified, and solutions will be found.

Activated sludge

So what has been accomplished so far? Shortly after the initial AMD studies, the activated sludge community (the next logical community with regard to complexity) proteomic studies were

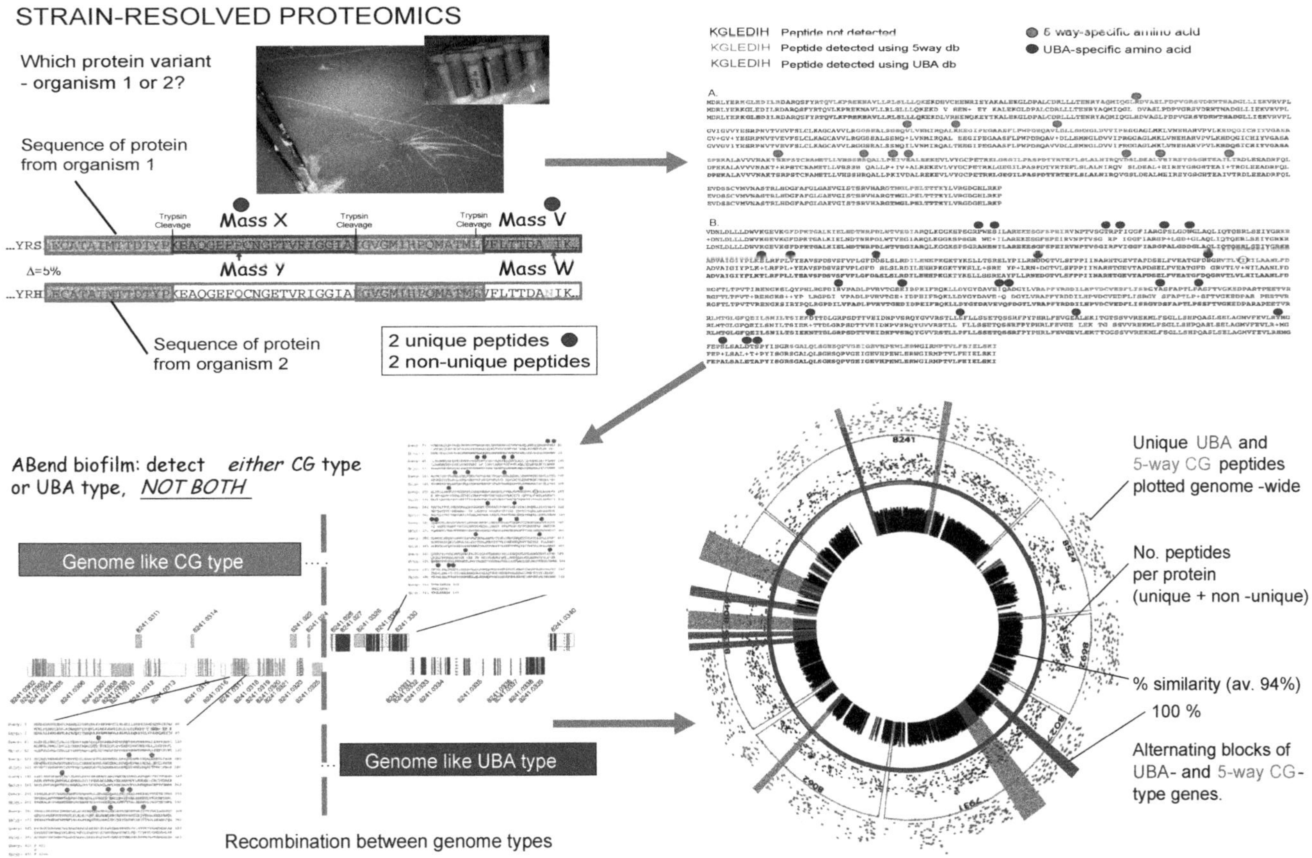

Figure 3.5 The PIGT (Proteomics-inferred genome typing) method. Proteomics is capable of strain-typing sampled bacteria by matching detected peptides to those belonging to a given variant or variants. Additionally, the genetic alignment of these variant peptides onto a genetic map can give information on genetic exchange including recombination between strains and phage-derived integration. First, peptides with unique sequences are identified between two strains from genomics-predicted sequences (top left panel). Next, mass spectra for peptides unique to one strain are matched to a database from one of the two strains (top right panel). Typing of unique peptides is related to parent proteins, which can then be mapped to the genome (lower left panel). Evidence for recombination events can be inferred from alternating blocks of proteins belonging to one strain type (bottom right panel). Adapted from Lo *et al*., 2007. A colour version of this figure is located in the plate section at the back of the book.

under way. Initial studies by Wilmes and Bond (2006b) applied 2D-PAGE to determine reference maps for anaerobic and aerobic phases and combined these data with fluorescence *in situ* hybridization (FISH) analyses. This study was followed by a more detailed study combining 2D-PAGE and MS to identify the proteins collected from the system against the recently released metagenome (García *et al.*, 2006), resulting in 46 protein identifications from 111 excised spots (Wilmes and Bond, 2008a). While this study showed potential, the primarily limitation was lack of dynamic range. Thus, similar samples from aerobic and anaerobic activated sludge communities were prepared and analysed by shotgun proteomics methods comparable to those discussed above for the AMD system (Wilmes *et al.*, 2008b). The goal of this study was not simply protein identification but also strain variation and quantitation via methods similar to Lo *et al.* (2007). Reference genomes from US and Australian samples (García *et al.*, 2006) were compared to the peptide MS/MS spectra from a UK sample. In total, over 5000 protein identifications were made, but, more importantly, 702 proteins from the dominant *A. phosphatis* population were identified with substantial differences in protein abundance for enzyme variants involved in both core metabolism and enhanced biological phosphorus removal (EPBR)-specific pathways. This study illustrates that shotgun MS techniques can be used to characterize systems with higher species diversity than the AMD system, but also provides insight into how strain variation in important biochemical pathways can 'fine-tune' the system for its current environmental niche.

Sargasso Sea

Moving up the scale of complexity, a recent study by Sowell *et al.* (2008) investigated the metaproteomes of the dominant members of the Sargasso Sea (see Chapter 8). The α-proteobacterial SAR11 clade, *Prochlorococcus*, and *Synechococcus* were analysed during a time of nutrient depletion. The annual cycle of the Sargasso Sea consists of a period of oligotrophy, an extreme condition in which there are large amounts of dissolved oxygen and a low abundance of plant nutrients, in particular phosphorus and nitrogen during the fall, followed by upwelling in the spring. Of particular interest is how these organisms adapt, compete, and flourish under the fluctuating and oftentimes harsh conditions when limited resources are available. This study focused on the measurement of proteins being expressed by the SAR11 clade during this period of oligotrophy and also measured the proteomes of *Prochlorococcus* and *Synechococcus*, which are present but not as abundant in the environment during this season.

The experimental approach used capillary liquid chromatography coupled with tandem mass spectrometry. Surface water samples were collected in the fall in order to measure microbial proteins during a time of limited nutrient availability. The proteins were tryptically digested, and the resulting peptides identified by spectral matches against databases containing genomes of SAR11, *Prochlorococcus*, and *Synechococcus* using the SEQUEST search algorithm. In this case, 239 SAR 11 proteins were identified (by 2279 unique peptides), as well as 402 *Prochlorococcus* and 404 *Synechococcus* proteins. The key finding in this study is the detection of large amounts of transporter proteins, especially in the SAR11 dataset, in which transporters comprised 67% of the total spectra. This is in relation to *in vitro* culture of SAR11, which only consists of 28–35% transport proteins. Specifically, proportionately high numbers of periplasmic substrate-binding proteins were expressed by the SAR11 cells, including those specific for acquiring phosphorus (PstS, PstB, PstA, PhoU, PhoR, PhoB) and phosphonate (PhnD, PhnE). These subunits of ABC transport systems are known to bind substrates with high affinity and specificity. The presence of components of these systems provides evidence supporting previous studies showing that SAR11 can uptake inorganic phosphate and is doing so in the environment. This is hypothesized to be a key reason this clade has evolved to become one of the dominant members in the nutrient-depleted area of the sea. In addition, proteins known to transport amino acids and sugars (DctP, PotD) as well as proteins involved in ammonium and reduced sulphur compound uptake (YhdWXYZ, OpuA) were present.

As expected, proteins from the two cyanobacteria were not as abundant as those from SAR11, but the proportion of transporter-associated proteins was again elevated. Also, as with SAR11, proteins associated with phosphate and phosphonate uptake were present, as well as those for carbon fixation, photosynthesis, and pigment proteins (chlorophyll and carotenoid for *Prochlorococcus* and phycobilisome for *Synechococcus*). According to the authors, these microbes need high-affinity uptake systems aimed at various substrates, which were found to be expressed.

Human microbiome

Another example of a complex community that was recently assessed using a metaproteomics approach is the human gut microbiota (VerBerkmoes *et al.*, 2008). The human microbiome has recently garnered great interest as it has become apparent that interactions of microbes and human cells are essential for normal human health. Recently, there has been greater effort to obtain metagenomic information from many different communities inhabiting the human body including the gastrointestinal tract (GI) (Gill *et al.*, 2006; Kurokawa *et al.*, 2007), mouth (Warburton *et al.*, 2008), skin, and vagina (see Chapters 2 and 9). With the completion of metagenomes, the possibility of proteomics studies became more tractable, and the metaproteome of the human gut microbiota was recently analysed (VerBerkmoes *et al.*, 2008). Working from predicted metagenome information, the expressed proteins from active members of the GI microbial community were measured by shotgun proteomics. The authors used existing metagenome sequence data (Gill *et al.*, 2006) as a reference database, as well as many known gut isolate genomes, a human protein database, and a list of distracters not associated with GI tract microbial communities. The samples collected for the VerBerkmoes *et al.* (2008) study were two healthy Swedish twins, while the metagenomes were from two Americans, so there were no direct matches between the metagenome and metaproteome datasets. As a representation of the colonic microbiota, faecal samples were analysed from the identical twins. The complex nature of this sample type, along with the numerous amounts of microbes in the sample (~10^{11} microbes/gram of faeces), yielded the most complex sample type analysed to date and offered many technical challenges to be overcome both experimentally and computationally.

Even with the lack of direct matches present in the database, over 2000 proteins were identified, many of which matched more closely to the Gill *et al.* (2006) metagenomes than the isolate databases. The statistical quality of the data (as estimated by false positive rates) indicates that searching proteome data with unmatched genomes is feasible, an important point that can be useful for future studies. In spite of these challenges, thousands of proteins were identified providing key insights to the functions of these microbes within the community. In comparison to the predicted metagenome, the distribution of expressed proteins was marked by a greater number of proteins involved in carbohydrate metabolism, translation, and energy production being expressed. In addition, human antimicrobial proteins and others human proteins involved in innate immunity were seen, providing a glimpse into the microbial–human interactions. This study laid the foundation for future work by illustrating that technology is available to make such measurements, even in such complex systems as the human gut (see Chapter 9).

Challenges in community proteomics

To date, most studies have focused on systems of minimal complexity, but the move to more complex systems, such as aquatic sediment and soil samples, is clearly on the horizon. One key to such studies will be the make up of the complexity. Are there a few dominant organism types with hundreds to thousands of low percentage components (i.e. a low evenness), or is the entire system roughly equally populated with most species at less than 1% (i.e. a high evenness)? Are there closely related reference metagenomes available, or is it essentially a genomically uncharacterized system? The answers to these questions will predict the likelihood of success with current metaproteomics techniques. Clearly, samples that are extremely complex but have a few dominant members will be relatively straightforward with current techniques for the dominant members but very challenging for the members that are more rare. True *in situ* analyses of highly diverse microbial systems that do not contain dominant members (all organisms at less than 1%) may be very difficult with current technologies without the use of enrichment techniques.

Currently, the clearest challenge is the availability of reference metagenomes with close similarities to the system of interest. Each of the microbial community studies discussed above used some form of reference metagenome that was not identical to the sample being analysed. Both the power and Achilles' heal of shotgun proteomics is the accuracy at which the mass and sequence information (MS/MS) of peptides are measured. A single amino acid change in a peptide (except Ile-Leu) prevents correctly identifying the peptide. The effects of genetic variation on peptide identification via MS/MS techniques was thoroughly studied for both isolate and community datasets (Denef *et al.*, 2007). From this study, a range of computational models set the boundaries at which half of the proteins in a proteomic experiment can be identified to 77–92% amino acid identity (AAI) between orthologues in the sample and database. Experimental data indicated loss of half of the identifiable proteins at 90% AAI. The results from this study suggested that shotgun proteomics is capable of cross-strain identifications but avoids most cross-species false positives.

Another serious challenge in environmental proteomics is the efficient and non-biased extraction of proteins from complex environmental matrices. In the initial studies of the AMD biofilm community, this was a clear focus. Researchers compared lysis and extraction methods via buffers of varying pH, as well as sonication versus other methods of lysis. While clear advantages in total number of protein identifications were found with some methods tested, biases were also found with other methods. These biases can sometimes be advantageous, as long as they are understood and acknowledged. For example, low level community members can sometimes be 'enriched' via biased lysis methods while still maintaining true *in situ* analyses.

The most difficult environmental matrices will be soil and sediment samples. These are potentially the most complex communities studied to date and have the greatest diversity from location to location, making the use of reference genomes less than practical. However, as most initial proteomics studies of soil have found (Schulze *et al.*, 2005; Benndorf *et al.*, 2007), the largest challenge will be the efficient lysis of microbial cells and the extraction of proteins. Soils and sediments are full of contaminating materials that adversely affect the typical proteomics methodology; furthermore, they can act as a mixed bed chromatographic column, sometimes irreversibly binding peptides and proteins. While a few studies have been undertaken in soil microcosms, only a handful of proteins have been identified or quantified (Schulze *et al.*, 2005; Benndorf *et al.*, 2007). Most studies have focused primarily on method development rather than biological insight into the system. Clearly, soils and sediments are the grand technical challenge for the emerging field of environmental microbial community proteomics.

Box 3.3 highlights some of the technical challenges for community proteomics and some possible analytical solutions on the horizon. As was clearly described by Wilmes and Bond (2006a), the greatest technical challenge in complex environmental proteomics will be dynamic range. Numerous scenarios can cause large dynamic range issues that will be difficult to overcome with current standard proteomics technologies, but the future should be bright. Over the last five years, standard proteomics LC-MS systems have increased dynamic range by 2–3 orders of magnitude (Fig. 3.2). This has resulted from the implementation of advanced chromatographic methods coupled with new generation trapping instruments that can hold much higher quantities of ions at any given point in time (Schwartz *et al.*, 2002; Hu *et al.*, 2005). The application of up-front rapid separation of cell types would greatly increase dynamic range but would have to be weighed against the stress effects on the individual proteomes over the time of enrichment. These methods move away from the pure *in situ* goal but provide the simplest way to increase dynamic range.

The availability of solid genomic data is key to community proteomics success. The best case scenario would always be a metagenome and metaproteome from the exact same sample, but usually that is impractical. As shown in the literature and discussed above, similar but distinct samples can be used for a metagenome

Box 3.3 Microbial community proteomics needs

1. Dynamic range (deeper and wider measurements).
2. Solid genomic foundation (accurately curated metagenomes).
3. Quantitative methods.
4. Accurate peptide identifications.
5. Characterization of strain variation at peptide and protein level.
6. Advanced sequence tagging/*de novo* sequencing techniques.
7. Miniaturization for fine scale resolution.

Potential MS-analytical solutions to address these needs

1. High-throughput LC-MS systems with sophisticated chromatographic separations.
2. High resolution and accurate mass measurements of both full scan and MS/MS spectra on liquid chromatography time scales.
3. Improvements in the collection of label-free quantitative data.
4. Alternative methods of MS/MS dissociation on LC time frames.
5. Examination of both peptides and intact proteins from the same sample.
6. Chip-based front ends that can perform cell lyses, protein digestions, peptide separation, and nanospray, all on one integrated chip.

and proteome and high quality results can still be obtained. Additionally, compiling all available genomes into a search database is impractical, due to search time considerations (even in this age of relatively cheap processing power, searching overly large databases is problematic) and an increase in false discovery rate (particularly as the number of proteins from species that are not present in a given sample increase beyond those actually present). One critical concept, though, is the quality of genome assembly and annotation. Quality proteomics results are heavily dependent on database quality and design. With the advent of new DNA sequencing technologies (reviewed in Holt and Jones, 2008; see Chapter 2), large quantities of DNA sequence can be acquired, but the smaller read lengths and qualities of the final assembled products are a concern for other functional genomics studies such as proteomics. The integration of new sequencing technologies with high-throughput proteomics technologies will therefore be necessary.

The field's second most pressing need is accurate quantitation of a wide diversity of proteins across a large dynamic range. This is even a challenge in isolate studies, but quantitation in microbial communities provides even greater dilemmas. In a simple binary comparison of two communities that are temporally resolved, not only are proteins changing in abundances within cells but the cell density of any given species is changing. If one cannot account for the approximate quantities of each species type, then the measurement only represents the global difference in quantity of the protein in the sample, not inside any given species. The clearest solution to this problem is the coupling of proteomics data with other measurements that give species/organism abundances such as FISH or rRNA profiling.

Accurate peptide identifications have always been the heart of shotgun or bottom-up proteomics methods. Typical 2D-LC-MS/MS experiments can create tens of thousands of spectra per day, many of which can be attributed to chemical noise. The search algorithms produce identifications for every spectra, thus filtering methods must be used to reduce false positives. The balance of false positives and false negatives is a complex process, with many computational techniques aimed at answering this problem. Minimizing false discovery rates is paramount to any quality published proteome dataset and has been reviewed thoroughly (Elias and Gygi, 2007; Nesvizhskii *et al.*, 2007; Tabb, 2008). In metaproteomics, where the proteome rarely matches the genome, the problem is greatly increased. If care is not taken in careful database construction, algorithm analyses, and data filtering, the results can be dominated by truly random or spurious identifications. There is a large effort to solve this problem through purely informatic techniques, but recent advances in MS instrumentation have changed the landscape. Original shotgun proteomics datasets were dominated by spectra collected on low resolution, fast-scanning mass spectrometers. In the last couple years, new hybrid mass spectrometers have been developed and commercialized, which combine high mass accuracy/high resolution data acquisition with rapid scanning MS systems. Specifically, the hybrid linear ion trap Orbitrap (Hu *et al.*, 2005) has become one of the instruments of choice for shotgun proteomics. As mentioned above, this instrument provides rapid, high resolution MS and MS/MS capabilities with a robust, simple MS system that operates with automated on-line LC systems. With this new instrument, it is possible to measure peptides masses at 1–10 p.p.m. mass accuracy (accuracy at the millidalton), as well as perform high resolution or low resolution MS/MS scans. The ability to accurately measure peptides masses is critical to keep false discovery rates low in very complex samples searched against large genomic databases. This next generation of hybrid, high resolution mass spectrometers will be critical for continued advances in metaproteomes of complex communities.

High mass accuracy will also be necessary when trying to *de novo* sequence or sequence tag peptides. Very often in complex and evolving communities, changes will occur between the original genomes and observed proteomes, ranging from simple amino acid variations to insertion of new genes. In either case, methods will be needed to identify peptides and thus proteins not present in the original database and, without high mass accuracy, the level of false positives with such analyses can be high. The development of *de novo*/sequence tagging techniques, as discussed above, will be critical for metaproteomics and will greatly benefit from new high resolution MS systems and new alternative methods for fragmenting peptides.

A final, grand challenge for metaproteomics will be the analyses of very finely spatially resolved environmental samples. Current techniques require bulk amounts of material for standard lysis and fractionation (hundreds of milligrams of wet cellular material) and 1–10 milligrams of wet cellular material for small-scale single-tube lysis (fractionation is not possible with these techniques). Clearly, reducing the material required by an order of magnitude would be advantageous for numerous studies. To achieve such gains in sensitivity, it would be necessary to abandon typical methods of lysis in microfuge tubes and chromatography in low flow columns. Instead, the entire process from cell lysis, protein digestion, and peptide separation would need to be accomplished on chip-based platforms directly connected to MS systems. While this is not likely to develop in the next few years, there is a large amount of fundamental research in this area that brings promise of commercially available robust chip-based platforms in the future.

Conclusions

Proteomics is a powerful and versatile tool that is currently emerging as a new approach that may play a significant role in unlocking the answers to long intractable mysteries in microbial ecology. While proteomics is largely a discovery-based approach that should be followed by other molecular or analytical techniques to provide a full picture of the subject system, it is not limited to simple cataloguing of proteins present in a sample. With the combined capabilities of 'top-down' and 'bottom-up' approaches, proteomics can pursue inquiries ranging from quantitation of protein expression between growth conditions (whether nutritional, spatial, temporal, or chemical) to protein structural information.

The power and utility of proteomics has grown alongside technological progress. The ability to ionize proteins/peptides via MALDI and ES, the development of rapid scanning tandem mass spectrometers along with informatics pipelines to handle the data, and the advent of low flow single- and multi-dimensional HPLC systems have allowed mass spectrometry to approach systems-level questions in proteomics. Several studies have laid the groundwork for using proteomics in model systems selected due to their limited biodiversity, including an acid mine drainage system (Ram *et al.*, 2005), an activated sludge community (Wilmes and Bond, 2006a,b), and the Sargasso Sea during oligotrophic conditions (Sowell *et al.*, 2008).

Despite these early successes, the future fate of the field of proteomics in ecological microbiology lies in overcoming a number of limitations and challenges. Some of these are unique to environmental microbiology studies, while others are shared with other applications of proteomics, such as human health, and thus can benefit from developments in those areas. As has been discussed, *in situ* measurements are an important goal, and unique and creative developments will be needed in community proteogenomics to properly handle samples derived from diverse origins such as soil, groundwater, oceans, and even the human microbiome. Development of statistical and bioinformatic approaches to quantitation using label as well as label-free methodology must be improved to allow greater confidence in conclusions and the adoption of universal standards across proteomics.

Improvements in dynamic range and sensitivity of proteomic measurements are paramount. These improvements will allow the move from select 'simple' microbial systems that are the exception rather than the rule, to other systems of widely ranging levels of microbial diversity. Achieving increased dynamic range will likely require the use of novel multi-dimensional (on- and off-line) separation techniques (at the cost of longer run times per sample) and/or even faster scanning mass spectrometers to sufficiently characterize more complex microbial communities. Increased sensitivity to delve deeper into the proteome and to obtain quality data from minimal sample sizes is also critical. The most promising avenue for improving sensitivity involves chip-based microfluidics for all sample preparation steps through ionization.

Finally, since a genomics-derived protein database underpins downstream protein identification, improved results can be obtained by improving the quality and scope of databases and/or reducing reliance on databases through *de novo* approaches. High-throughput genome sequencing has enabled the move from reference genomes derived from isolates to construction of metagenomes directly from the same sample to be analysed by proteomics. Also, development of *de novo* bioinformatic techniques promises to supplant total reliance on databases, minimizing issues arising from strain variation.

Future trends

Targeted nanoscale proteomics for ecological monitoring

Traditionally in microbial ecology, reverse transcription of mRNA followed by PCR amplification is the method of choice for monitoring expression of biomarkers across time and space directly from environmental samples. This approach is not without weaknesses including (1) the rapid turnover of mRNA requires rapid and careful sample handling, which is not a trivial concern in the field; (2) the general instability of mRNA, which limits the technique in some hostile environments; and (3) mRNA levels often do not directly correlate with protein expression. If monitoring of environmental biomarkers can be done at the protein level instead of the mRNA level, the three main problems above are eliminated, and, furthermore, the measurement is addressing the actual functional units of interest, the proteins.

The generation and use of antibodies for detecting biomarkers is one option for analysing protein content in a sample, but this also has some shortcoming including the need to generate quality antibodies, the inability of antibodies to detect small changes in protein sequence, and cross-reactivity between antibodies and non-target background proteins. Recently, new methodologies have emerged in targeted mass spectrometry-based proteomics and have become more widespread especially in the medical and pharmaceutical fields, where analyses of biomarkers from hundreds to thousands of samples is paramount (Anderson and LaBaer, 2005). Targeted proteomics focuses on the selective identification and quantitation of a protein or proteins of interest from an array of proteins. These methodologies may have the ability to overcome the shortcomings of PCR and antibody-based approaches, as well as provide accurate quantitation of hundreds of biomarkers per hour; targeted proteomics provides a considerable improvement in precision (Kuhn *et al.*, 2004). Furthermore, targeted proteomics addresses one of the weaknesses of MS, its inability to detect low abundance proteins in a complex sample, due to the presence of high abundance proteins (Anderson and LaBaer, 2005).

In developing a targeted proteomics approach, studies should initially focus on method development to plot where peptide biomarkers elute from in the chromatography system. For targeted proteomics, an elution tag and a mass tag are needed for every biomarker. To maintain elution profiles, the same LC and method should be used on all MS systems. A triple quadrupole mass spectrometer is an ideal instrument for monitoring large numbers of biomarkers from complex matrices and provides unparalleled quantitation. The method of choice for this is selected reaction monitoring (SRM); SRM can trigger normal MS/MS, thus peptide sequences can then be confirmed by MS/MS, validating as well as quantifying the biomarker targets.

Protein-SIP

The use of stable isotope labelling for the purpose of relative quantitation has become routine in mass spectrometry studies via comparison of peak intensities between the heavy and light versions of a given peptide. Isotope labelling can also be used to provide information linking metabolic activities with individual species within a mixed bacterial culture by a technique termed protein-based stable isotope probing (protein-SIP). Following culture with ^{13}C- or ^{15}N-labelled substrates, incorporation of these labels into proteins can be traced to specific species within the culture, thus providing insight into functional relationships and role partitioning within the microbial community. Specifically, incorporation of a given labelled substrate indicates a given species is capable of utilization of that substrate, and, subsequently, incorporation into proteins belonging to other species can be tracked to elucidate the flow of carbon within a microbial system.

Jehmlich *et al.* (2008) demonstrated the efficacy of this approach in a co-culture experiment with the aromatic-metabolizing *Aromatoleum aromaticum* grown in a mixed culture with an enriched pond sludge sample; the latter was unable to utilize toluene as a carbon source. ^{13}C-labelled toluene was added to the mixed culture and mass spectrometry analysis of 2D-PAGE-separated proteins identified a mass shift attributable to ^{13}C in only *A. aromaticum* proteins.

While DNA (Friedrich, 2006) and RNA-based (Manefield *et al.*, 2002; Whiteley *et al.*, 2006) approaches to SIP have been used previously (see Chapter 4), these methods rely on differential ultracentrifugation to separate the labelled and unlabelled species. Because protein-SIP utilizes mass spectrometry to detect mass shifts, detection of labelled species is considerably more sensitive. While DNA/RNA-based SIP requires ~30% labelling, 1–2% labelling is sufficient for protein-SIP. Neither protein nor nucleic acid-based SIP is capable of true *in situ* analysis due to the problems inherent in providing a labelled substrate; however, representative microcosms can substitute.

Acknowledgments

Funding was provided by the DOE Genomics: GTL Program. Oak Ridge National Laboratory is managed by University of Tennessee-Battelle LLC for the Department of Energy under contract DOE-AC05-00OR22725.

References

Anderson, K.S., and LaBaer, J. (2005). The sentinel within: exploiting the immune system for cancer biomarkers. J. Proteome Res. *4*, 1123–33.

Baker, B.J., and Banfield, J.F. (2003). Microbial communities in acid mine drainage. FEMS Microbiol. Ecol. *44*, 139–152.

Banfield, J.F., VerBerkmoes, N.C., Hettich, R L., and Thelen, M.P. (2005). Proteogenomic approaches for the molecular characterization of natural microbial communities. OMICS 9, 301–333.

Benndorf, D., Balcke, G.U., Harms, H., and von Bergen, M. (2007). Functional metaproteome analysis of protein extracts from contaminated soil and groundwater. ISME J. *1*, 224–234.

Biemann, K. (1986). Mass spectrometric methods for protein sequencing. Anal. Chem. *58*, 1288A-1300A.

Biemann, K. (1988). Contributions of mass spectrometry to peptide and protein structure. Biomed. Environ. Mass Spectrom. *16*, 99–111.

Binnewies, T.T., Motro, Y., Hallin, P.F., Lund, O., Dunn, D., La, T., Hampson, D.J., Bellgard, M., Wassenaar, T.M., and Ussery, D.W. (2006). Ten years of bacterial genome sequencing: comparative-genomics-based discoveries. Funct. Integr. Genomics *6*, 165–185.

Brown, S.D., Thompson, M.R., Verberkmoes, N.C., Chourey, K., Shah, M., Zhou, J., Hettich, R.L., and Thompson, D.K. (2006). Molecular dynamics of the *Shewanella oneidensis* response to chromate stress. Mol. Cell Proteomics 5, 1054–1071.

Callister, S.J., Dominguez, M.A., Nicora, C.D., Zeng, X., Tavano, C.L., Kaplan, S., Donohue, T.J., Smith, R.D., and Lipton, M.S. (2006). Application of the accurate mass and time tag approach to the proteome analysis of sub-cellular fractions obtained from *Rhodobacter sphaeroides* 2.4.1. Aerobic and photosynthetic cell cultures. J. Proteome Res. *5*, 1940–1947.

Chourey, K., Thompson, M.R., Morrell-Falvey, J., Verberkmoes, N.C., Brown, S.D., Shah, M., Zhou, J., Doktycz, M., Hettich, R.L., and Thompson, D.K. (2006). Global molecular and morphological effects of 24-h chromium exposure on *Shewanella oneidensis* MR-1. Appl. Environ. Microbiol. *72*, 6331–6344.

Corbin, R.W., Paliy, O., Yang, F., Shabanowitz, J., Platt, M., Lyons, C.E. Jr, Root, K., McAuliffe, J., Jordan, M.I., Kustu, S., Soupene, E., and Hunt, D.F. (2003). Toward a protein profile of *Escherichia coli*: comparison to its transcription profile. Proc. Natl. Acad. Sci. U.S.A. *100*, 9232–9237.

Craig, R., and Beavis, R.C. (2004). TANDEM: matching proteins with mass spectra, Bioinformatics *20*, 1466–1467.

DeLong, E.F. (1997). Marine microbial diversity: the tip of the iceberg. Trends Biotechnol. *15*, 203–207.

DeLong, E.F. (2005). Microbial community genomics in the ocean. Nat. Rev. Microbiol. *3*, 459–69.

Denef, V.J., Shah, M.B., VerBerkmoes, N.C., Hettich, R.L., and Banfield, J.F. (2007). Implications of strain- and species-level sequence divergence for community and isolate shotgun proteomic analysis. J. Proteome Res. *6*, 3152–3161.

Denef, V.J., Verberkmoes, N.C., Shah, M.B., Abraham, P., Lefsrud, M., Hettich, R.L., and Banfield, J.F. (2008). Proteomics-inferred genome typing (PIGT) demonstrates interpopulation recombination as a strategy for environmental adaptation. Environ Microbiol. *11*, 3113–3125.

Elias, J.E., and Gygi, S.P. (2007). Target-decoy search strategy for increased confidence in large-scale protein identifications by mass spectrometry. Nat. Methods, *4*, 207–214.

Eng, J.K., McCormack, A.L., and Yates, III, J.R. (1994). An approach to correlate tandem mass spectral data of peptides with amino acid sequences in a protein database. J. Am. Scoc. Mass Spectrom. *5*, 976–989.

Fenn, J.B., Mann, M., Meng, C.K., Wong, S.F., and Whitehouse, C.M. (1989). Electrospray ionization for mass spectrometry of large biomolecules. Science *246*, 64–71.

Florens, L., Carozza, M.J., Swanson, S.K., Fournier, M., Coleman, M.K., Workman, J.L., and Washburn, M.P. (2006). Analyzing chromatin remodeling complexes using shotgun proteomics and normalized spectral abundance factors. Methods *40*, 303–311.

Frank, A., and Pevzner, P. (2005). PepNovo: *de novo* peptide sequencing via probabilistic network modeling. Anal. Chem. *77*, 964–973.

Friedrich, M.W. (2006). Stable-isotope probing of DNA: insights into the function of uncultivated microorganisms from isotopically labeled metagenomes. Curr. Opin. Biotechnol. *17*, 59–66.

García, M.H., Ivanova, N., Kunin, V., Warnecke, F., Barry, K.W., McHardy, A.C., Yeates, C., He, S., Salamov, A.A., Szeto, E., *et al* (2006). Metagenomic analysis of two enhanced biological phosphorus removal (EBPR) sludge communities. Nat. Biotechnol. *24*, 1263–1269.

Gill, S.R., Pop, M., Deboy, R.T., Eckburg, P.B., Turnbaugh, P.J., Samuel, B.S., Gordon, J.I., Relman, D.A., Fraser-Liggett, C.M., and Nelson, K.E. (2006). Metagenomic analysis of the human distal gut microbiome. Science *312*, 1355–1359.

Hillenkamp, F., Karas, M., Beavis, R.C., and Chait, B.T. (1991). Matrix-assisted laser desorption/ionization mass spectrometry of biopolymers. Anal. Chem. *63*, 1193A-1203A.

Holt, R.A., and Jones, S.J. (2008). The new paradigm of flow cell sequencing. Genome Res. *18*, 839–846.

Hu, Q., Noll, R.J., Li, H., Makarov, A., Hardman, M., and Cooks, R.G. (2005). The Orbitrap: a new mass spectrometer. J. Mass Spectrom. *40*, 430–443.

Hunt, D.F., Buko, A.M., Ballard, J.M., Shabanowitz, J., and Giordani, A.B. (1981). Sequence analysis of polypeptides by collision activated dissociation on a triple quadrupole mass spectrometer. Biomed. Mass Spectrom. *8*, 397–408.

Hunt, D.F., Yates III, J.R., Shabanowitz, J., Winston, S., and Hauer, C.R. (1986). Protein sequencing by tandem mass spectrometry. Proc. Natl. Acad. Sci. U.S.A. *83*, 6233–6237.

Ishihama, Y., Oda, Y., Tabata, T., Sato, T., Nagasu, T., Rappsilber, J., and Mann, M. (2005). Exponentially modified protein abundance index (emPAI) for estimation of absolute protein amount in proteomics by the number of sequenced peptides per protein. Mol. Cell Proteomics *4*, 1265–1272.

Jeans C., Singer, S.W., Chan, C.S., Verberkmoes, N.C., Shah, M., Hettich, R.L., Banfield, J.F., and Thelen, M.P. (2008). Cytochrome 572 is a conspicuous membrane protein with iron oxidation activity purified directly from a natural acidophilic microbial community. ISME *2*, 542–550.

Jehmlich, N., Schmidt, F., von Bergen, M., Richnow, H.H., and Vogt, C. (2008). Protein-based stable isotope probing (Protein-SIP) reveals active species within anoxic mixed cultures. ISME *2*, 1122–1133.

Kent, A.D., and Triplett, E.W. (2002). Microbial communities and their interactions in soil and rhizosphere ecosystems. Annu. Rev. Microbiol. *56*, 211–36.

Klose, J. (1975). Protein mapping by combined isoelectric focusing and electrophoresis of mouse tissues. A novel approach to testing for induced point mutations in mammals. Humangenetik *26*, 231–243.

Kuhn, E., Wu, J., Karl, J., Liao, H., Zolg, W., and Guild, B. (2004). Quantification of C-reactive protein in the serum of patients with rheumatoid arthritis using multiple reaction monitoring mass spectrometry and ^{13}C-labeled peptide standards. Proteomics *4*, 1175–1186.

Kurokawa, K., Itoh, T., Kuwahara, T., Oshima, K., Toh, H., Toyoda, A., Takami, H., Morita, H., Sharma, V.K., Srivastava, T.P., *et al.* (2007). Comparative metagenomics revealed commonly enriched gene sets in human gut microbiomes. DNA Res. *14*, 169–181.

Lipton, M.S., Pasa-Tolic', L., Anderson, G.A., Anderson, D.J., Auberry, D.L., Battista, J.R., Daly, M.J., Fredrickson, J., Hixson, K.K., Kostandarithes, H., *et al.* (2002). Global analysis of *Deinococcus* radiodurans proteome by using accurate mass tags. Proc. Natl. Acad. Sci. U.S.A. *99*, 11049–11054.

Liu, H., Lin. D., and Yates III, J.R. (2002). Multidimensional separations for protein/peptide analysis in the post-genomic era. Biotechnology *32*, 898–902.

Liu, H., Sadygov, R.G., and Yates III, J.R. (2004). A model for random sampling and estimation of relative protein abundance in shotgun proteomics. Anal. Chem. *76*, 4193–4201.

Lo, I., Denef, V.J., Verberkmoes, N.C., Shah, M.B., Goltsman, D., DiBartolo, G., Tyson, G.W., Allen, E.E., Ram, R.J., Detter, J.C., *et al* (2007). Strain-resolved community proteomics reveals recombining genomes of acidophilic bacteria. Nature *446*, 537–541.

Lu, P., Vogel, C., Wang, R., Yao, X., and Marcotte, E.M. (2007). Absolute protein expression profiling estimates the relative contributions of transcriptional and translational regulation. Nat. Biotechnol. *25*, 117–1124.

Luo, Q., Hixson, K.K., Callister, S.J., Lipton, M.S., Morris, B.E., and Krumholz, L.R. (2007). Proteome analysis of *Desulfovibrio desulfuricans* G20 mutants using the accurate mass and time (AMT) tag approach. J Proteome Res. *8*, 3042–3053.

Ma, B., Zhang, K., Hendrie, C., Liang, C., Li, M., Doherty-Kirby, A., and Laioie, G. (2003). PEAKS: powerful software for peptide *de novo* sequencing by tandem mass spectrometry. Rapid Commun. Mass Spectrom. *17*, 2337–2342.

Manefield, M., Whiteley, A.S., Griffiths, R.I., and Bailey, M.J. (2002). RNA stable isotope probing, a novel means of linking microbial community function to phylogeny. Appl. Environ. Microbiol. *68*, 5367–5373.

Mann, M., Hendrickson, R.C., and Pandey, A. (2001). Analysis of proteins and proteomes by mass spectrometry. Annu. Rev. Biochem. *70*, 437–473.

McCormack, A.L., Schieltz, D.M., Goode, B., Yang, S., Barnes, G., Drubin, D., and Yates III, J.R. (1997). Direct analysis and identification of proteins in mixtures by LC-MS/MS and database searching at the low-femtomole level. Anal. Chem. *69*, 767–776.

McDonald, H., Ohi, R., Miyamoto D.T., Mitchison T.J., and Yates, J.R. (2002). Comparison of three directly coupled HPLC MS/MS strategies for identification of proteins from complex mixtures: single-dimension LC-MS/MS, 2-phase MudPIT, and 3-phase MudPIT. Int. J. Mass Spetrom. *219*, 245–251.

Nakanishi, T., Okamoto, N., Tanaka, K., and Shimizu, A. (1994). Laser desorption time-of-flight mass spectrometric analysis of transferrin precipitated with antiserum: a unique simple method to identify molecular weight variants. Biol. Mass. Spectrom. *23*, 230–233.

Nesvizhskii, A.I., Vitek, O., and Aebersold, R. (2007). Analysis and validation of proteomic data generated by tandem mass spectrometry. Nat. Methods *4*, 787–797.

O'Farrell, P.H. (1974). High resolution two-dimensional electrophoresis of proteins. J. Biol. Chem. *250*, 4007–4021.

Old, W.M., Meyer-Arendt, K., Aveline-Wolf, L., Pierce, K.G., Mendoza, A., Sevinsky, J.R., Resing, K.A., and Ahn, N.G. (2005). Comparison of label-free methods for quantifying human proteins by shotgun proteomics. Mol. Cell Proteomics *4*, 1487–1502.

Ong, S.E., and Mann, M. (2005). Mass spectrometry-based proteomics turns quantitative. Nat. Chem. Methods *1*, 252–262.

Peng, J., and Gygi, S.P. (2001). Proteomics: the move to mixtures. J. Mass Spectrom. *36*, 1083–1091.

Peng, J., Elias, J.E., Thoreen, C.C., Licklider, L.J., and Gygi, S.P. (2003). Evaluation of multi-dimensional chromatography coupled with tandem mass spectrometry (LC/LC-MS/MS) for large-scale protein analysis: The yeast proteome. J. Proteome Res. *2*, 43–50.

Perkins, D.N., Pappin, D.J., Creasy, D.M., and Cottrell, J.S. (1999). Probability-based protein identification by searching sequence databases using mass spectrometry data. Electrophoresis *20*, 3551–3567.

Ram R.J., Verberkmoes, N.C., Thelen, M.P., Tyson, G.W., Baker, B.J., Blake, II, R.C., Shah, M., Hettich, R.L., and Banfield, J.F. (2005). Community proteomics of a natural microbial biofilm. Science *308*, 1915–1920.

Sadygov, R.G., Cociorva, D., and Yates, J.R. 3rd. (2004). Large-scale database searching using tandem mass spectra: looking up the answer in the back of the book. Nat. Methods *3*, 95–202.

Schwartz, J.C., Senko, M., and Syka, J.E.P. (2002). A two-dimensional quadrupole ion trap mass spectrometer. J. Am. Soc. Mass Spectrom. *13*, 659–669.

Schulze, W.X., Gleixner, G., Kaiser, K., Guggenberger, G., Mann, M., and Schulze, E.D. (2005). A proteomics fingerprint of dissolved organic carbon and of soil particles. Oecologia *142*, 335–343.

Shevchenko, A., Jensen, O.N., Podtelejnikov, A.V., Sagliocco, F., Wilm, M., Vorm, O., Mortensen, P., Shevchenko, A., Boucherie, H., and Mann, M. (1996). Linking genome and proteome by mass spectrometry: large-scale identification of yeast proteins from two dimensional gels. Proc. Natl. Acad. Sci. U.S.A. *93*, 14440–14445.

Singer, S.W., Chan, C.S., Zemla, A., VerBerkmoes, N.C., Hwang, M., Hettich, R.L., Banfield, J.F., and Thelen, M.P. (2008). Characterization of Cytochrome 579, an unusual cytochrome isolated from an iron oxidizing microbial community. Appl. Environ. Microbiol. *74*, 4454–4462.

Sowell, S.M., Wilhelm, L.J., Norbeck, A.D., Lipton, M.S., Nicora, C.D., Barofsky, D.F., Carlson, C.A., Smith, R.D., and Giovanonni, S.J. (2008). Transport functions dominate the SAR11 metaproteome at low-nutrient extremes in the Sargasso Sea. ISME J. *3*, 93–105.

Strader, M.B., Verberkmoes, N.C., Tabb, D.L., Connelly, H.M., Barton, J.W., Bruce, B.D., Pelletier, D.A., Davison, B.H., Hettich, R.L., Larimer, F.W., and Hurst, G.B. (2004). Characterization of the 70S Ribosome from *Rhodopseudomonas palustris* using an integrated 'top-down' and 'bottom-up' mass spectrometric approach. J. Proteome Res. *5*, 965–78.

Tabb, D.L. (2008). What's driving false discovery rates? J. Proteome Res. *7*, 45–46.

Tabb, D.L., McDonald, W.H., and Yates, J.R. III (2002). DTASelect and contrast: tools for assembling and comparing protein identifications from shotgun proteomics. J. Proteome Res. *1*, 21–26.

Thompson, M.R., VerBerkmoes, N.C., Chourey, K., Shah, M., Thompson, D.K., and Hettich, R.L. (2007). Dosage-dependent proteome response of *Shewanella oneidensis* MR-1 to acute chromate challenge. J. Proteome Res. *6*, 1745–1757.

Thompson, M.R., Chourey, K., Froelich, J.M., Erickson, B.K., Verberkmoes, N.C., and Hettich, R.L. (2008). Experimental approach for deep proteome measurements from small-scale microbial biomass samples. Anal. Chem. *80*, 9517–9525.

Tringe, S.G., von Mering, C., Kobayashi, A., Salamov, A.A., Chen, K., Chang, H.W., Podar, M., Short, J.M., Mathur, E.J., Detter, J.C., Bork, P., Hugenholtz, P., and Rubin, E.M. (2005). Comparative metagenomics of microbial communities. Science *308*, 554–557.

Turnbaugh, P.J., Ley, R.E., Hamady, M., Fraser-Liggett, C.M., Knight, R., and Gordon, J.I. (2007). The human microbiome project. Nature *449*, 804–810.

Tyson, G.W., Chapman, J., Hugenholtz, P., Allen, E.E., Ram, R.J., Richardson, P.M., Solovyev, V.V., Rubin, E.M., Rokhsar, D.S., and Banfield, J.F. (2004). Community structure and metabolism through reconstruction of microbial genomes from the environment. Nature *428*, 37–43.

Venter, J.C., Remington, K., Heidelberg, J.F., Halpern, A.L., Rusch, D., Eisen, J.A., Wu, D., Paulsen, I., Nelson, K.E., Nelson, W., Fouts, *et al.* (2004). Environmental genome shotgun sequencing of the Sargasso Sea. Science *304*, 66–74.

VerBerkmoes, N.C., Bundy, J.L., Hauser, L., Asano, K.G., Razumovskaya, J., Larimer, F., Hettich, R.L., and Stephenson, J.L. Jr (2002). Integrating 'top-down' and 'bottom-up' mass spectrometric approaches for proteomic analysis of *Shewanella oneidensis*. J. Proteome Res. 3, 239–52.

VerBerkmoes, N.C., Connelly, H.M., Pan, C., and Hettich, R.L. (2004). Mass spectrometric approaches for characterizing bacterial proteomes. Expert Rev Proteomics *1*, 433–47.

VerBerkmoes, N.C., Shah, M.B., Lankford, P.K., Pelletier, D.A., Strader, M.B., Tabb, D.L., McDonald, W.H., Barton, J.W., Hurst, G.B., Hauser L., *et al* (2006). Determination and comparison of the baseline proteomes of the versatile microbe *Rhodopseudomonas palustris* under its major metabolic states. J. Proteome Res. *5*, 287–298.

VerBerkmoes, N.C., Russell, A.L., Shah, M., Godzik, A., Rosenquist, M., Halfvarson, J., Lefsrud, M.G., Apajalahti, J., Tysk, C., Hettich, R.L., and Jansson, J.K. (2008). Shotgun metaproteomics of the human distal gut microbiota. ISME J. *3*, 179–189.

Washburn, M.P., Wolters, D.A., Yates III, J.R. (2001). Large-scale analysis of the yeast proteome by multidimensional protein identification technology. Nature Biotechnology *19*, 242–247.

Warburton, P., Roberts, A.P., Allan, E., Seville, L., and Mullany, L.J. (2008). Characterization of the genes from the oral metagenome. Antimicrob Agents Chemother. *53*, 273–276.

Whiteley, A.S., Manefield, M., and Lueders, T. (2006). Unlocking the 'microbial black box' using RNA-based stable isotope probing techniologies. Curr. Opin. Biotechnol. *17*, 67–71.

Wilmes, P., and Bond, P.L. (2006a). Metaproteomics: studying functional gene expression in microbial ecosystems. Trends Microbiol. *14*, 92–97.

Wilmes, P., and Bond, P.L. (2006b). Towards exposure of elusive metabolic mixed-culture processes: the application of metaproteomic analyses to activated sludge. Water Sci. Technol. *54*, 217–226

Wilmes, P., Wexler, M., and Bond, P.L. (2008a). Metaproteomics provides functional insight into activated sludge wastewater treatment. PLoS ONE 3, e1778.

Wilmes, P., Andersson, A.F., Lefsrud, M.G., Wexler, M., Shah, M., Zhang, B., Hettich, R.L., Bond, P.L., VerBerkmoes, N.C., and Banfield, J.F. (2008b). Community proteogenomics highlights microbial strain-variant protein expression within activated sludge performing enhanced biological phosphorus removal. ISME 2, 853–864.

Zhang, B., VerBerkmoes, N.C., Langston, M.A., Uberbacher, E., Hettich, R.L., and Samatova, N.F. (2006). Detecting differential and correlated protein expression in label-free shotgun proteomics. J. Proteome Res. 5, 2909–2918.

Nucleic-acid-based Characterization of Community Structure and Function

4

James Prosser, Janet K. Jansson and Wen-Tso Liu

Abstract

Nucleic acid-based techniques were first used to characterize natural microbial communities in the early 1990s and are now used routinely. The ability to characterize communities without the requirement for cultivation has led to enormous advances in our ability to describe microbial communities and to determine the factors that influence their structure. New generations of molecular techniques provide even greater descriptive power and can be used to assess the physiological potential and ecosystem function of communities. They also enable microbial ecologists to address fundamental questions in population and community ecology, including investigation of the links between diversity and function. This chapter describes methods currently used to analyse nucleic acids extracted from environmental samples, and shows how they can be used to characterize communities. It also looks ahead to exciting new technologies that are likely to increase greatly our ability to explore and understand the complex functions and interactions of microbial communities in natural environments.

Introduction

Molecular approaches have proved invaluable for characterization of the structure of microbial communities in natural environments. In particular they have been used to determine the identities of microorganisms that have never been cultivated, and in some cases to predict their functional roles. Most DNA-based approaches for predicting identities of microorganisms have utilized the small subunit (SSU) rRNA genes as phylogenetic markers. Increasingly, genes encoding specific enzymatic processes are being used to gain information on the diversity, composition and abundance of communities performing important, microbially mediated functions. The advantages and disadvantages of these approaches are detailed in Chapter 1. Currently, one of the biggest restrictions is the reliance of these approaches on DNA sequences that are deposited in databases. For 16S rRNA gene sequences, however, the number of deposited sequences has been increasing exponentially and this increase is likely to continue through the application of high-throughput sequencing methods. There remain concerns regarding the quality of sequences that have been deposited in databases and thus there is a need for curation and validation of existing databases. Despite these caveats, the amount of qualitative and quantitative information that can be obtained about microorganisms in the environment, without the necessity for cultivation, is enormous and continually growing. The increasing number of microbial genome sequences and metagenome sequence data should also improve the ability to place sequences from natural communities within an increasingly rich phylogenetic framework, based on both SSU rRNA genes and functional genes. Molecular analysis of cultivated organisms and metagenomic studies are also enabling increasing analyses of the links between phylogenies created from rRNA genes and functional genes, thus facilitating identification of microorganisms performing particular ecosystem functions. In this chapter we will explore the use of nucleic acid-based approaches that are currently used to characterize the structure, activity and function of complex microbial communities and methods used to characterize specific members of those communities.

Assessment of community structure 'microbiomics'

When discussing the structure of a microbial community we most often refer to the identities of the different microorganisms that are present. Traditional techniques required isolation and phenotypic characterisation of pure cultures for identification. Owing to the inability to cultivate the majority of microorganisms in nature, nucleic acid-based techniques have now become the method of choice for determining the composition of microbial communities, largely based on characterization of SSU rRNA genes. The PCR-amplified product of the gene encoding rRNA is sometimes referred to as rDNA; however we strongly argue against this abbreviation, as ribosomes do not contain DNA and rDNA is also used as an abbreviation for recombinant DNA. Recently the term 'microbiomics' has been proposed as a catch-all term describing the different molecular fingerprinting techniques used to characterize complex microbial communities; www.microbiomics.se (Fig. 4.1).

PCR amplification

As described in Chapter 1, microbial species can be defined at different phylogenetic levels by targeting genes that are evolutionarily conserved, such as the 16S rRNA gene for bacteria and archaea, or other phylogenetic genes. Target specificity is determined by the primer sequences used during PCR amplification. For example, when using the 16S rRNA gene as a target sequence, primers can be designed for amplification of most bacteria, or for amplification of specific bacterial groups. Sometimes it is possible to design primers that can detect a specific genus or (rarely) species, but PCR primers generally target higher level phylogenetic groups of bacteria, archaea or fungi. For fungi, the target region is often the intergenic

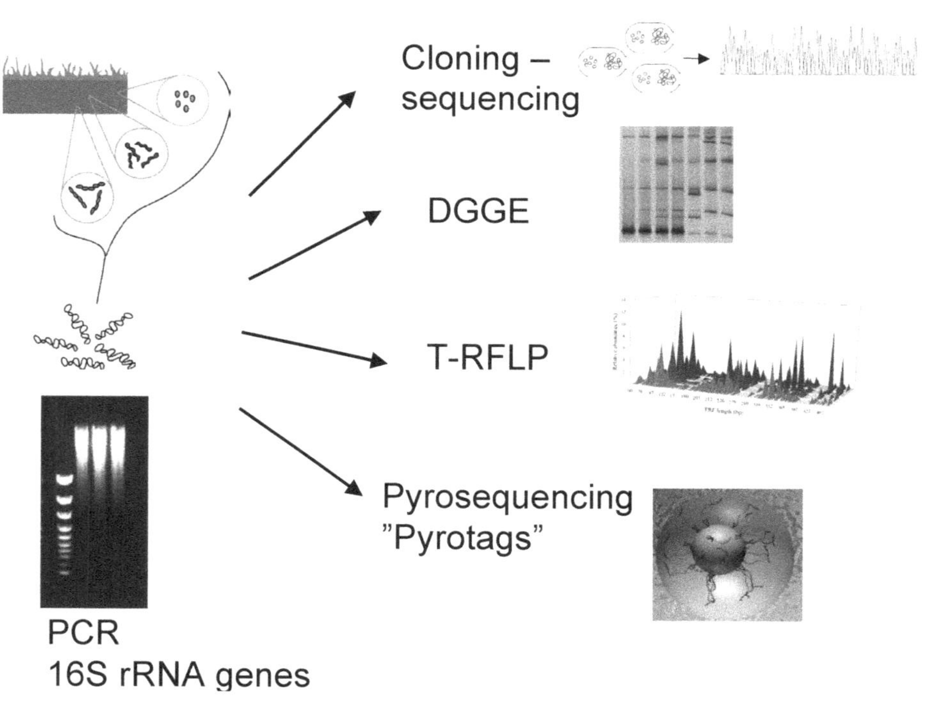

Figure 4.1 An overview of fingerprinting techniques to study complex microbial community structures in environmental samples 'Microbiomics' based on differentiation and/or identification of 16S rRNA genes amplified from the community. A colour version of this figure is located in the plate section at the back of the book.

transcribed spacer region (ITS) of the 18S rRNA gene. The choice of primer can introduce bias both with respect to the annealing region and the resultant amplification efficiency and specificity. The extent of bias is rarely determined and can be difficult to assess reliably. It is therefore difficult to generalize or even speculate on the extent to which different primers introduce bias in community structure analysis. This approach is, however, considered to introduce less bias than, e.g. use of different growth media and cultivation conditions in traditional approaches.

The use of molecular approaches that target DNA sequences is dependent on the successful extraction of DNA from the environmental sample of interest and significant technical advances have been made over the past few decades. Extraction of DNA from soil is particularly challenging and modern methods, frequently based on commercial kits, are reviewed in Chapter 6. Each DNA extraction approach has three key steps that determine the DNA yield and purity: (1) cell lysis, (2) DNA purification and (3) DNA concentration. Cell lysis is usually accomplished by enzymatic digestion or physical disruption (e.g. bead beating) or by a combination of these approaches. Bead beating is currently the most rigorous approach for cell lysis and is capable of lysing high GC Gram-positive bacteria. Physical disruption does lead to damage of nucleic acids. A balance is therefore required between applying bead beating for sufficient time to enable lysis of all cells and preventing DNA damage and degradation of DNA released from cells that are lysed rapidly. This represents a further source of bias.

Sequence analysis

Two types of approach have been primarily used to determine the microbial community structure in environmental samples based on genes amplified from extracted nucleic acids: (1) sequencing of genes and (2) resolution of SSU rRNA genes without sequencing; i.e. 'fingerprinting'. Until recently, sequencing was mainly performed after first cloning PCR amplification products to produce a clone library of the target gene amplicons and then sequencing representatives of the libraries. For example, this approach has been widely used for characterisation of both community 16S rRNA genes and functional genes and is now the standard method for fine resolution characterization of natural communities. The sequence data can be used to identify which organisms are present, by comparison with sequences in databases. The power of this approach has increased considerably over the past two decades as 16S rRNA gene databases have increased in size. The information content depends on the primers used and the length of sequence amplified. For short sequence lengths (< 100 bp), identification is possible, but only to a gross phylogenetic resolution (see Chapter 1). With increased sequence length, the ability to perform rigorous phylogenetic analysis improves. Sequencing of SSU rRNA genes, in particular, has been invaluable in identifying new groups of microorganisms with no cultivated representatives.

Comparison of sequences with those in databases enables putative identification of organisms present in the environment from which the sequences were derived. The value of this approach depends on several factors. Sequencing accuracy is important, particularly with newly developed, high-throughput approaches. However, as mentioned above, a bigger concern is the nature and reliability of sequences deposited in publicly available databases. Databases contain sequences of cultivated organisms but many (now the majority) of sequences are from environmental clones, amplified directly from extracted nucleic acids. The lengths of deposited sequences and the region amplified vary considerably, due to the use of different primers, many of which are specific for particular phylogenetic groups. The 16S rRNA gene databases are now extensive but the accuracy of the data cannot be assured, as there is no quality control of material entered into databases. For example, Ashelford *et al.* (2005) suggest that 5% contain anomalies and 3% are chimeric and software is available for detecting chimeric sequences (Ashelford *et al.*, 2006; DeSantis *et al.*, 2006). Nevertheless, this is the method of choice for phylogenetic analysis and identification of natural communities and is of tremendous value. Identification is achieved by sequence comparisons, such as Blast searches, or by phylogenetic analysis. The latter is more rigorous and, when performed for many sequences from the same environment, can lead to the discovery of novel phylogenetic groups.

Database coverage for functional genes is less comprehensive and more variable but is rapidly increasing. For many functional genes, phylogenies of 16S rRNA genes and functional genes of cultivated organisms have been compared. Where these phylogenies are congruent, functional gene sequences (through Blast searches) can provide information on identity. Again, reliability of the accuracy of database sequences is not assured and lengths and regions of genes sequenced vary.

Analysis of clone library sequences has demonstrated enormous diversity within most natural microbial communities (McCaig *et al.*, 1999). Early studies, involving analysis of 16S rRNA gene clone libraries, were limited by technical and resource limitations and were generally restricted to analysis of < 100–200 sequences. In clone libraries of this size, identical sequences are rarely obtained in complex communities, such as soil, even when using primers that amplify members of low level taxonomic groups, such as genera and species. Complete coverage of the diversity of terrestrial and marine environments cannot be achieved with such small sequencing efforts. If sequences are grouped into operational taxonomic units (OTUs) at specified similarity levels, for example corresponding to genera or species, it is possible to construct collector's curves in which the number of different OTUs identified is plotted against the number of sequences examined. This approach, coupled to rarefaction analysis, has been used to estimate richness of microbial communities (see below).

Recently, another approach has been introduced that has promise to replace sequence analysis of clone libraries with respect to speed and data accumulation. This technique is most commonly termed 454 sequencing, but is also referred to as 'pyrosequencing'. It is based on sequencing by synthesis (Ronaghi *et al.*, 1998) and relies on attaching adapters to the PCR products (either before or after PCR amplification). The adapters are used to attach the PCR products to microscopic beads in a water/oil emulsion. The beads with bound DNA are then placed into a 44-μm well on a PicoTiterPlate with 1.6 million wells. Normally, only one PCR product is attached per bead and only one bead is distributed to each well. The PCR amplification can take place directly on the bead or in an oil emulsion prior to distribution onto the PicoTiterPlate. A mix of enzymes such as DNA polymerase, ATP sulphurylase and luciferase are also added to each well of the plate. During the sequencing stage, the four nucleotides are washed in series over the PicoTiterPlate. If a nucleotide complementary to the template strand is added to a well, the polymerase extends the existing DNA strand by adding that nucleotide. Addition of a nucleotide results in release of pyrophosphate that initiates an enzymatic reaction that generates a light signal that is recorded by the CCD camera in the instrument. The signal strength is proportional to the number of nucleotides that have been incorporated. Pyrosequencing costs can be greatly reduced by using bar codes or sample specific identifiers that are incorporated into one of the primers during PCR amplification. This allows several samples to be loaded simultaneously to the same PicoTiterPlate. The sequences are resolved according to sample origin by identification of the sample specific sequence tag (Sogin *et al.*, 2006; Andersson *et al.*, 2008).

To date, pyrosequencing has primarily been used to complement genome sequencing projects but has recently been modified for sequencing of 16S rRNA genes of microbial communities. One advantage of 454 sequencing is that a tremendous amount of data is obtained in a relatively short time, since current machines are capable of sequencing 20 megabases per 4.5-hour run, and technical developments are increasing sequencing rates. This method does not require cloning and therefore overcomes cloning bias. Until recently, the GS20 could only read a maximum of 100 bp at a time. However, the new FLX system is able to sequence 200–300 bp regions that are suitable for 16S rRNA phylogeny and the recently launched 'Titanium' system enables 400–500 bp reads. There are now several published examples of use of these systems for analysis of natural microbial communities (Sogin *et al.*, 2006; Roesch *et al.*, 2007; Andersson *et al.*, 2008; Krause *et al.*, 2008).

Fingerprinting techniques

Different nucleic acid based methods have been developed to provide a holistic view of natural microbial communities. These approaches, termed 'fingerprinting' are generally based on methods for distinguishing characteristics of the

SSU rRNA or functional genes, of populations comprising those communities. One of the first fingerprinting approaches to become widely adopted in microbial ecology was denaturing gradient gel electrophoresis (DGGE) (Muyzer *et al.*, 1993). This relies on separation of PCR-amplified genes on the basis of heterogeneities in their GC content and sequence, leading to differences in their melting properties as they migrate through a gel matrix containing a gradient of a denaturing chemical. A variation of this approach is temperature gradient gel electrophoresis (TGGE) that incorporates a temperature gradient into the gel to separate molecules according to their melting characteristics. Double-stranded DNA molecules have different migration properties and mobilities through a gel. When the strands start to separate due to melting, migration is retarded and denatured molecules have a much slower gel migration. Complete separation of strands is prevented by using primers with a GC-clamp. For mixed communities containing genes from different populations the result is a series of bands on a gel, with each band theoretically corresponding to one population in the original community. The relative intensity of each band is a quantitative measure of the relative abundance of that population within the target community. The resulting bands can be excised and sequenced to obtain information about the identities of the respective populations. A miniaturized version of TGGE is now available for profiling of genomic DNA in a matter of minutes and this technology should have application for typing of strains isolated from environmental or clinical samples (Biyani and Nishigaki, 2001).

Another widely used approach is terminal-restriction fragment length polymorphism (T-RFLP), which separates PCR products based on sequence polymorphisms in the amplified region (Liu *et al.*, 1997). During PCR amplification one or both of the primers is fluorescently labelled (most often the 5′ primer). The amplified product is digested with one or more restriction enzymes that cleave at different locations depending on the product sequence. The digested products are then loaded onto a gel or capillary in an automated sequencer. The fluorescent, terminal restriction product is the only DNA fragment that is detected by the laser detector in the automated sequencer. The resulting output is in the form of an electropherogram in which each terminal restriction fragment (TRF) is separated according to size (bp) and represented as a peak with peak height corresponding to the fluorescence intensity and therefore the amount of that particular DNA fragment in the DNA mixture. Theoretically, each TRF corresponds to a specific microbial population in the original community. In order to determine the relative abundance of each population (TRF), all of the peak areas over a given fluorescence intensity threshold value, depending on background, are summed and the relative peak area is determined by dividing individual peak area by total peak area (Edlund *et al.*, 2006). Often the data are presented as relative abundance (%).

The major features, advantages and disadvantages of the four major sequence-based approaches are presented in Table 4.1, although others exist, and an additional fingerprinting approach, using the Phylochip method, is explained in detail in Chapter 5 and will not be further discussed in this chapter.

Drawbacks and limitations of current methods

In many studies, the relative abundance of phylogenetic groups is inferred from, e.g. relative abundances of phylotypes within clone libraries, relative intensities of DGGE bands and relative peak areas in T-RFLP electropherograms. A number of problems inherent in this approach are specific to particular methods. For example, DGGE bands or T-RFLP peaks may contain more than one sequence. Another fundamental problem is bias introduced during PCR amplification of the target gene. This is unavoidable for all PCR-amplification-based methods but can be checked by alternative techniques, such as direct microscopic enumeration of fluorescently labelled cells by fluorescence *in situ* hybridization (FISH), although this technique also has bias. In general, comparative analysis and seeding experiments with known amounts of added microbes suggest that sequencing of clone libraries provides a good measure of relative abundance of different phylotypes. The reliability of clone library analysis for assessment of relative abundance does, however, rely on good coverage which is often difficult to

Table 4.1 Features, advantages and disadvantages of the major sequence-based approaches to analysis of microbial community structure based on differentiation of SSU rRNA genes

Fingerprinting approach	Key feature	Advantages	Disadvantages
DGGE/TGGE	Separation of PCR products based on their melting characteristics in a chemical or temperature denaturation gradient	Relatively cheap and easy to perform Highly reproducible Bands of interest can be cloned and sequenced to identify corresponding populations	Only the most dominant members of the community are detected One band can represent several sequences Quantification of relative band intensity depends on gel quality
T-RFLP	Separation of PCR products based on the size of their fluorescently labelled terminal restriction fragments	Relatively cheap and easy to perform Highly reproducible Suitable for comparing large number of samples over time or according to treatment	Only the most dominant members of the community are detected One TRF can represent several sequences Only putative identities of populations corresponding to a given TRF can be obtained by comparison to sequences deposited in databases that are digested in silico. This problem is reduced by using multiple restriction enzymes in parallel
Clone libraries	Sequence analysis of clones following PCR amplification of SSU rRNA or functional genes	Provides information for phylogeny and identification	Many clones need to be sequenced to obtained required coverage of diversity
454 pyrosequencing	Pyrosequencing of PCR products attached to beads in a PicoTiterPlate	Very large amounts of sequences (>100,000) can be obtained to generate deep coverage of a community Sequencing is rapid Several samples can be loaded simultaneously and distinguished using sample specific barcodes to reduce costs	Currently limited by availability of 454 FLX sequencers The technology is relatively expensive The read lengths vary depending on the platform used (GS20 versus FLX, versus Titanium) There are discrepancies in the use of primer binding sites making comparisons between studies difficult

achieve for complex environmental samples (see below).

Some problems are associated with specific techniques. For example, defining T-RFLP peaks and DGGE bands is subjective and carryover reduces discriminative ability. These issues are greater when analysing complex communities and complex fingerprint patterns. For example, bacterial 16S rRNA gene primers typically generate 50–100 bands on DGGE gels or peaks on T-RFLP electropherograms, whereas 16S rRNA gene analysis of ammonia oxidising bacteria generates less than ten bands. In many cases, qualitative comparisons are sufficient, determining the appearance or disappearance of bands, peaks, or clone types, or the changes in relative intensity, peak area or abundance over time or according to treatment. In other cases more rigorous, quantitative comparisons and a range of techniques are required, many of which are based on multivariate analysis and cluster analysis techniques.

For all comparisons, qualitative or quantitative, it is essential that replicates are analysed. This is normally feasible for fingerprint analyses, as many samples can be run relatively cheaply. However, the need for replication when comparing community composition determined by sequencing representatives of clone libraries is equally essential, despite the additional resources required. Thus, it is meaningless to compare relative abundances or presence/absence of clone sequences of different phylotypes in clone libraries from different treatments or environments if only a single clone library is available for each treatment. Clone libraries from replicate samples of each treatment must be obtained and analysed for valid statistical comparisons.

Richness, evenness and diversity indices

In most environments, molecular techniques detect only the most abundant phylotypes. In some ways, the analysis of a bacterial community in a single gram of soil is equivalent to analysis of the global human community. The total number of individuals in each is in the order of 10^9 and the distribution of humans among different nationalities follows a curve that is similar to species distribution curves for the bacterial 16S rRNA gene. Typical clone libraries analyse approximately 100 sequences, which is equivalent to characterizing all humans on earth by sampling only 100 individuals. Fingerprinting techniques access a greater proportion of the total community, as many representatives of a single phylotype will contribute to a DGGE band or T-RFLP peak. These techniques are considered to detect phylotypes with relative abundance down to 1% and even moderately abundant organisms will not be analysed. The greatly increased power of high-throughput sequencing techniques increases significantly the ability to detect and identify rare organisms, but even generation of 10^5 16S rRNA gene sequences will only give moderate coverage of total diversity in environments with low total abundance.

It is conceivable that further development of sequencing techniques will overcome these problems but a more fundamental issue is definition of the basic phylogenetic unit on which we base measurement of richness. For higher organisms, richness generally refers to the number of different species present in an environment. Although a number of definitions of species exist for eukaryotes, the biological species concept (Mayr, 1957) is used most frequently. This defines species as 'groups of actually or potentially interbreeding natural populations which are reproductively isolated from each other', sexual reproduction thereby acting as a barrier to recombination. This definition is of little use for bacteria, archaea and some eukaryotes, which reproduce asexually, and alternative definitions are required. Those used are subjective, operational and generally have no evolutionary or mechanistic basis, although ecological species concepts attempt to overcome some of these limitations. Whole genome sequencing of cultivated organisms has also led to the concept of a core genome, common to all members of a 'species', and an accessory genome, which varies in composition between different members of the species. Unfortunately, the core genome tends to decrease in size with each new species sequenced. Traditional prokaryotic species descriptions frequently depend on physiological characteristics encoded by accessory genes and accessory genes may encode important ecological characteristics. In addition, all species definitions must consider lateral gene transfer, through plasmid transfer, transduction or transformation.

As a consequence, species definitions are subjective and operational. The standard technique required for acceptance of a cultivated organism as a new species is DNA–DNA hybridization, but this is not applicable for analysis of large numbers of strains or, obviously, for uncultivated organisms. Molecular taxonomy, based largely on sequences of SSU rRNA genes, provides some consistency and can be used to analyse large numbers of sequences representing community composition (see Chapter 1). Typically a nucleotide sequence similarity of 97% is used to define species, but there are suggestions that higher levels (99%) are required. Techniques such as multilocus sequence typing, in which organisms are grouped on the basis of sequences of several genes, demonstrate significant diversity within species defined by SSU rRNA gene sequences alone. This additional 'subspecies' diversity can be associated with considerable ecological importance and cannot be ignored when attempting to determine mechanisms driving diversity and influence of community structure on ecosystem function.

Many ecological questions require an estimate of the richness of a community, which for plants and animals equates to the number of species. Early estimates of bacterial richness were based on DNA–DNA hybridization of environmental DNA and clone library sequence data. For example, Torsvik estimated up to several thousand 'distinct genomes' per gram of soil (Torsvik *et al.*, 1990). Estimation of richness from clone library sequence data using rarefaction analysis also indicated high levels of richness and collector's curves rarely reach plateaux without analysis of very high numbers of sequences, or use of low similarity levels for grouping of sequences.

Improved theoretical approaches to analysis of richness (Curtis *et al.*, 2002; Gans *et al.*, 2005) using both DNA–DNA hybridization and sequence data have led to even higher estimates of richness. For example, Gans *et al.* (2005) used a number of theoretical species distribution patterns to determine richness from DNA:DNA hybridisation data, leading to estimates of up to 10^6 genomes per gram of soil. The assumptions on which species distribution curves are based arise from those found in higher organism communities and are difficult to determine, with confidence, for bacteria and archaea. Nevertheless, they indicate enormous diversity and predict the level of sequencing necessary for coverage. For example, based on the estimates of Gans *et al.* (2005) sequencing of 10^6 clones is required to detect 80% of the species present in 1 g soil. We note that this number of sequences can now be obtained using the 454 pyrosequencing approach with bar coded 'pyrotags' (see below) and we anticipate the increased adaptation of this method for profiling of microbial community compositions.

That being said, it is clear that the other fingerprinting approaches commonly in use are not able to estimate true bacterial and archaeal richness in soil (Bent *et al.*, 2007) and that total richness is not equal to the number of bands on a gel or peaks on a T-RFLP trace, which typically resolve in the order of 100 bands or peaks. At best, fingerprinting techniques will only detect approximately 0.01–10% of the most dominant or readily amplifiable phylotypes present, the amount depending on sample complexity. They may, however, provide information on evenness and dominance. Thus, a reduction in the number of bands on a DGGE gel does not necessarily indicate a reduction in richness of the community, as there will potentially be many thousands of phylotypes that are below the detection limit. Rather, the reduction in band number indicates decreased evenness within the community and increased dominance by a few phylotypes.

The value of diversity indices, such as the Shannon diversity index, depends on the degree to which coverage is approached. So, for example, if high level taxonomic groups are being characterized, coverage may be high and profiles or relative abundances of different phylogenetic groups within clone libraries may provide a good indication of true relative abundance. If coverage is low, true richness will not be reflected in profiles and diversity indices have little value.

Community function

A wide range of techniques are available that contribute information on community function and several are discussed in other chapters. These include functional gene microarrays (Chapter 5), proteomics (Chapter 3) and metagenomics (Chapter 2). Discussion here will focus on nucleic acid-based techniques and those involving detection of nutrient assimilation by active organisms.

RNA-targeted 16S rRNA-gene analysis

DNA-targeted techniques provide information on presence/absence, abundance and relative abundance of organisms belonging to particular groups, i.e. they indicate which organisms are present, irrespective of their physiological status. For example, 16S rRNA genes amplified from DNA extracted from environmental samples could originate not only from viable and active cells, but also from dead, inactive or dormant cells. To determine which phylogenetic groups are active or which are functioning, alternative methods are required. One approach is to target rRNA, rather than DNA, by reverse transcription (RT) of rRNA, creating cDNA, and subsequent PCR amplification (RT-PCR amplification). The rationale behind this approach is that active or growing cells will contain more ribosomes than inactive or non-growing cells. The evidence for this comes from physiological studies of a relatively limited range of cultivated organisms which demonstrate increasing ribosome content with increasing growth rate (Kerkhof and Kemp, 1999). These studies also indicate that actively growing cells may contain many ribosomes (e.g. in *E. coli* tens of thousands of ribosomes per cell). This approach is therefore similar to BrdU immunocapture (see below) in targeting all of the active organisms, rather than just those performing a specific activity.

While the increase in ribosome content with increasing growth rate and activity may be a widespread phenomenon, the generality of these results must be, at least, questioned. Studies have been performed on cultivated strains, growing at relatively high specific growth rates. These organisms are unlikely to be representative or typical of the majority of soil microorganisms and high growth rates are unlikely to be achieved in nutrient-limited environments. Nevertheless, RNA-copy (ribosome) numbers per cell are likely to be greater than 16S rRNA gene copy number and RNA-targeted analysis theoretically provides much greater sensitivity than DNA-targeted techniques. A second concern is that, while an increase in growth rate or cell activity may lead to a rapid increase in ribosome number, the converse may not be the case. There is evidence for some strains of environmental importance (for example, ammonia oxidizers) that cellular rRNA levels do not decrease rapidly when cells are starved. Indeed, maintenance of high cellular ribosome levels may be a sensible strategy for organisms with transient nutrient supply to enable rapid recovery and growth following periods of starvation. RNA-based techniques may therefore reflect a history of activity rather than indicate which organisms are currently active.

Fig. 4.2 illustrates application of rRNA-targeted approaches in a microcosm study designed to determine the influence of pentachlorophenol (PCP) on soil bacterial communities (Mahmood *et al.*, 2005). Soil was incubated for 63 days after addition of PCP and of ^{13}C-labelled PCP. (This study also investigated PCP-effects using stable isotope probing, which will be described below). Samples were removed, nucleic acids extracted and changes in the bacterial community were followed by DNA-targeted amplification of 16S rRNA genes and RNA-targeted amplification of 16S rRNA using universal bacterial primers. Fig. 4.2 shows changes in DGGE-fingerprints obtained following DNA- and rRNA-targeted amplification after incubation with PCP for 35 days. Degradation of PCP was accompanied by changes in bacterial community structure, with the appearance of seven bands (labelled 1–7 in Fig. 4.2) in both DNA- and RNA-based DGGE profiles. Sequencing of these bands showed phylogenetic relationships either with known hydrocarbon degraders or with sequences of uncultured organisms from polluted soils. Changes in RNA fingerprints were apparent earlier than changes in DNA fingerprints, as changes in ribosome content will precede growth and DNA synthesis. In addition, as indicated by bands a–c in Fig. 4.2b, targeting RNA rather than DNA detected more community changes and provided clearer evidence of selection for PCP degraders.

Quantification of rRNA by sequence-specific enzymatic cleavage

rRNA can also be quantified using a sequence-specific enzymatic cleavage method (Uyeno *et al.*, 2004; *Suenaga et al.*, 2005). In this method, high-quality rRNA is extracted and enzymatically cleaved, and the size and abundance of the

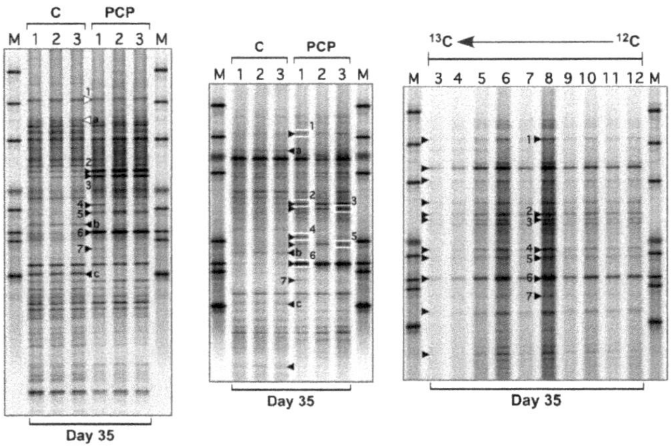

Figure 4.2 Denaturing gradient gel electrophoresis (DGGE) banding profiles of bacterial 16S rRNA genes PCR-amplified from **A** DNA and **B** RNA extracted from soil microcosms amended with water (control, C) or ^{12}C- and ^{13}C-labelled pentachlorophenol (PCP). Lanes 1–3 relate to samples from triplicate microcosms of control or PCP treatments harvested after incubation for 35 days. Black arrowheads indicate bands that increased in relative intensity (right-directing arrowheads) or disappeared (left-directing arrowheads) during incubation with PCP. White arrowheads designate bands that did not change significantly in relative intensity during incubation with PCP. Bands 1–6 were excised and sequenced. Markers in lane M consisted of seven unrelated bacteria. **C** Stable isotope probing (SIP)-based DGGE banding profiles of bacterial 16S rRNA genes RT-PCR-amplified from PCP-amended microcosms after incubation for 35 days. Extracted RNA was subjected to density gradient centrifugation to separate ^{12}C- and ^{13}C-labelled RNA. Lane numbers 3–12 correspond to density gradient fractions (heavy ^{13}C- to light ^{12}C-labelled RNA). Black arrowheads indicate the bands that appeared in the heavy fractions. Markers in lane M in all gels consisted of seven unrelated bacteria. (From Mahmood *et al.*, 2005, with permission).

cleaved fragments analysed using a chip-based electrophoresis system. The abundance of a single targeted 16S rRNA is determined by comparing the peak area (intensity) of cleaved 16S rRNA to that of total 16S rRNA. Either RNase H or DNAzyme can be used in the enzyme cleavage step. DNAzyme is reported to be more specific than RNAase H (Suenaga *et al.*, 2005). Both enzymes can use target-binding probe sequences designed at different levels of phylogenetic specificity (e.g. domain, group, and species). They can quantify bacterial and/or archaeal fractions in model communities and can detect specific microbial populations (> 1–2% of total rRNA). Sensitivity depends on (i) the use of high-quality, intact, extracted rRNA, (ii) the mismatched position of the non-target rRNA molecules and (iii) the resolution of electrophoresis systems for small-sized cleaved rRNA fragments (Uyeno *et al.*, 2004; Suenaga *et al.*, 2005).

Analysis of expressed functional genes

Although information on some functional groups can be obtained by analysis of their 16S rRNA genes, for example all known betaproteobacterial autotrophic ammonia oxidizers fall within a single monophyletic group, this situation is quite rare, and functional groups are usually defined by functional genes that encode key enzymes in particular processes. The value and efficiency of these functional genes vary considerably. For example, primers are available for the ammonia monooxygenase gene (*amoA*) that detects the gene in all known bacterial ammonia oxidizers. *AmoA* gene sequences are well represented in sequence databases and *amoA* gene phylogeny is congruent with that of 16S rRNA genes (Purkhold *et al.*, 2000). Primers are also available which detect methane monooxygenase (*pmoA*), an equivalent enzyme which also catalyses ammonia oxidation. However, neither of these primer sets detects *amoA* genes in the recently discovered crenarchaeal ammonia oxidizers (Schleper *et al.*, 2005) and universal *amoA* gene primers that amplify genes from all three groups are not available. It is therefore not possible to characterize all genes involved in the first step of the ammonia oxidation process using the same primers.

The situation is more complex for other processes. Denitrification involves four key reduction steps, and four functional genes: nitrate reductase, nitrite reductase, nitric oxide reductase and nitrous oxide reductase. This process is distributed widely through soil bacteria and archaea, such that approximately 50% of phylogenetic groups with cultivated representatives include denitrifiers. This distribution has resulted from widespread lateral gene transfer. Consequently, functional gene- and 16S rRNA gene-based phylogenies are generally not congruent and functional gene sequences provide little information on identity or physiological characteristics and diversity (Philippot and Hallin, 2005). In addition, some steps are catalysed by different enzymes. Nitrite reductase genes exist in two forms, *nirK* and *nirS*, which are never found in the same organism and with no apparent pattern to their distribution.

DNA-targeted studies of functional genes provide information only on the potential ability of that function to operate within the environment, as cells may be inactive or dormant, the gene may be silent and never expressed and cell physiology may be unfavourable for transcription of the gene and/or translation to a protein. Studies of gene expression therefore provide a much better indication of the ability of a cell to perform a particular function under the prevailing environmental conditions. Table 4.2 lists some of the genes that have been used to target specific ecosystem functions and to determine the influence of environmental conditions or other

Table 4.2 Examples of functional genes used in environmental studies to characterize organisms performing particular ecosystem functions

Ecosystem function	Functional gene	Enzyme	Example reference
Carbon dioxide fixation	*cbbL*	Ribulose-1,5-bisphosphate carboxylase/oxygenase (RubisCO)	Selesi *et al.*, 2005
Chitin degradation	*chiA*	Chitinase	Hjort *et al.*, 2007
Ammonia oxidation	*amoA* *hao*	Ammonia monooxygenase, Hydroxylamine oxidoreductase	Bruns *et al.*, 1998; Nicol *et al.*, 2008
Methane oxidation	*pmoA* *mmoX* *mxaF*	Particulate methane monooxygenase Soluble methane monooxygenase Methanol dehydrogenase	Morris *et al.*, 2002
Denitrification	*narG*, *napA* *nirK*, *nirS* *norB*, *norZ* *nosZ*	Nitrate reductase Nitrite reductase Nitric oxide reductase Nitrous oxide reductase	Philippot and Hallin, 2005
Sulphate reduction	*dsrAB* *aprBA*	Dissimilatory sulphite reductase Dissimilatory adenosine-5′-phosphosulphate reductase	Muyzer and Stams, 2008
Nitrogen fixation	*nifH*	Nitrogenase	Moisander *et al.*, 2008
Polyaromatic hydrocarbon (PAH) degradation	*nahAc* *phnAc*	Naphthalene dioxygenase PAH dioxygenase	Edlund and Jansson, 2008
Methanogenesis	*mcrA*	Methyl coenzyme M reductase	Lueders *et al.*, 2001
Ureolysis	*ureC*	Urease	Gresham *et al.*, 2007
Oxygenic photosynthesis	*psbA*	D1 protein	Zeidner *et al.*, 2003
Antibiotic resistance	*ermB*	Erythromycin resistance methylase	Jernberg *et al.*, 2005
Mercury resistance	*merA*	Mercury reductase	Barkay and Wagner-Dobler, 2005

factors on the abundance, relative abundance and community composition of the responsible organisms.

The ability to measure gene expression in environmental samples has been limited by difficulties in extracting mRNA and maintaining its integrity and considerable effort has been expended in devising RNA extraction and purification techniques that preserve mRNA. Methods for extraction of mRNA, for example from soil, are now established and are reviewed (Saleh-Lakha *et al.*, 2005). Assessment of transcriptional activity is therefore increasingly being used to determine *in situ* ecosystem activity in the soil and to identify the organisms performing specific functions. Analysis of expressed genes, from mRNA, essentially uses methods used for DNA, and is subject to the same benefits and limitations. In addition, there are issues and considerations specific to mRNA analysis and gene expression that will be illustrated in two examples of this approach, both involving ammonia-oxidizing bacteria.

The first involves field and microcosm studies designed to determine the relative roles of bacterial and archaeal ammonia oxidation in soil (Nicol *et al.*, 2008). DNA-targeted analysis of 16S rRNA and *amoA* genes demonstrated significant pH-associated differences in both bacterial and archaeal ammonia oxidizer communities in soils maintained at pH values in the range 4.9–7.5 for 46 years. For both groups, DGGE analysis demonstrated distinct populations in acid and neutral soils. This analysis, however, only indicates differences in the potentials for both groups to contribute to ammonia oxidation, and not their actual activities. To assess transcriptional activity, both archaeal and bacterial *amoA* gene copies and mRNA transcripts were quantified by real-time qPCR (see below) across the pH gradient. At all pH values, archaeal *amoA* genes and gene transcripts were more abundant than those for bacteria. However, contrasting patterns were seen with respect to pH. Archaeal *amo*A gene and transcript abundances decreased with increasing soil pH, while bacterial gene abundance showed little change, and transcripts increased with increasing pH. These data suggest that archaea may play a greater role in ammonia oxidation in acid soils, while bacteria may have greater relative importance in neutral soils. To investigate short-term transcriptional activity of archaeal ammonia oxidizers at different pH values, microcosms were constructed containing acid soil, neutral soil or mixtures of both with pH readjusted to a pH of either 4.9 or 7.5. DGGE analysis of gene transcripts (Fig. 4.3) shows the extent of differences in active archaeal communities at the different pH soils. It also demonstrates that transcriptional activity of organisms selected by low pH was greater in mixed soil microcosms adjusted to pH 4.9, while neutral pH adapted communities showed greater transcriptional activity at pH 7.5.

This study demonstrates a number of features of gene expression studies and the need for careful analysis. For example, differences in abundance estimates for bacterial and archaeal

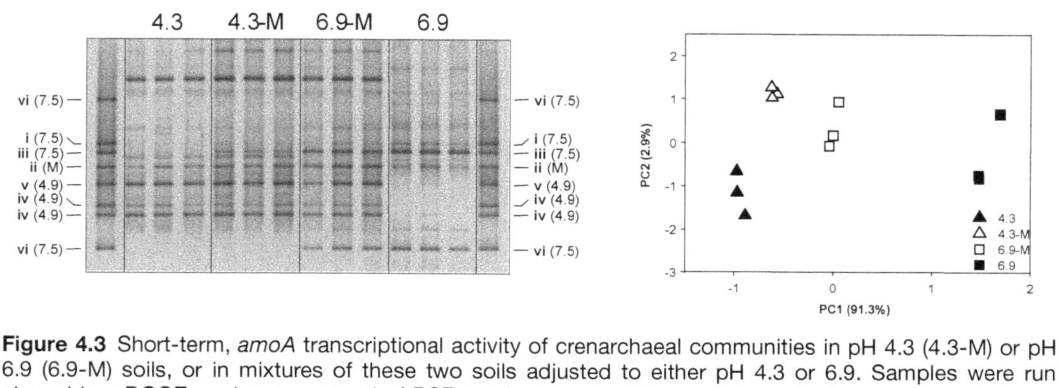

Figure 4.3 Short-term, *amoA* transcriptional activity of crenarchaeal communities in pH 4.3 (4.3-M) or pH 6.9 (6.9-M) soils, or in mixtures of these two soils adjusted to either pH 4.3 or 6.9. Samples were run alongside a DGGE marker composed of PCR products derived from eight *amoA* clones representative of major clades of sequences identified in clone libraries from the same soil. DGGE and principal component analysis of crenarchaeal amoA transcripts from microcosms after incubation for 12 days. (From Nicol *et al.*, 2008, with permission.)

ammonia oxidizers based on qPCR of *amoA* may result from real differences and/or from biases in cell lysis efficiency, nucleic acid extraction efficiency or primer bias (as *amoA* primers for the two groups differ). The *amoA* gene product may have functions other than ammonia oxidation and the abundance of *amoA* might result from processes other than ammonia oxidation. For example, although there is no evidence for heterotrophic growth of bacterial ammonia oxidizers, heterotrophic or mixotrophic growth of archaeal ammonia oxidizers might lead to greater abundance of *amoA*, independent of any role in ammonia oxidation. In addition, we need to consider the links between gene expression (transcript abundance) and ammonia oxidising activity. This will depend on several factors, such as the stability and rate of turnover of *amoA* mRNA, the stability of the ammonia monooxygenase protein and on whether the environment within the cell and external environmental conditions are sufficient to maintain ammonia oxidizer activity. Fig. 4.3 also indicates a common feature of transcript abundance studies, in that there are significantly fewer gene transcripts than genes. This may reflect technical difficulties in extracting mRNA and in reverse transcription. Alternatively, it suggests that the majority of potential ammonia oxidizers (as indicated by gene abundance) are not transcriptionally active.

In the second study, bacterial and archaeal *amoA* genes and mRNA transcripts were quantified in a range of soils (Leininger *et al.*, 2006). In three soils tested, archaeal *amoA* gene transcripts were detected and were generally more abundant than bacterial *amoA* transcripts. In addition, the metatranscriptome was obtained by 454 pyrosequencing of cDNA created from extracted RNA and was found to contain two archaeal *amoA* transcripts but no bacterial *amoA* transcripts. One drawback of this approach was the dominance of rRNA genes in the transcripts that masked detection of many functional genes. Detection of *amoA* and other functional genes would have benefited from techniques that are now available for elimination of such rRNA genes prior to sequencing. One large advantage of this type of metatranscriptomic approach is that it eliminates cloning bias, which may occur in traditional clone library analysis, and provides an indication of functional importance rather than mere presence. This approach has recently been applied by Frias-Lopez *et al.* (2008) to measure bacterial and archaeal gene expression in ocean waters. In that study small amounts of RNA (ng) were polyadenlyated, linearly amplified with T7 RNA polymerase and converted to cDNA, which was directly sequenced by 454 pyrosequencing (Frias-Lopez *et al.*, 2008).

Bromodeoxyuridine (BrdU) incorporation

In the late 1990s Borneman *et al.* (1999) and Urbach *et al.* (1999) proposed use of bromodeoxyuridine (BrdU), a thymidine analogue that is incorporated into the DNA of replicating cells, to distinguish actively growing bacteria in environmental samples (Borneman, 1999; Urbach *et al.*, 1999). Urbach *et al.* (1999) also used immunocapture of BrdU-labelled DNA as a simple and elegant approach for separation of the BrdU-labelled DNA from unlabelled DNA. The ability of this approach to assess the identities of growing bacteria in soil has since been demonstrated in soil and sediment. For example, BrdU-immunocapture was used to identify active bacteria in association with arbuscular mycorrhizal fungi (Artursson and Jansson, 2003) and those stimulated by chitin, to identify potential chitinase-producing bacteria in suppressive soil (Hjort *et al.*, 2007). In the latter study, a high correlation was found between bacteria identified using the BrdU approach and chitinase producers isolated on selective medium. Fig. 4.4 shows an illustration of the use of fingerprinting techniques to determine the community structure of soil bacterial populations that are activated upon chitin addition. Close examination of this figure reveals that fewer TRFs were detected that represent actively growing populations compared to the number of TRFs found in the 'total' bacterial community, including presumably inactive, dormant cells. In another study, BrdU immunocapture was used to identify bacterial populations that were active before and after dredging of PAH-contaminated Baltic Sea surface sediments (Edlund and Jansson, 2006) and along a eutrophication gradient in the Baltic Sea (Edlund *et al.*, 2006). In both studies, the bacterial populations that were identified from the total DNA extracts were significantly different

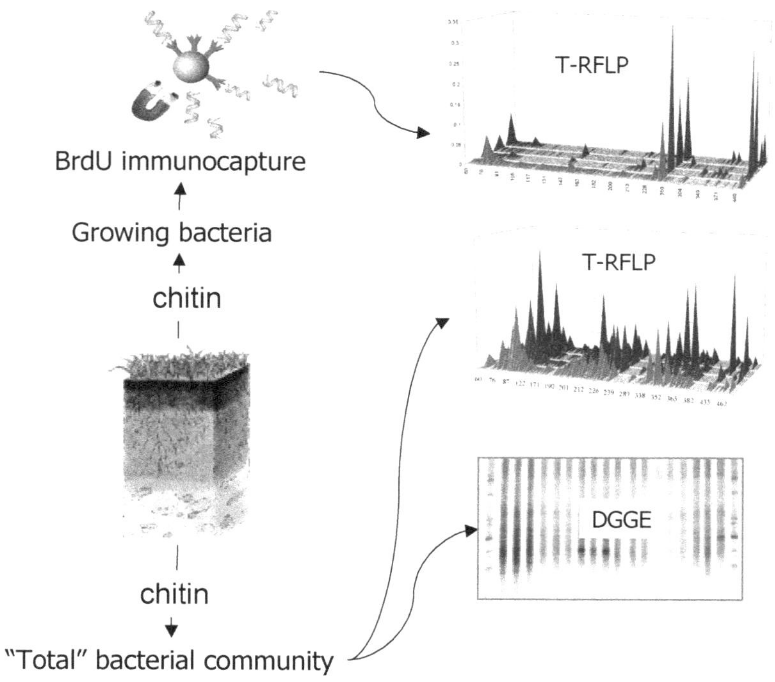

Figure 4.4 Illustration of the use of bromodeoxyuridine (BrdU) immunocapture to fingerprint actively growing bacterial populations upon chitin addition, in comparison to the 'total' microbial community represented in directly extracted DNA (adapted from results described in Hjort *et al.*, 2007). A colour version of this figure is located in the plate section at the back of the book.

from those that had incorporated BrdU, indicating that the dominant populations identified in the total extracts do not necessarily represent the growing community members and vice versa.

In a later study of the same PAH-polluted sediments, phenanthrene-degrading populations were enriched by spiking with phenanthrene, and the emergence of active potential phenanthrene-degrading community members was monitored by identification of sequences captured by BrdU immunocapture during the enrichment (Edlund and Jansson, 2008). A high correlation was found between these sequences and those of organisms previously cultured on phenanthrene-selective medium from the same polluted sediments (Edlund *et al.*, 2006; Edlund and Jansson, 2008). Recently, BrdU immunocapture was compared with analysis of reverse transcribed 16S rRNA from total RNA extracts from samples taken at different redox depths in Baltic Sea sediments (Fig. 4.5; Edlund *et al.*, 2008). Several populations were detected by both approaches, showing a high correlation between these different approaches for detection of active bacteria. However, analysis of reverse transcribed RNA indicated greater dominance of *Gammaproteobacteria* than analysis of BrdU-labelled DNA (Edlund *et al.*, 2008). This discrepancy may reflect different intracellular levels of activity (i.e. DNA replication and transcription). Also, RT-PCR is a snapshot approach, i.e. the community members that are active during sampling are analysed, whereas BrdU-immunocapture requires a period of incubation, increasing the variety of potentially active cells. The dominance of *Gammaproteobacteria* indicated by RT-PCR indicates that these community members were transcribing 16S RNA genes more actively than other bacterial groups at this redox depth at the time of sampling.

BrdU incorporation has also been combined with fluorescence activated cell sorting (FACS) to sort active cells from marine samples (Mou *et al.*, 2007). The BrdU molecule was tagged with a fluorescent marker that was taken up and

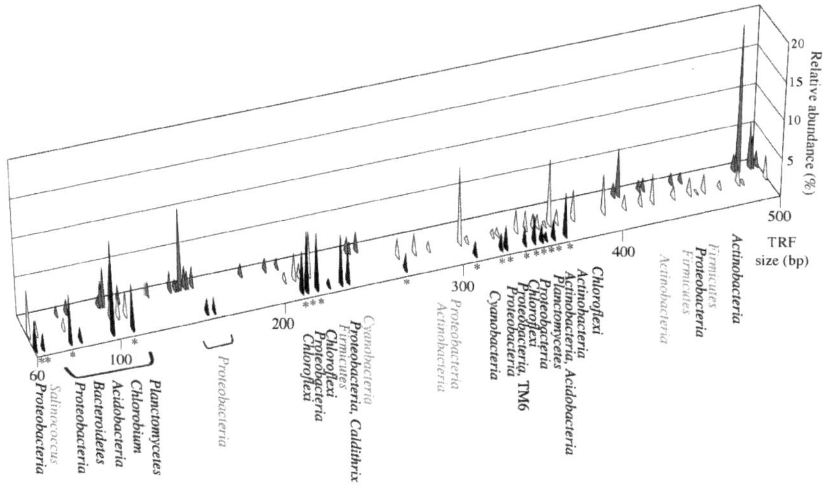

Figure 4.5 Comparison of phylotypes putatively identified by matching TRF fragment lengths (base pairs) to sequences deposited n ribosomal databases and their relative abundances. Black TRFs = 16S rRNA gene fragments amplified from BrdU-labelled DNA; White TRFs= cDNA from reverse transcribed RNA, Grey TRFs = 16S rRNA gene fragments amplified from directly extracted community DNA. Genera listed below the x-axis correspond to sequence identities of clones amplified from cDNA and BrdU-labelled DNA (black text) and TRF identities from the RDP-TAP database (grey text). TRFs that were detected using both RT-PCR and BrdU-labelled DNA are indicated with an asterix below the x-axis. Modified from Edlund *et al.*, 2008.

incorporated into actively growing bacterial cells. Fluorescent cells were then detected by flow cytometry, sorted into a chamber and identified by 16S rRNA gene sequencing. Although this method holds promise, it has similar drawbacks to FISH with respect to background fluorescence, including that from naturally fluorescent cells; low fluorescence yields from cells that have taken up BrdU; and the unknown ability of different cell types to take up and incorporate the fluorescently labelled BrdU molecule. There are also concerns about difficulties in removing unlabelled DNA from the magnetic beads during the immunocapture procedure (Walters and Field, 2006), such that amplified products could represent DNA from both actively growing and inactive populations. Appropriate controls are therefore essential throughout the procedure. A further problem is the unknown ability of different bacteria to assimilate BrdU and rates of incorporation, but Edlund *et al.* (2008) identified representatives of Gram-negative and Gram-positive bacteria, aerobes and anaerobes and phylogenetically diverse bacteria using this approach. Nevertheless, rates of incorporation will vary and the approach should be used with caution when assessing the diversity of active communities. Rather, it is a suitable approach for detecting populations that are active under specific conditions, such as following substrate addition (Edlund and Jansson, 2008; Hjort *et al.*, 2007). It is therefore ideal as a first step in identifying potential degraders or other populations of interest, such as mycorrhiza-helper bacteria, that can then be selectively targeted for isolation on selective medium (Artursson and Jansson, 2003).

Stable isotope probing (SIP)

Stable isotope probing (SIP) provides a means of linking phylogeny with function, i.e. determining who is doing what in soil. Specifically, it identifies which members of a microbial community actively assimilate compounds or groups of compounds labelled with a stable isotope. In theory, SIP can be applied to study assimilation of compounds containing any stable isotope. However, the vast majority of applications have involved assimilation of ^{13}C-labelled compounds, with recent development of ^{15}N-SIP to investigate nitrogen fixing organisms. Most of this section will describe the techniques in terms of ^{13}C-SIP studies and several reviews (Dumont and Murrell, 2005;

Prosser *et al.*, 2006) of this approach provide references to specific studies.

SIP requires the supply of a ^{13}C-labelled substrate to the community, which may be a laboratory culture, an environmental microcosm or a field site and supply is usually achieved by direct addition of the labelled compound. Such compounds have included CO_2, methane, methanol, pesticides and pollutants. Labelled substrates may also be supplied indirectly. For example, organisms assimilating root exudates may be assessed following their release from plants incubated (in pots or in the field) with ^{13}C-labelled CO_2.

During incubation, organic cell components of actively assimilating organisms will be labelled with ^{13}C. Subsequent analysis of ^{13}C- and ^{12}C-labelled compounds depends on the cell components chosen. Initial SIP approaches focused on analysis of phospholipid fatty acids (PLFA), which are major components of cell membranes and whose composition and structure vary between different phylogenetic groups. PLFAs are extracted from soil during incubation and are analysed by transmethylation to fatty acid methyl esters (FAME). FAME concentrations are determined by GC-flame ionization detection and $\delta^{13}C$ values are determined by GC-IRMS (gas chromatography isotope ratio mass spectrometry). The identities of labelled and unlabelled PLFAs are then determined by comparison of retention times with reference compounds and by mass spectral analysis.

PLFA discriminates between high level-phylogenetic groups but nucleic acid sequences provide much greater taxonomic resolution and, particularly for functional gene analysis, provide information on functional group diversity. DNA-SIP involves extraction of DNA from soil during incubation. ^{12}C- and ^{13}C-DNA are separated by caesium chloride density gradient centrifugation for 16–24 h. In some applications, samples are removed from two regions of the fractionated DNA that are considered to be dominated by ^{12}C- or ^{13}C-labelled DNA. More rigorous analysis requires removal of samples from fractions taken along the ^{12}C–^{13}C DNA gradient. Molecular analysis is then carried out on each DNA fraction. Typically, this analysis involves amplification of 16S or 18S rRNA genes or functional genes, as described above, with analysis of amplified genes by cloning and sequencing or by fingerprinting techniques. In addition, DNA fractions can be subjected to microarray or metagenomics analysis.

DNA-SIP is less sensitive that PLFA-SIP, but sensitivity can be increased by analysis of RNA, rather than DNA. In addition, incorporation of label into rRNA, through increases in ribosome contents of active cells, is likely to occur sooner than incorporation into DNA, which occurs during DNA replication. In this approach, extracted ^{12}C-RNA and ^{13}C-RNA are fractionated using isopycnic density gradient centrifugation in CsTFA (caesium trifluoroacetate) for 40 h. Molecular analysis is currently limited to amplification and analysis of 16S or 18S rRNA genes, as described above. RNA-SIP analysis of transcribed genes is theoretically possible but has not been reported, presumably because mRNA yields are not sufficiently high for detection. In addition, to an extent, mRNA analysis and SIP are complementary techniques. mRNA-SIP would, however, enable determination of which substrates lead to transcription of a particular gene.

There are several limitations to this method, such as sensitivity that was discussed above, depending on the target molecule. Separation of unlabelled and labelled DNA is possible following incorporation of as little as 20% of a ^{13}C-labelled substrate, while RNA-SIP can analyse as little as 500 ng of rRNA. One seeding experiment, in which ^{13}C-labelled *Pseudomonas fluorescens* cells were added to soil, demonstrated a lower detection limit for RNA-SIP of approximately 10^6 cells g^{-1} soil of *Pseudomonas fluorescens* fully labelled with ^{13}C. Sensitivity will obviously be less for partially labelled cells and separation of ^{12}C- and ^{13}C-RNA is more difficult in complex communities. Sensitivity will also vary with the degree of labelling of the substrate, the rate of substrate assimilation and the level of turnover. The concentration of labelled substrate may actually be reduced by turnover with unlabelled substrate within the soil. For example, respiration of existing organic carbon will reduce the concentration of added $^{13}CO_2$. In addition, since SIP depends on assimilation of label it is of limited value for assessing extracellular enzyme activity (unless products are rapidly assimilated) or processes in which substrate is not assimilated. For example, N-SIP may be useful for analysis of nitrogen fixation but not ammonia oxidation, in which ammonium is converted to nitrite, which then diffuses out of the cell.

All techniques that involve the addition of substrate (including the BrdU immunocapture method above) must consider potential bias through disturbance of soil conditions, release of other nutrients (if the substrate is added in solution) and unrealistically high substrate concentrations, although this disturbance is less for gaseous substrates, such as methane and carbon dioxide. The influence of disturbance can be assessed by use of appropriate controls. Ideally substrates will be added at low concentration but the degree to which this can be achieved is severely restricted by the relatively low sensitivity of SIP (particularly for nucleic-acid-based SIP) and the requirement for assimilation of significant amounts of ^{13}C to facilitate separation of labelled and unlabelled material.

SIP analysis may also be complicated by production of ^{13}C-labelled products, which are then assimilated by secondary utilizers. This problem, often referred to as cross-feeding, may be alleviated by reducing incubation times, but this will reduce levels of ^{13}C assimilation, reducing sensitivity. The appearance of label in secondary utilizers is advantageous when the aim is to determine which organisms are involved during sequential degradation or utilisation of a compound. This issue does, however, demonstrate the need for time-course studies, rather than end-point analysis only, to distinguish primary and secondary utilizers. Interactions between different functional groups and multi-stage substrate conversion can be investigated using dual-labelled compounds, by addition of labelled pathway intermediates and by addition of labelled compounds after initial incubation with unlabelled compounds.

The method used to fractionate labelled and unlabelled nucleic acids also leads to fractionation on the basis of GC content. This is particularly important when using DNA-SIP and when analysing high-level taxonomic groups, within which GC content may vary significantly. The problem may be less for RNA-SIP as rRNA GC content is believed to vary less because of regions of highly conserved sequence. The potential for GC-bias necessitates control experiments in which unlabelled substrate is added.

SIP is based on the assumptions that strains do not discriminate between ^{13}C- and ^{12}C-labelled substrates and that growth and activity are equivalent on labelled and unlabelled substrates. High concentrations of ^{13}C-labelled organic compounds do inhibit growth, but there is no evidence that the levels of stable isotope used for SIP studies cause significant bias. In addition, DNA-SIP requires chromosome replication and will therefore not detect activity, and assimilation, of resting, non-dividing cells. Such activity will be detected by RNA-SIP if cells are turning over ribosomes.

SIP analysis using stable isotopes of nitrogen is more difficult than carbon for two main reasons. The first is reduced discrimination between labelled and unlabelled nucleic acids, as the differential in mass between ^{14}N- and ^{15}N-labelled DNA is less than that between ^{12}C- and ^{13}C-labelled materials. Secondly, the N-content of nucleic acids is significantly less than C-content, reducing sensitivity. Both of these factors also increase the likelihood of 'contamination' with high-GC DNA. To reduce these effects, Buckley *et al* (2007) developed a two-stage density gradient centrifugation method (Buckley *et al.*, 2007). The first stage utilizes CsCl density gradients to recover ^{15}N-labelled DNA followed by purification of ^{15}N-labelled DNA from unlabelled high GC-DNA by a second density gradient centrifugation step with CsCl gradients containing bis-benzimide. This approach has been used to characterize free-living soil nitrogen fixing bacteria by analysis of both 16S rRNA and *nifH* genes in ^{15}N-labelled DNA following incubation with ^{15}N-labelled dinitrogen (Buckley *et al.*, 2008). Results suggested that nitrogen fixation was carried out by members of cultivated groups within the Alphaproteobacteria, including the Rhizobiales, but also divergent 16S rRNA lineages within the Betaproteobacteria and Actinobacteria and within *nifH* cluster 1 with no cultivated representatives.

SIP has now been used to address a wide range of ecological questions. The method was developed for investigation of methane oxidizers and both PLFA- and DNA-SIP have demonstrated which organisms contribute to methane oxidation in soil and which are active under high and low methane concentrations. Assimilation of a range of organic compounds has been studied, and growth on carbon dioxide by autotrophs. SIP has also been used to determine which organisms utilize plant root exudates, following growth of plants on ^{13}C-CO_2. This approach has determined how liming of soil influences microbial communities using grassland root exudates, in the field, and to determine which methanogens contributed to hydrogen- and acetate-dependent methanogenesis in the rice rhizosphere.

SIP is illustrated here with regard to degradation of pentachlorophenol (PCP). In the experiment described in Fig. 4.3, soil microcosms were supplied with both ^{12}C- and ^{13}C-labelled PCP and bacteria assimilating PCP were determined by amplification of 16S rRNA genes from fractionated RNA. This study provides a direct comparison of DNA- and RNA-DGGE and RNA-based SIP-DGGE. The three approaches distinguished, respectively, between changes in relative abundance of different bacterial groups in response to PCP amendment, changes in relative activity within communities and bacterial groups which were assimilating PCP. The last of these was evidence from DGGE bands appearing in ^{13}C-RNA fractions during incubation. In this particular study, all of the approaches indicated similar changes in DGGE profiles. This suggests that changes in community structure were due to growth and assimilation of PCP, rather than extracellular enzyme activity or reduction in relative abundance due to inhibition.

Quantification of nucleic acid sequences

In comparison to community fingerprinting methods, fewer methods are currently available for the quantitative analysis of targeted microbial populations and functional activity. Methods currently available include whole-cell fluorescence *in situ* hybridisation (FISH), membrane hybridisation, quantitative PCR and hierarchical oligonucleotide primer extension (HOPE) (Wu and Liu, 2007). Here, we will restrict the examples and discussion to quantitative PCR and HOPE approaches.

Quantitative PCR

Quantitative PCR (typically abbreviated Q-PCR, qPCR or real-time PCR) quantifies the copy number of target DNA present initially in a mixture of DNA molecules by comparing the observed amplified signal intensity with a standard curve (Fig. 4.6). This standard curve is constructed using a standard DNA template, which is diluted in series at 10- or 5-fold, e.g. ranging from 10^7 down to 10^2 DNA copies. During PCR amplification, the signal intensities of amplified DNA products under different template concentrations at different cycles are recorded using a selected fluorescent-reporting system, and then normalized. Common fluorescence reporting chemistries include *Taq*Man probes, molecular beacons and DNA intercalating dyes such as SYBR Green I (Giulietti *et al.*, 2001). By selecting an arbitrary threshold, usually set at a level that is 10 times the standard deviation of the baseline signal observed between cycles 3 and 15, the corresponding threshold cycle (Ct) at each reference template concentration can be defined (see inset in Fig. 4.6). It is possible to quantify up to four different targets simultaneously down to a concentration theoretically close to 1–2 copies of DNA template with a dynamic range of 7 orders of magnitude (Giulietti *et al.*, 2001). Q-RT-PCR is mainly used for the quantification of RNA. For this, total RNA is extracted and then reverse transcribed to cDNA, which is then used as template for Q-PCR analysis. To eliminate

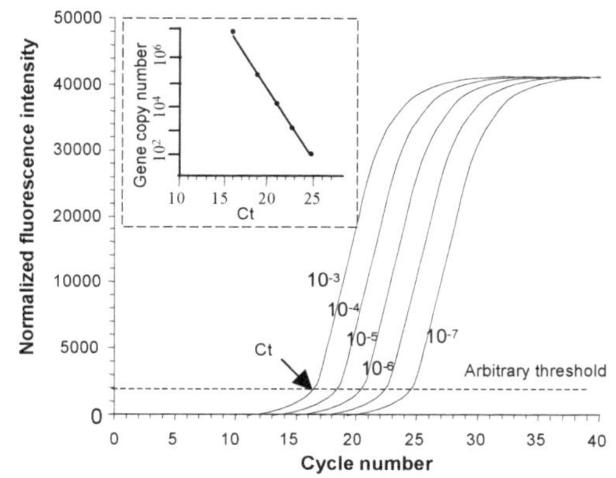

Figure 4.6 Fluorescent intensities of quantitative PCR obtained at different template dilution concentration and the Ct values for each of the amplified standards. The inset illustrates the relation between the Ct value and the standard gene copy number.

the possible co-amplification of genomic DNA, extracted RNA should be treated with DNase.

Although Q-PCR and Q-RT-PCR have been widely applied in environmental microbiology studies, these two methods require rigorous optimization before application. The accuracy of Q-PCR methods can be compromised by the standard curves constructed using genomic DNA, plasmids, PCR amplicons, or oligonucleotides as the standard DNA template. This is mainly because the gene copy number per genome, for example of the SSU rRNA gene, is not constant across different microbial species (Klappenbach *et al.*, 2001). When quantifying uncultivated organisms, gene copy number is usually obtained by estimation, and the standard curve is constructed using a PCR amplicon instead of genomic DNA. As a result, variation can also be introduced due to the effect of target size (i.e. PCR amplicon vs. genomic DNA) on amplification efficiency, or during the comparison of results obtained from multiple targets. To resolve the latter issue, Holmes *et al.* (2006) used a single plasmid containing multiple targeted genes as the reference DNA template. To eliminate false-positive signals at low copy number of targets, a no-template control (NTC) can be introduced in the construction of the standard curve. It has been demonstrated that the lowest Ct value obtained from a sample at a given dilution rate should be 3.3 cycles greater than the Ct value of the NTC and, based on this, the detection limit or sensitivity can be determined (Smith *et al.*, 2006). Nevertheless, environmental template concentration can vary due to different DNA extraction methods used and subsequently affect absolute quantification. Possible inter-plate variation can also occur with the use of different DNA extraction protocols or during analysis of a large set of samples, and thus additional control and replication reactions are needed (Smith *et al.*, 2006). In Q-RT-PCR experiments, it is difficult to normalize the number of transcripts from a specific gene to that of a housekeeping gene, as there are no stably expressed housekeeping genes in bacterial cells. Furthermore, during the reverse-transcription step, additional variation can be introduced, as methods used can vary in RT efficiency. Thus, careful selection of methods to generate cDNA is very important, and will need to be systematically investigated. As described in early sections, these optimized Q-PCR reactions can then be used to investigate the spatial and temporal changes of those targeted organisms in different microbial ecosystems, or applied to the detection of functionally important organisms in various environments (Table 4.2).

HOPE

As a method to complement the strength of T-RFLP or DGGE, HOPE determines the relative abundance of targeted 16S rRNA genes among total PCR-amplified bacterial or archaeal 16S rRNA genes within a given microbial ecosystem (Fig. 4.7). This method employs multiple oligonucleotide primers, which differ in length or are designed at different lengths by the addition of poly-A(s), to target specific 16S rRNA gene sequences at different taxonomical levels (i.e. domain, phylum, order ... and species). The first step of the method involves 20–25 thermal cycles of denaturation, annealing and single base extension with fluorescently labelled ddNTPs. At the end of thermal cycle, the number of individual extended primers is amplified n-fold. These extended primers are then separated and identified in a DNA auto-sequencer, and their peak areas are quantified based on the fluorescence intensity detected. By calculating a relative ratio between a lower ranked primer and the highest ranked primer, the relative abundance of specific bacterial targets in a given microbial environment can be calculated. HOPE can discriminate sequences with single base-paired mismatch occurring near the 3′ end of the primers and attain detection sensitivity down to 0.01–0.05% of total PCR amplicons for model bacterial mixtures and at least 0.1% for environmental samples. HOPE has been successfully demonstrated for rapid and specific determination of *Bacteroides* spp. present in faeces and wastewaters by firstly developing a total of ten primers targeting 16S rRNA genes of different *Bacteroides* spp. at different taxonomical levels. Results indicate *B. vulgatus*, *B. thetaiotaomicron*, *B. caccae*, *B. uniformis*, *B. fragilis*, *B. eggerthii* and *B. massiliensis* were individually detected in human faeces at the species-level. Furthermore, Bac303-related targets (i.e. most *Bacteroidales*) in human faecal microbiota were observed to account for 28–44% of PCR-amplified 16S rRNA

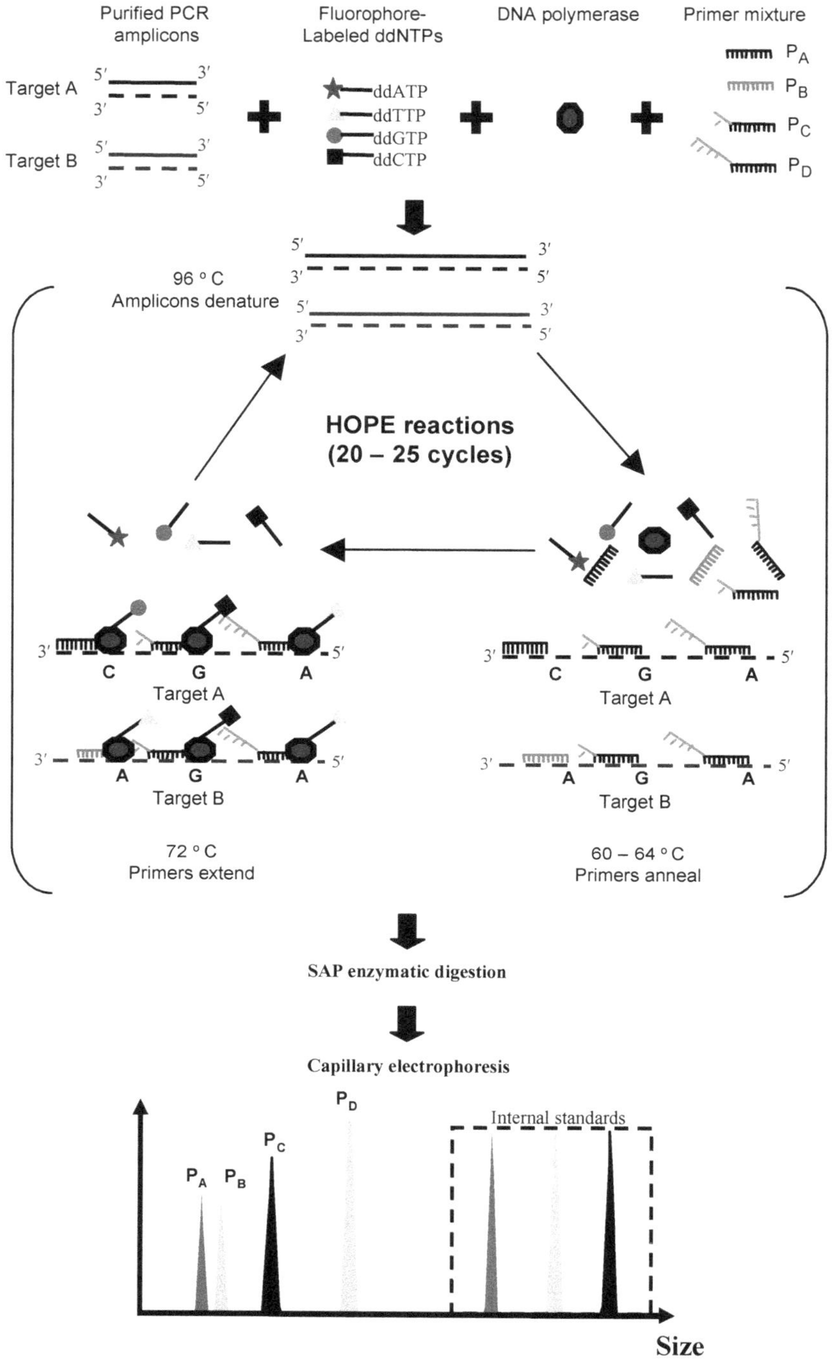

Figure 4.7 A schematic illustration of HOPE. PCR amplicons, fluorophore-labelled ddNTPs, DNA polymerase and oligonucleotide primers were mixed together to form the HOPE reaction mixture. The mixture undergoes a thermal program involving 20–25 cycles of denaturation (96 °C), annealing (60–64 °C) and single-base extension (72 °C). The products were purified and the excess ddNTP residual was removed by shrimp alkaline phosphatase (SAP) digestion. An appropriate amount of the product mixture was mixed with synthetic oligonucleotide standards, and then subjected to capillary electrophoresis. From the electropherograms, the extended primers were recognized by means of consistent fragment sizes and the colours of the extended ddNTP. (Reprinted with permission from Hong *et al.*, copyright 2008 ASM.)

genes and their abundances were two-fold higher than that observed in wastewaters. This study was further extended to detect and rapidly measure the relative abundance of a total of 14 functionally important *Bacteroides-Parabacterioides* spp. at species, group and domain levels in faecal microbiota and wastewaters. These results were comparable to those obtained by quantitative PCR or reported previously using FISH and 16S rRNA clone library methods, supporting HOPE as a sensitive, specific and rapid method to determine relative abundance of *Bacteroides* spp. predominant in faecal samples. HOPE can serve as a rapid, medium-throughput (in comparison to DNA microarray) detection method that examines the relative abundances of bacterial targets at various taxonomical levels, and its simplicity and adaptability can facilitate attempts to further understand microbial diversity in different ecosystems (Hong *et al.*, 2008). As HOPE is a newer method, its limitation remains to be further explored. One likely constraint is the need for reference 16S rRNA gene clones or reference microbial strains for the optimization on the primer specificity for targeted microbial groups. Others may include the complexity in extracting the information from an electropherogram containing multiple peaks, and the higher start-up cost in comparison with microbial fingerprinting methods. It is necessary that more rigorous statistical methods can be further developed for analysis and comparison of data obtained from different environments. Like other microbial fingerprinting methods, HOPE after a few more improvements can be a very useful tool in the quantitative study of microbial ecology.

Future aspects

It is anticipated that next-generation molecular tools will be developed from advances in 'lab-on-a-chip (LOC) devices' or 'micro total analysis systems'. This platform employs the most recent technologies in micro/nanofabrication and micro/nanofluidics to produce micro-scale components including wells, interconnected fluid reservoirs and pathways, heaters, mixers, valves and pumps (Liu and Zhu, 2005). One example, discussed above, is the high-throughput pyrosequencing approach that is performed simultaneously in a plate containing hundreds of thousands of microtitre wells. Another example is the use of microfluidic chambers to carry out high-throughput parallel 'digital PCR' reactions possibly from a single cell (Ottesen *et al.*, 2006). This concept has been demonstrated in a commercially available microfluidic device that precisely disperses microbial cells together with PCR solution into 1176 independent nanolitre volume compartments using micromechanical valves. This device is then placed on the top of a conventional flat block thermocycler to carry out PCR reactions, and the PCR amplification signal can be monitored using a microarray scanner to detect fluorescence signals of amplified PCR products. Using this approach, Ottesen and co-workers successfully co-amplified a phylogenetic marker (i.e. 16S rRNA gene) and a functional gene (formyl-tetrahydrofolate synthetase, FTHFS) involved in the acetyl-CoA pathway in individual wells and, based on their results, the microbial function could be correlated with bacterial populations present in the termite hindgut community. PCR products could be further retrieved from individual wells, cloned and sequenced to provide the diversity of the targeted FTHFS gene within a specific microbial group or the diversity of microbial populations involved in homoacetogenesis in the termite hindgut.

Concluding remarks

The continued discovery of many abundant groups of microorganisms with no cultivated representatives has necessitated development of cultivation-independent nucleic acid-based techniques for the discovery and assessment of their physiological characteristics and ecosystem function. This chapter has given an overview of currently available techniques and examples of their power in demonstrating who is doing what in the environment. These techniques are also proving invaluable for assessment of *in situ* activities of microbial groups for which cultivated representatives are available. Indeed, we are now moving away from our traditional reliance on extrapolation of information gained from laboratory culture to explain and understand the activities of microorganisms in natural environments. There is now real potential to determine the '*in situ* physiology' of microorganisms, even in complex environments such as soil and marine sediments.

The new generation of molecular techniques for assessing function described in this chapter have already led to major advances in linking microbial diversity and ecosystem function. However, this chapter represents a snapshot in a rapidly and continuously developing field. The previous section has given some indication of the enormous potential of the next generation of techniques. In applying these techniques, and prioritising technical developments, it is important to focus attention on the important questions in microbial ecology and on the ultimate value of the information gained. In this respect, techniques are required which provide quantitative data, determining not just who is doing what, but how quickly. Quantitative analysis of *in situ* growth and activity of different microbial groups, of their contributions to biogeochemical cycling processes and of their interactions with other groups will significantly advance our understanding of microbial ecology. They will also contribute more fully to predictive models of environmental change and to other applications where lack of information leads to treatment of microbial communities as a black-box.

Web source

www.microbiomics.se

References

Andersson, A.F., Lindberg, M., Jakobsson, H., Backhed, F., Nyren, P., and Engstrand, L. (2008). Comparative analysis of human gut microbiota by barcoded pyrosequencing. PLoS ONE *3*, e2836.

Artursson, V., and Jansson, J.K. (2003). Use of bromodeoxyuridine immunocapture to identify active bacteria associated with arbuscular mycorrhizal hyphae. Appl. Environ. Microbiol. *69*, 6208–6215.

Ashelford, K.E., Chuzhanova, N.A., Fry, J.C., Jones, A.J., and Weightman, A.J. (2005). At least 1 in 20 16S rRNA sequence records currently held in public repositories is estimated to contain substantial anomalies. Appl. Environ. Microbiol. *71*, 7724–7736.

Ashelford, K.E., Chuzhanova, N.A., Fry, J.C., Jones, A.J., and Weightman, A.J. (2006). New screening software shows that most recent large 16S rRNA gene clone libraries contain chimeras. Appl. Environ. Microbiol. *72*, 5734–5741.

Barkay, T., and Wagner-Dobler, I. (2005). Microbial transformations of mercury: potentials, challenges, and achievements in controlling mercury toxicity in the environment. Adv Appl Microbiol *57*, 1–52.

Bent, S.J., Pierson, J.D., Forney, L.J., Danovaro, R., Luna, G.M., Dell'anno, A., and Pietrangeli, B. (2007). Measuring species richness based on microbial community fingerprints: the emperor has no clothes. Appl. Environ. Microbiol. 73, 2399–2401; author reply 2399–2401.

Biyani, M., and Nishigaki, K. (2001). Hundredfold productivity of genome analysis by introduction of microtemperature-gradient gel electrophoresis. Electrophoresis *22*, 23–28.

Borneman, J. (1999). Culture-independent identification of microorganisms that respond to specified stimuli. Appl. Environ. Microbiol. *65*, 3398–3400.

Bruns, M.A., Fries, M.R., Tiedje, J.M., and Paul, E.A. (1998). Functional gene hybridization patterns of terrestrial ammonia-oxidizing bacteria. Microb Ecol *36*, 293–302.

Buckley, D.H., Huangyutitham, V., Hsu, S.F., and Nelson, T.A. (2007). Stable isotope probing with $^{15}N_2$ reveals novel noncultivated diazotrophs in soil. Appl. Environ. Microbiol. *73*, 3196–3204.

Buckley, D.H., Huangyutitham, V., Hsu, S.-F., and Nelson, T.A. (2008). $^{15}N_2$-DNA-stable isotope probing of diazotrophic methanotrophs in soil. Soil Biol. Biochem. *40*, 1272–1283.

Curtis, T.P., Sloan, W.T., and Scannell, J.W. (2002). Estimating prokaryotic diversity and its limits. Proc. Natl. Acad. Sci. U.S.A. *99*, 10494–10499.

DeSantis, T.Z., Hugenholtz, P., Larsen, N., Rojas, M., Brodie, E.L., Keller, K., Huber, T., Dalevi, D., Hu, P., and Andersen, G.L. (2006). Greengenes, a chimera-checked 16S rRNA gene database and workbench compatible with ARB. Appl. Environ. Microbiol. *72*, 5069–5072.

Dumont, M.G., and Murrell, J.C. (2005). Stable isotope probing – linking microbial identity to function. Nat. Rev. Microbiol. *3*, 499–504.

Edlund, A., and Jansson, J.K. (2006). Changes in active bacterial communities before and after dredging of highly polluted Baltic Sea sediments. Appl. Environ. Microbiol. *72*, 6800–6807.

Edlund, A., and Jansson, J.K. (2008). Use of bromodeoxyuridine immunocapture to identify psychrotolerant phenanthrene-degrading bacteria in phenanthrene-enriched polluted Baltic Sea sediments. FEMS Microbiol. Ecol. *65*, 513–525.

Edlund, A., Soule, T., Sjoling, S., and Jansson, J.K. (2006). Microbial community structure in polluted Baltic Sea sediments. Environ. Microbiol. *8*, 223–32.

Edlund, A., Hardeman, F., Jansson, J.K., and Sjoling, S. (2008). Active bacterial community structure along vertical redox gradients in Baltic Sea sediment. Environ. Microbiol. *10*, 2051–2063.

Frias-Lopez, J., Shi, Y., Tyson, G.W., Coleman, M.L., Schuster, S.C., Chisholm, S.W., and Delong, E.F. (2008). Microbial community gene expression in ocean surface waters. Proc. Natl. Acad. Sci. U.S.A. *105*, 3805–10.

Gans, J., Wolinsky, M., and Dunbar, J. (2005). Computational improvements reveal great bacterial diversity and high metal toxicity in soil. Science *309*, 1387–1390.

Giulietti, A., Overbergh, L., Valckx, D., Decallonne, B., Bouillon, R., and Mathieu, C. (2001). An overview of real-time quantitative PCR: Applications to quantify cytokine gene expression. Methods 25, 386–401.

Gresham, T.L.T., Sheridan, P.P., Watwood, M.E., Fujita, Y., and Colwell, F.S. (2007). Design and Validation of *ure*C-based Primers for groundwater detection of urea-hydrolyzing bacteria. Geomicrobiol. J. *24*, 353–364.

Hjort, K., Lembke, A., Speksnijder, A., Smalla, K., and Jansson, J.K. (2007). Community structure of actively growing bacterial populations in plant pathogen suppressive soil. Microb. Ecol. *53*, 399–413.

Holmes, V.F., He, J.Z., Lee, P.K.H., and Alvarez-Cohen, L. (2006). Discrimination of multiple *Dehalococcoides* strains in a trichloroethene enrichment by quantification of their reductive dehalogenase genes. Appl. Environ. Microbiol. *72*, 5877–5883.

Hong, P.Y., Wu, J.H., and Liu, W.T. (2008). Relative abundance of *Bacteroides* spp. in stools and wastewaters as determined by hierarchical oligonucleotide primer extension. Appl. Environ. Microbiol. *74*, 2882–2893.

Jernberg, C., Sullivan, A., Edlund, C., and Jansson, J.K. (2005). Monitoring of antibiotic-induced alterations in the human intestinal microflora and detection of probiotic strains by use of terminal restriction fragment length polymorphism. Appl. Environ. Microbiol. *71*, 501–506.

Kerkhof, L., and Kemp, P. (1999). Small ribosomal RNA content in marine Proteobacteria during non-steady-state growth. FEMS Microbiol Ecol *30*, 253–260.

Klappenbach, J.A., Saxman, P.R., Cole, J.R., and Schmidt, T.M. (2001). rrndb: the Ribosomal RNA Operon Copy Number Database. Nucleic Acids Research *29*, 181–184.

Krause, L., Diaz, N.N., Goesmann, A., Kelley, S., Nattkemper, T.W., Rohwer, F., Edwards, R.A., and Stoye, J. (2008). Phylogenetic classification of short environmental DNA fragments. Nucleic Acids Res. *36*, 2230–2239.

Leininger, S., Urich, T., Schloter, M., Schwark, L., Qi, J., Nicol, G.W., Prosser, J.I., Schuster, S.C., and Schleper, C. (2006). Archaea predominate among ammonia-oxidizing prokaryotes in soils. Nature *442*, 806–9.

Liu, W.T., and Zhu, L. (2005). Environmental microbiology-on-a-chip and its future impacts. Trends Biotechnol. 23, 174–179.

Liu, W.T., Marsh, T.L., Cheng, H.,and Forney, L.J. (1997). Characterization of microbial diversity by determining terminal restriction fragment length polymorphisms of genes encoding 16S rRNA. Appl. Environ. Microbiol. *63*, 4516–4522.

Lueders, T., Chin, K.J., Conrad, R., and Friedrich, M. (2001). Molecular analyses of methyl-coenzyme M reductase alpha-subunit (mcrA). genes in rice field soil and enrichment cultures reveal the methanogenic phenotype of a novel archaeal lineage. Environ. Microbiol. 3, 194–204.

Mahmood, S., Paton, G.I., and Prosser, J.I. (2005). Cultivation-independent *in situ* molecular analysis of bacteria involved in degradation of pentachlorophenol in soil. Environ. Microbiol. 7, 1349–60.

Mayr, E. (1957). The species problem. In: Mayr, E. (ed.). American Association for the Advancement of Science, Washington, DC, p. 1–22.

McCaig, A.E., Glover, L.A., and Prosser, J.I. (1999). Molecular analysis of bacterial community structure and diversity in unimproved and improved upland grass pastures. Appl. Environ. Microbiol. *65*, 1721–1730.

Moisander, P.H., Beinart, R.A., Voss, M., and Zehr, J.P. (2008). Diversity and abundance of diazotrophic microorganisms in the South China Sea during intermonsoon. ISME J. *2*, 954–967.

Morris, S.A., Radajewski, S., Willison, T.W., and Murrell, J.C. (2002). Identification of the functionally active methanotroph population in a peat soil microcosm by stable-isotope probing. Appl. Environ. Microbiol. *68*, 1446–1453.

Mou, X., Hodson, R.E., and Moran, M.A. (2007). Bacterioplankton assemblages transforming dissolved organic compounds in coastal seawater. Environ. Microbiol. *9*, 2025–2037.

Muyzer, G., and Stams, A.J. (2008). The ecology and biotechnology of sulphate-reducing bacteria. Nat Rev Microbiol *6*, 441–454.

Muyzer, G., de Waal, E.C., and Uitterlinden, A.G. (1993). Profiling of complex microbial populations by denaturing gradient gel electrophoresis analysis of polymerase chain reaction-amplified genes coding for 16S rRNA. Appl. Environ. Microbiol. *59*, 695–700.

Nicol, G.W., Leininger, S., Schleper, C., and Prosser, J.I. (2008). The influence of soil pH on the diversity, abundance and transcriptional activity of ammonia oxidizing archaea and bacteria. Environ. Microbiol. *10*, 2966–2978.

Ottesen, E.A., Hong, J.W., Quake, S.R., and Leadbetter, J.R. (2006). Microfluidic digital PCR enables multigene analysis of individual environmental bacteria. Science *314*, 1464–1467.

Philippot, L., and Hallin, S. (2005). Finding the missing link between diversity and activity using denitrifying bacteria as a model functional community. Curr. Opin. Microbiol. *8*, 234–239.

Prosser, J.I., Rangel-Castro, J.I., and Killham, K. (2006). Studying plant–microbe interactions using stable isotope technologies. Curr. Opin. Biotechnol. *17*, 98–102.

Purkhold, U., Pommerening-Roser, A., Juretschko, S., Schmid, M.C., Koops, H.P., and Wagner, M. (2000). Phylogeny of all recognized species of ammonia oxidizers based on comparative 16S rRNA and amoA sequence analysis: implications for molecular diversity surveys. Appl. Environ. Microbiol. *66*, 5368–5382.

Roesch, L.F., Fulthorpe, R.R., Riva, A., Casella, G., Hadwin, A.K., Kent, A.D., Daroub, S.H., Camargo, F.A., Farmerie, W.G., and Triplett, E.W. (2007). Pyrosequencing enumerates and contrasts soil microbial diversity. ISME J. *1*, 283–290.

Ronaghi, M., Uhlen, M., and Nyren, P. (1998). A sequencing method based on real-time pyrophosphate. Science 281, 363, 365.

Saleh-Lakha, S., Miller, M., Campbell, R.G., Schneider, K., Elahimanesh, P., Hart, M.M., and Trevors, J.T. (2005).

Microbial gene expression in soil: methods, applications and challenges. J. Microbiol. Methods *63*, 1–19.

Schleper, C., Jurgens, G., and Jonuscheit, M. (2005). Genomic studies of uncultivated archaea. Nat. Rev. Microbiol. *3*, 479–488.

Selesi, D., Schmid, M., and Hartmann, A. (2005). Diversity of green-like and red-like ribulose-1,5-bisphosphate carboxylase/oxygenase large-subunit genes (cbbL) in differently managed agricultural soils. Appl. Environ. Microbiol. *71*, 175–184.

Smith, C.J., Nedwell, D.B., Dong, L.F., and Osborn, A.M. (2006). Evaluation of quantitative polymerase chain reaction-based approaches for determining gene copy and gene transcript numbers in environmental samples. Environ. Microbiol. *8*, 804–815.

Sogin, M.L., Morrison, H.G., Huber, J.A., Mark Welch, D., Huse, S.M., Neal, P.R., Arrieta, J.M., and Herndl, G.J. (2006). Microbial diversity in the deep sea and the underexplored 'rare biosphere'. Proc. Natl. Acad. Sci. U.S.A. *103*, 12115–12120.

Suenaga, H., Liu, R., Shiramasa, Y., and Kanagawa, T. (2005). Novel approach to quantitative detection of specific rRNA in a microbial community, using catalytic DNA. Appl. Environ. Microbiol. *71*, 4879–4884.

Torsvik, V., Goksoyr, J., and Daae, F.L. (1990). High diversity in DNA of soil bacteria. Appl. Environ. Microbiol. *56*, 782–787.

Urbach, E., Vergin, K.L., and Giovannoni, S.J. (1999). Immunochemical detection and isolation of DNA from metabolically active bacteria. Appl. Environ. Microbiol. *65*, 1207–1213.

Uyeno, Y., Sekiguchi, Y., Sunaga, A., Yoshida, H., and Kamagata, Y. (2004). Sequence-specific cleavage of small-subunit (SSU) rRNA with oligonucleotides and RNase H: a rapid and simple approach to SSU rRNA-based quantitative detection of microorganisms. Appl. Environ. Microbiol. *70*, 3650–3663.

Walters, S.P., and Field, K.G. (2006). Persistence and growth of fecal Bacteroidales assessed by bromodeoxyuridine immunocapture. Appl. Environ. Microbiol. *72*, 4532–4539.

Wu, J.H., and Liu, W.T. (2007). Quantitative multiplexing analysis of PCR-amplified ribosomal RNA genes by hierarchical oligonucleotide primer extension reaction. Nucleic Acids Res. *35*, e82.

Zeidner, G., Preston, C.M., Delong, E.F., Massana, R., Post, A.F., Scanlan, D.J., and Beja, O. (2003). Molecular diversity among marine picophytoplankton as revealed by psbA analyses. Environ. Microbiol. *5*, 212–6.

The Use of Microarrays in Microbial Ecology

5

Gary L. Andersen, Zhili He, Todd Z. DeSantis, Eoin L. Brodie and Jizhong Zhou

Abstract

Microarrays have proven to be a useful and high-throughput method to provide targeted DNA sequence information for up to many thousands of specific genetic regions in a single test. A microarray consists of multiple DNA oligonucleotide probes that, under high stringency conditions, hybridize only to specific complementary nucleic acid sequences (targets). A fluorescent signal indicates the presence and, in many cases, the abundance of genetic regions of interest. In this chapter we will look at how microarrays are used in microbial ecology, especially with the recent increase in microbial community DNA sequence data. Of particular interest to microbial ecologists, phylogenetic microarrays are used for the analysis of phylotypes in a community and functional gene arrays are used for the analysis of functional genes, and, by inference, phylotypes in environmental samples. A phylogenetic microarray that has been developed by the Andersen laboratory, the PhyloChip, will be discussed as an example of a microarray that targets the known diversity within the 16S rRNA gene to determine microbial community composition. Using multiple, confirmatory probes to increase the confidence of detection and a mismatch probe for every perfect match probe to minimize the effect of cross-hybridization by non-target regions, the PhyloChip is able to simultaneously identify any of thousands of taxa present in an environmental sample. The PhyloChip is shown to reveal greater diversity within a community than rRNA gene sequencing due to the placement of the entire gene product on the microarray compared with the analysis of up to thousands of individual molecules by traditional sequencing methods. A functional gene array that has been developed by the Zhou laboratory, the GeoChip, will be discussed as an example of a microarray that dynamically identifies functional activities of multiple members within a community. The recent version of GeoChip contains more than 24,000 50mer oligonucleotide probes and covers more than 10,000 gene sequences in 150 gene categories involved in carbon, nitrogen, sulphur, and phosphorus cycling, metal resistance and reduction, and organic contaminant degradation. GeoChip can be used as a generic tool for microbial community analysis, and also link microbial community structure to ecosystem functioning. Examples of the application of both arrays in different environmental samples will be described in the two subsequent sections.

Introduction

The use of microarray technologies for microbial ecology studies

Microarrays have become an increasingly popular method in the microbial ecologist's toolkit for analysing microbial communities. This trend has also been observed in other disciplines resulting in an increase in PubMed entries mentioning microarrays from 20 in 1998 to over 34,300 by November 2008 (Loring, 2006). Within the scope of this chapter, microarrays can be simply thought of as a high-throughput way to detect the presence and concentration of multiple nucleotide sequences within an environmental sample. Sequence complementarity between single-stranded nucleic acid molecules, one of which is typically immobilized on the microarray, leads to the hybridization of specific target sequences from a sample. The advent of large genomic sequencing centres and the continuing decrease in sequencing costs have lead to an explosion of DNA sequence data, including data from environmental microorganisms. This has created an even greater need for high-throughput methods such as microarrays to make efficient use of the expanding sequence databases.

In this chapter we will discuss two specific examples of microarrays that are being used to answer questions of interest to microbial ecologists. The first example targets the 16S rRNA gene of bacteria and archaea (PhyloChip) to identify specific members within a complex microbial community. We will discuss how this microarray has been used to characterize the microbial diversity of selected environments. The second example targets known functional gene markers to study functional gene diversity and activities of microorganisms in the environment. We will demonstrate how functional gene arrays (GeoChips) have been used to analyse microbial communities, and provide linkages of microbial genes/populations to ecosystem processes and functions.

DNA microarrays

Based on the target of molecular markers, two types of major microarrays have been used for microbial community analysis. The first are phylogenetic microarrays that target phylogenetic genes, such as the 16S rRNA gene and *gyrB* gene. Phylogenetic microarrays allow us to identify microorganisms and their phylogenetic relationships in a community of interest. The second are functional gene arrays that target key functional gene markers that are indicative of a specific physiological or metabolic process, such as *nirK* and *nirS* encoding nitrite reductases, key enzymes of the denitrification process (Braker *et al.*, 2000), and *amoA* encoding ammonia monooxygenase, a key enzyme for ammonia oxidization (Rotthauwe *et al.*, 1997). Functional gene arrays allow us to study functional gene diversity and activities of microbial communities. In this section, both microarrays are introduced below.

Phylogenetic microarrays

Analysis of microbial communities with 16S rRNA targeted microarrays

Molecular methods for detecting and monitoring bacteria and archaea routinely rely upon classifying heterogeneous 16S rRNA molecules, either as RNA or as gene fragments encoding RNA that are amplified by universal PCR primers. The general method of sampling sequence types has been to clone and sequence PCR products derived from these biomarkers. However, the number of clones required to adequately catalogue the majority of taxa in a sample is typically unwieldy assuming a log-normal distribution (Curtis and Sloan, 2005). For example, in a typical soil sample with one billion bacterial cells and 10,000 different species, one would need to sample at least one million sequences (Gans *et al.*, 2005). Pyrosequencing has provided a alternative to cloning by producing a greater sample size at a lower cost (Huse *et al.*, 2008). But, as they are currently practiced, neither method is able to cover environmental microbial populations spanning multiple orders of magnitude within a single sample. Sequencing community members is an essential but inefficient process, where the biomarkers representing the most abundant phylotypes or species mask less abundant but potentially significant members.

As an alternative approach to biomarker sampling by cloning, hybridization of target sequences to an array of probes, permits much greater numbers of molecules to be sampled compared with the hundreds or thousands that usually comprise an environmental clone library. Because of the small, oftentimes single nucleotide differences within the probed regions of the biomarker genes for the differentiation of microbial taxa, a high level of sequence specificity is desired. Oligonucleotide DNA microarrays often consist of large numbers of individual short, 15- to 30-nucleotide capture probes to offer the highest level of specificity for the identification of specific target sequences, particularly in a background of closely related sequences. As stated earlier in Chapter 1, the 16S rRNA gene sequence provides a number of advantages in its use as a biomarker for the identification of individual bacterial components in complex environmental samples. Rather than translating its genetic code into protein, rRNA acts directly in the protein assembly machinery as a functional molecule. Due to structural constraints of this molecule, specific regions throughout the 16S rRNA gene have a highly conserved nucleotide sequence while non-structural segments may have a high degree of variability (Woese *et al.*, 1975). Probing the regions of high variability can be used to identify microorganisms at the species level while regions of less variability are used for group-level identification. With only one to a few nucleotides of sequence variability, at best, within any 15- to 30-bp region that may be targeted by a probe for discrimination between related microbial species, it is imperative to maximize the probe–target sequence specificity in the microarray system.

One example of a microarray that has been successfully used to discriminate bacterial species uses a hierarchical set of oligonucleotide probes to target organisms at different levels of taxonomic specificity on a matrix of acrylamide gel pads on a glass slide (Liu *et al.*, 2001). Developed by Liu and Stahl, this method uses the 3-dimensional nature of the gel matrix to allow solution-based probe kinetics with a non-equilibrium dissociation approach for high levels of discrimination between target and non-target 16S rRNA gene sequences. Amplified 16S rRNA gene sequences are placed on the gel-pad microarray and allowed to hybridize under low-stringency conditions. Increasing the hybridization temperature (increased stringency) results in the preferential dissociation of non-homologous probe–target complexes. The simultaneously generated melting curves for the perfect match (PM) and mismatch (MM) duplexes are used to define the temperature at which 50% of the starting duplex remains intact (dissociation temperature, T_d.) It was found that, for the most part, a probe–target duplex with one MM has a greater than two-fold level of discrimination from a PM duplex at T_d, thus allowing greater specificity of detection by differentiating between PM and MM complexes.

By contrast, other 16S rRNA gene-targeted or phylogenetic microarrays use short oligonucleotide probes bound to a two-dimensional surface (e.g. glass) with specific sequences located at defined two dimensional coordinates. Many studies have successfully used these 16S rRNA gene-targeted microarrays to differentiate bacteria in specific groups, such as *Enterococcus* (Lehner *et al.*, 2005), *Cyanobacteria* (Castiglioni *et al.*, 2004), nitrifying bacteria (Kelly *et al.*, 2005) and fish pathogens (Warsen *et al.*, 2004) and for quantitative tuning of sampling protocols to enhance detection of desired taxonomic groups.

A comprehensive view of microbial diversity

Instead of targeting specific groups or classes of organisms, another strategy is to perform a comprehensive screen for all known bacterial or archaeal taxa on a single microarray. This relies, initially, on obtaining all known 16S rRNA gene sequences from the major sequence repositories, including Lawrence Berkeley National Laboratory's Greengenes (greengenes.lbl.gov) (DeSantis *et al.*, 2006b), Michigan State University's ribosomal database project (RDP; rdp.cme.msu.edu) (Cole *et al.*, 2005), the Max Planck Institute for Marine Microbiology's Silva database (http://www.arb-silva.de/), and the National Institute of Health's NCBI (www.ncbi.nlm.nih.gov). There are currently over 700,000 individual sequences housed in these repositories but due to the lack of peer-review before inputting the sequences from individual submitters there are a number of quality control issues that may not have been addressed, producing inaccurate data. To reduce this problem, a series of filters on the sequence data may be employed to reduce the possibility that an assay is created for a non-existing target. The Greengenes database of 16S rRNA gene sequences compiles information from other databases to produce a set of sequences that are compatible with a comprehensive phylogenetic microarray design. Among the issues addressed by this database are (i) standardized taxonomic placement of individual 16S rRNA gene sequences, (ii) removal of chimeric sequences, (iii) removal of poor quality (ambiguous) sequences, and (iv) distribution of data in a consistently aligned sequence format.

Because the discovery rate of 16S sequence records from uncultured organisms now exceeds that from their cultured counterparts, taxonomic placement of sequences lags behind. In fact, over one-third of full-length 16S records in GenBank are presented without taxonomic nomenclature and are simply annotated as 'environmental samples' or 'unclassified'. In contrast, records in the Greengenes dataset are annotated with taxonomy proposed by five independent curators – NCBI, RDP, based on *Bergey's Manual*?; Cole *et al.* (2005); Wolfgang Ludwig (Ludwig *et al.*, 2004); Phil Hugenholtz (Hugenholtz, 2002); and Norm Pace (Pace, 1997) – collectively covering over 95% of the database. Incongruent taxonomic nomenclature exists among curators even at the phylum level, yet each is tracked to promote user awareness of several estimations of phylogenetic descent allowing a balanced approach to node and operational taxon unit (OTU) nomenclature when labelling probe specificity.

Since 16S rRNA genes from environmental DNA are usually PCR amplified, it has been suspected that many chimeric sequences are present in the public repositories. A PCR-generated chimeric sequence usually comprises two phylogenetically distinct parent sequences and occurs when a prematurely terminated amplicon re-anneals to a foreign DNA strand and is copied to completion in the following PCR cycles. The point at which the chimeric sequence changes from reflecting one parent to the next (breakpoint) can be misconstrued as novel biomarker for a unique organism (Fig. 5.1). The trend has been observed in 3% of the sequences from uncultured organisms and 0.2% of sequences annotated as pure cultures (Ashelford *et al.*, 2005; DeSantis *et al.*, 2006b). Very recently, large 16S rRNA clone libraries have been deposited to GenBank with an inferred chimeric content of up to 8% of the sequences (Ashelford *et al.*, 2006). The Bellerophon (Huber *et al.*, 2004) chimera test result is available for each Greengenes 16S record allowing putative chimeras to be avoided. This step helps to ensure that all probes in a microarray design are complementary to sequences from true organisms.

Public DNA data repositories rarely quality-check the primary data (chromatograms) of the

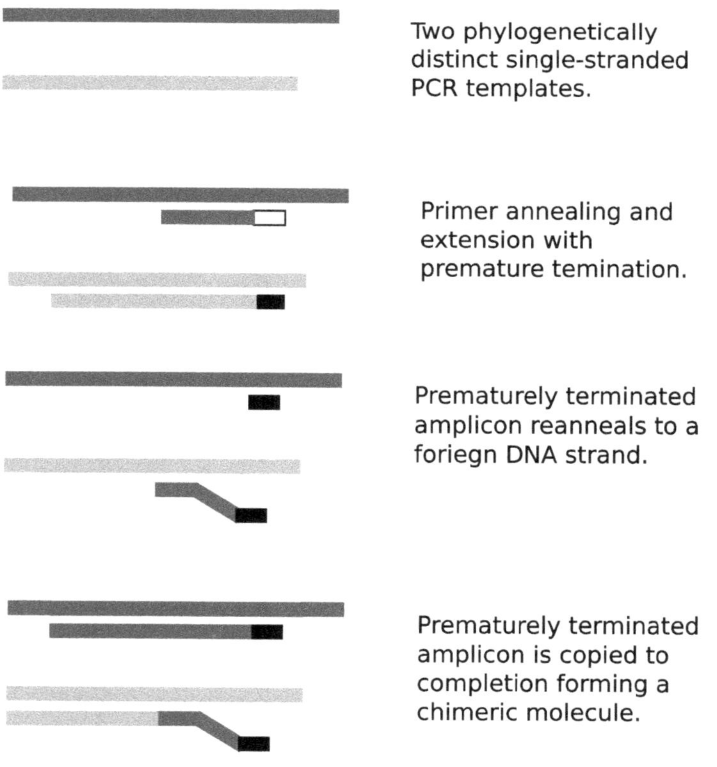

Figure 5.1 An example of a chimeric artifact generated during PCR amplification of a mixed population using broad-specificity 16S rRNA gene primers. Partial amplicons may form hybrids with dissimilar templates because conserved regions exist at positions medial to the PCR primer targets. The partial amplicon can be extended using the dissimilar 16S gene as a template.

sequences they distribute. Unfortunately, the majority of 16S rRNA gene sequences in the public databases are decoded from single coverage sequencing reads typically yielding 'non-ambiguous' yet non-reproducible base calls in 2 of 400 chromatogram peaks (0.5%) even under ideal conditions where data is collected in small batches by experienced research groups (Fields *et al.*, 2006). The problem is magnified in high-throughput efforts. Assuming chromatograms or quality scores are not obtainable for the sequence collection, various poor quality records can still be identified. Obvious sequencing reaction failures for the majority of the gene or a just a region can be easily discarded or trimmed based on the distribution of ambiguous base-calls (non-'ACGT' characters). Interpreted chromatograms should also have zero or few long homopolymeric regions (i.e. eight or more sequential adenines reported) since these have rarely been confirmed in isolates and base calls downstream of the homopolymer commonly have increased error rates. Furthermore, an emphasis should be placed on reads covering a large portion of the gene of interest so that multiple probes can be picked from throughout the domains. Greengenes enables de-selection of records with an undesirable percentage of ambiguous calls, one or more homopolymeric runs, and low gene coverage (short reads).

Functional gene arrays

As described in Chapter 6, 1 gram of soil contains more than 5000 microorganisms, and a majority of them (> 99%) have not yet been cultivated (Whitman *et al.*, 1998), which presents enormous difficulties for microbiologists to study microbial composition, structure, function, and dynamics in natural and/or contaminated environments. Using functional gene markers to investigate such microbial communities is therefore necessary. Schadt *et al.* (2004) listed some of those functional genes currently used in environmental studies that allow us to study functional gene diversity and activities of microorganisms in the environment. Currently, conventional molecular methods, such as PCR-based cloning, and *in situ* hybridization are very useful in providing snap shots for microbial diversity, structure and function, but they fail to provide a full picture of microbial activities and dynamics in a rapid and high-through-put fashion. Although microarray technology has been used successfully to analyse global gene expression in pure culture studies, adapting microarray technology for use in environmental studies presents numerous challenges in terms of probe design, the coverage of gene sequences, specificity, sensitivity and quantitative capability. To overcome such obstacles for studying microbial communities in natural settings, a particular type of microarrays, called functional gene arrays (FGAs), has been developed and used. This type of microarray contains probes from genes involved in key microbially mediated biogeochemical processes, such as C, N, P, and S cycling and utilization, organic contaminant degradation, and metal reduction and resistance. FGAs that mainly target geochemical processes are also called GeoChips (He *et al.*, 2007). FGAs are powerful tools to address some fundamental questions in microbial ecology, biogeochemistry, and environmental biology, such as: (i) What functional genes/microorganisms are in a microbial community? (ii) What biological or geochemical processes dominate in a microbial community? (iii) What are the dynamics of activity for a given gene or process? (iv) How does microbial community structure link to its function? and (v) What are the relationships between functional gene activities/abundance and geochemical parameters?

For almost a decade, microbiologists, especially microbial ecologists, have tried to answer a central question whether microarray-based FGA technologies can provide specific, sensitive and quantitative detection of microbial populations and activities within the context of environmental applications. The technology has evolved several generations in terms of gene/probe coverage and related computational techniques for functional gene sequence retrieval, probe selection, data analysis, and information storage. The first generation of FGAs was prototype microarrays with a focus on proof of the concept. For example, such a prototype microarray was constructed with 89 PCR-amplicon probes targeting functional genes involved in nitrogen cycling using pure cultures and laboratory clones (Tiquia *et al.*, 2004; Wu *et al.*, 2001). It presents a great challenge to obtain PCR amplicon probes from a diversity of environmental clones and bacterial strains from various sources, so constructing a comprehensive FGA is very difficult. Therefore, oligonucleotide-based FGAs have become more popular due to their high specificity, ease of construction, and low cost. Oligonucleotide FGAs are fabricated with synthetic oligonucleotide probes with 50mer (Rhee, 2004) or 70mer (Taroncher-Oldenburg *et al.*, 2003) in length, and they can be spotted on glass slides (Rhee *et al.*, 2004; Taroncher-Oldenburg *et al.*, 2003; Tiquia *et al.*, 2004; Wu *et al.*, 2001), or nylon membranes (Steward *et al.*, 2004). Different FGAs have been systematically evaluated in terms of their sensitivity, specificity, and quantitation, and the results demonstrate that FGA technology holds promise for the analysis of microbial communities. In addition, an FGA may target both functional and phylogenetic markers. For example, Loy *et al.* (Loy *et al.*, 2004) constructed an array containing both phylogenetic (16S rRNA gene) and functional (*dsrA/B*) markers and those example FGAs were classified as GeoChip 1.0. The current version, GeoChip 2.0 has been designed and used for a comprehensive analysis of microbial community structure, function, and dynamics in a natural or contaminated environment. The coverage of gene sequences/probes has greatly increased from hundreds to tens of thousands, and related computational techniques, such as sequence retrieval, probe design, and data analysis have been greatly improved. GeoChip 2.0 contains 24,243 oligonucleotide (50mer) probes and covers > 10,000 genes in > 150 functional groups involved in nitrogen, carbon, sulphur and phosphorus cycling, metal reduction and resistance, and organic contaminant degradation (Table 5.1; He *et al.*, 2007). This array uses experimentally established probe design criteria (He *et al.*, 2005b; Liebich *et al.*, 2006) and a new computational software tool, CommOligo (Li *et al.*, 2005) for oligonucleotide probe selection. In addition, an analysis of subnanogram quantities of microbial community DNA has been achieved by whole-community genome amplification (Wu *et al.*, 2006). This approach made it possible to analyse microbial communities with low biomass. The next generation, GeoChip 3.0, is currently being developed. It is expected to be more comprehensive, covering > 45,000 gene variants (sequences) in > 290 gene categories. Many new features will be implemented.

General experimental procedures

Sample preparation

For more detailed description on how microarrays are used to identify and quantify specific genetic sequences for biological research in general please refer to the excellent recent reviews by R. B. Stoughton and J. W. Edwards (Allison *et al.*, 2007; Stoughton, 2005). One of the main factors distinguishing the use of microarrays within microbial ecology from other areas of study is that the genetic sequences are derived from multiple organisms in what is usually a complex microbial community. Because the constituents of a microbial community typically have very different types of cell walls it is important to find a target isolation and purification method that is suitable for a wide range of conditions. Equally important for microbial ecology applications is to remove environmental inhibitors of nucleic acid amplification. An example of a robust nucleic acid extraction and purification protocol is the

Table 5.1 List of major functional markers on the GeoChip 2.0

Gene category	Example of key enzyme (gene)	Total_probe#
Nitrogen cycling		5310
Nitrogen fixation	Nitrogenase (nifH)	1225
Denitrification	Nitrate reductase (*narG*, *napA*, *nasA*), nitrite reductase (*nirS*, *nirK*), nitric oxide reductase (*norB*), nitrous oxide reductase (*nosZ*)	2086
Nitrogen reduction	Nitrate reductase (*napA*, *nasA*)	220

Table 5.1 Continued

Gene category	Example of key enzyme (gene)	Total_probe#
Nitrification	Ammonium monooxygenase (amoA), hydroxylamine oxidoreductase (hao)	347
Nitrogen mineralization	Urease (*ureC*), glutamate dehydrogenase (*gdh*)	1432
Carbon cycling		4599
Carbon fixation	Rubisco (cbbL, rbcL), Acl (aclB), CODH, FTHFS	1018
Cellulose degradation	Cellulase, endoglucanase	1285
Lignin degradation	Laccase, mannanase	513
Chitin degradation	Endochitinase (chiA), exochitinase	744
Methane production	Methyl coenzyme M reductase (mcrA)	437
Methane oxidation	Methane monooxygenase (pmoA)	336
Others	Lignin peroxidase (*lip*), pectinase, cellobiase	266
Sulphate reduction		1615
	Sulphite reductase (dsrA/B), APS (apsA)	1615
Phosphorus utilization		145
	Exopolyphosphatase (*ppx*), phytase	145
Metal reduction and resistance		4546
Arsenic resistance	Arsenate reductase (arsC, arsB, arsC)	877
Cadmium resistance	Cadmium transporter (cadA, cadB, cadC)	282
Chromium resistance	Chromium/chromate transporter (chrA)	319
Mercury resistance/reduction	Mercuric ion reductase/transporter (mer, merA, merB)	548
Nickel resistance	Nickel transporter (nccA), permease (nreB)	140
Zinc resistance	Zinc resistance protein (zntA)	128
Other metal resistance/reduction	cobalt resistance proteins, selenium reductase, etc.	2252
Contaminant degradation		8028
Benzene, toluene, ethylbenzene, and xylene (BTEX) and related aromatics	Benzene 1,2-dioxygenase (ben), ethylbenzene dehydrogenase (ebd), benzylsuccinate synthase (bss), xylene monooxygenase (*xyl*), benzoyl-CoA reductase (*bad*), and catechol 1,2-dioxygenase (*cat*, *tfd*).	4176
Chlorinated aromatics	Chlorophenol reductive dehalogenase (cpr)	90
Nitroaromatics	Nitrobenzene nitroreductase (nbz), 4-nitrobenzaldehyde dehydrogenase (ntn), p-nitrobenzoate reductase (pnb)	152
Polycyclic aromatic hydrocarbons (PAHs)	Naphthalene dioxygenase (nah), PAH ring-hydroxylating dioxygenase (pdo)	741
Polychlorinated biphenyls (PCBs)	Biphenyl dioxygenase (bph)	388
Chlorinated solvents (e.g. PCE)	PCE/TCE reductive dehalogenase (rdh, pceA, tecA)	232
Other organic compounds/by-products	Alkane hydroxylase (alk), homogentisate 1,2-dioxygenase (hmg), vanillate O-demethylase oxygenase (van), etc.	2249
Total	19959	24243

method established by Zhou *et al.* (1996) or using other approaches described in Chapters 4 and 6. The method used normally requires modification depending on the experimental goals and environmental sample type, such as soils, sediments, and groundwater (Hurt *et al.*, 2001). Many DNA extraction and purification kits are commercially available, and these were recently compared in terms of DNA extraction efficiency from different types of samples (Klerks *et al.*, 2006). Purified DNA samples should have A_{260}/A_{280} > 1.80, and A_{260}/A_{230} > 1.70. Since a typical hybridization for GeoChip analysis requires 2–5 µg purified DNA, samples with lower than 2 µg require amplification using the currently developed WCGA (whole community genome amplification) method (Wu *et al.*, 2006).

Obtaining purified mRNA from environmental samples is an even greater challenge than DNA, especially for low-abundance mRNAs. This is still often desirable since mRNA is an ideal indictor of microbial activity. Total RNA can be isolated and purified using the approach described by Hurt *et al.* (2001). This method can isolate DNA and RNA simultaneously with the same sample. Recently, a new gel electrophoresis method to isolate community RNA was developed (McGrath *et al.*, 2008). Normally, the ratios of A_{260}/A_{280} and A_{260}/A_{230} for a purified RNA are expected to be > 1.90, and > 1.70, respectively. For analysis of a microbial community, a normal hybridization requires 10–20 µg purified RNA, samples with lower than 5 µg will again require amplification, as may be performed with the novel whole community RNA amplification approach (WCRA) (Gao *et al.*, 2007) to obtain cDNA. With such a method, 1200- to 1800-fold amplification can be obtained with 10–100 ng of RNA as templates (Gao *et al.*, 2007).

Microarray fabrication

There are several different styles of microarrays used for microbial community analysis. The Affymetrix (Santa Clara, CA, USA) platform DNA arrays that are used on phylogenetic arrays such as the PhyloChip have the short oligonucleotide probes (~25mer) synthesized directly on the glass surface by a photolithography method at an approximate density of 10,000 molecules per µm^2 (Chee *et al.*, 1996). Spotted DNA arrays use oligonucleotides that are synthesized individually at a predefined concentration and are applied to a chemically activated glass surface. Oligonucleotide length can range from a few nucleotides to hundreds of bases in length but are typically in the 50mer range for functional arrays such as the GeoChip.

Target labelling

The nucleic acid targets are labelled so that a laser scanner tuned to a specific wavelength can measure the number of fluorescent molecules that hybridized to a specific DNA probe. For photolithography arrays such as Affymetrix, the nucleic acid targets are fragmented to between 50 and 100 bp in size and a biotinylated nucleotide is added to the end of the fragment by terminal DNA transferase. At a later stage, the biotinylated fragments that hybridize to the oligonucleotide probes are used as a substrate for the addition of multiple phycoerythrin fluorophores by a sandwich antibody (Streptavidin) method.

For spotted arrays such as the GeoChip, the purified community DNA can be fluorescently labelled by random priming using the Klenow fragment of DNA polymerase as described previously (Wu *et al.*, 2006) and more than one fluorescent moiety can be used (e.g. controls could be labelled with Cy3, and experimental samples labelled with Cy5 for direct comparison by hybridization to a single microarray). Total community RNA (e.g. 5–10 µg) can be labelled using Cy5 or Cy3 with Superscript^{TM} II/III RNase H$^-$ reverse transcriptase (Invitrogen Life Technologies, CA, USA) as described by He *et al.* (2005a). The labelled cDNA target is then purified and concentrated.

Hybridization

Microarray hybridizations are then carried out under stringent conditions described previously (Rhee *et al.*, 2004; Wu *et al.*, 2006). The temperature can be lowered to reduce stringency and allow the detection of more divergent sequences. Robotic hybridization and stringency wash stations can be used to give more consistent results. The photolithography arrays use an automated Affymetrix hybridization and fluidics station for all washes as well as fluorescent staining with antibody and phycoerythrin. Spotted arrays can use the TECAN HS4800 (Tecan US, Inc., Durham, NC, USA) to replace manual hybridization, which allows 48 hybridizations to be completed in 6 hours.

Signal quantification and analysis

After hybridization the arrays are scanned using a microarray scanner (e.g. GeneChip Scanner 3000, Affymetrix, Santa Clara, CA for PhyloChip, or ProScan Array, Perkin Elmer, Boston, MA, USA, for GeoChip) equipped with lasers at a resolution of 10 µm or finer. The scanned image displays are saved and analysed by quantifying the pixel density (intensity) of each spot using image quantification software (e.g. GeneChip Microarray Analysis Suite, version 5.1 Affymetrix, Santa Clara, CA for PhyloChip, or ImaGene 6.0, Biodiscovery Inc. Los Angeles, CA, USA, for GeoChip).

PhyloChip, a phylogenetic microarray

PhyloChip design

The key considerations that must be taken into account in designing a 16S rRNA gene-based microarray to identify individual organisms in a complex environmental mixture are (1) natural sequence diversity and (2) potential cross-hybridization. Sequence diversity is an issue as we sample new and distinctive environments such as bioaerosols. There may be many undocumented organisms with 16S rRNA gene sequences that are similar, but not identical to the sequences that were used for array design. Microarrays based upon single sequence-specific hybridizations (single probes) per OTU may be ineffective in detecting such environmental sequences with one or several polymorphisms. To overcome this obstacle, an Affymetrix-style photolithography chip was designed with a minimum of 11 different, short oligonucleotide probes for each taxonomic grouping, allowing for the failure of one or more probes. Also important is non-specific cross hybridization, especially when an abundant 16S rRNA gene shares sufficient sequence similarity to non-targeted probes, such that a weak but detectable signal is obtained. It has been found that the perfect match-mismatch (PM-MM) probe pair approach effectively minimizes the influence of cross-hybridization. Widely used on expression arrays as a control for non-specific binding (Chee *et al.*, 1996), the central nucleotide is replaced with any of the three non-matching bases so that the increased hybridization intensity signal of the PM over the paired MM indicates a sequence-specific, positive hybridization. By requiring multiple PM-MM probe-pairs to have a positive interaction, we substantially increase the chance that the hybridization signal is due to a predicted target sequence.

Once a reference set of valid genes is established it is important to create a multiple sequence alignment (MSA). The MSA allows confident comparisons between sequences when selecting probes. For instance, when a candidate probe does not complement a sequence it is practical to determine if sequence data is available at the expected probe position. Filtered, aligned 16S rRNA sequence records can be exported directly from Greengenes. Since the Greengenes database consistently spreads the gene into an alignment of 7682 characters in width, then private, in-house sequences can be formatted into the same MSA using the NAST (Nearest Alignment Space Termination) web tool (DeSantis *et al.*, 2006a). Other strategies exist for compiling MSAs from various sequence sources and the choice is governed by the size of the project. In general, alignment can be done with clustalw (Thompson *et al.*, 1994) for small MSAs (< 500 sequences), MUSCLE (Edgar, 2004) for mid-size MSAs, (500–10,000 sequences) or NAST (DeSantis *et al.*, 2006a) for large MSAs (> 10,000 genes).

Simultaneous clustering of genes and probes.

The objective of the probe selection strategy for comprehensive prokaryotic identification is to obtain an effective set of probes capable of correctly categorizing mixed amplicons into their proper operational taxonomic unit (OTU) designations. Each OTU is formed from sequences, which have common oligomer targets, considering only those targets that meet the G+C, melting temperature, and secondary structure constraints of the design. Whereas clustering implemented to infer phylogeny may utilize similarities along the entire gene, creating OTUs relies on finding shared attributes that can be assayed. A supervised clustering procedure is used which consists first of generating numerous candidate OTUs by unsupervised hierarchical clustering using pairwise gene distances measured by megaBLAST (Zhang *et al.*, 2000) or counts of unique targets. Then, each candidate OTU is evaluated to determine the count of targets which are simultaneously prevalent across the genes of the candidate OTU and also incapable of hybridization to genes outside the OTU.

In designing the G2 PhyloChip, probes presumed to have the capacity to correctly hybridize were those unique 25mers that also contain a central 17mer not matching any sequences outside the OTU (Urakawa *et al.*, 2002). Thus, probes that were unique to an OTU solely due to a distinctive base in one of the outer four bases were avoided. For each OTU harvested from the hierarchical trees, a set of 11 or more specific 25mers (probes) was sought. Three classes of probe sets resulted from the clustering procedure. The resulting 8741 OTUs, each containing an average of 3% sequence divergence, represented all 121 demarcated bacterial and archaeal orders. In most cases, as expected, the OTUs contained sequences that were previously identified as related using the phylogenic tree approaches. For a majority of the OTUs represented on the PhyloChip (5737; 65%), probes were designed from regions of gene sequences that have been identified only within a given taxon. For 1198 taxa (14%), no probe-level sequence could be identified that was not shared with other groups of 16S rRNA gene sequences, although the gene sequence as a whole was distinctive. For these taxonomic groupings, an average of 24 probes (but no less than 11 probes) was designed to a combination of regions on the 16S rRNA gene that taken together as a whole did not exist in any other taxa. For the remaining 1806 taxa (21%), a set of probes were selected to minimize the number of putative cross-reactive taxa. Although more than half of the probes in this group have a hybridization potential to one outside sequence, this sequence was typically from a phylogenetically similar taxon. For all three probe set classes, the advantage of the hybridization approach used was that multiple OTUs could be identified simultaneously by targeting unique regions or combinations or regions from genes.

Analysing PhyloChip data

The G2 PhyloChip consists of 506,944 probe features, arranged as a grid of 712 rows and columns. Of these features, 297,851 are oligonucleotide PM or MM probes with exact or inexact complementarity, respectively, to 16S rRNA genes. The remaining probes are used for image orientation, normalization controls, or for pathogen-specific signature amplicon detection using additional targeted regions of the chromosome (Wilson *et al.*, 2002). Each high-density 16S rRNA gene microarray is designed with additional probes that (1) target amplicons of prokaryotic metabolic genes spiked into the 16S rRNA gene amplicon mix in defined quantities just prior to fragmentation and (2) are complementary to pre-labelled oligonucleotides added into the hybridization mix. The first control collectively tests the target fragmentation, labelling by biotinylation, array hybridization, and staining/scanning efficiency. It also allows the overall fluorescent intensity to be normalized across all the arrays in an experiment. The second control directly assays the hybridization, staining and scanning.

Complementary targets to the probe sequences hybridize to the array and fluorescent signals are captured as pixel images using standard Affymetrix software (GeneChip Microarray Analysis Suite, version 5.1) that reduces the data to an individual signal value for each probe and is typically exported as a human readable 'CEL' file. Background probes are identified from the CEL file as those producing intensities in the lowest 2% of all intensities. The average intensity of the background probes is subtracted from the fluorescence intensity of all probes. The noise value (N) is the variation in pixel intensity signals observed by the scanner as it reads the array surface. The standard deviation of the pixel intensities within each of the identified background probe intensities is divided by the square root of the number of pixels comprising that feature. The average of the resulting quotients is used for N in the calculations described below.

Probe pairs scored as positive are those that meet two criteria: (i) the fluorescence intensity from the perfectly matched probe (PM) is at least 1.3 times greater than the intensity from the mismatched control (MM), and (ii) the difference in intensity, PM minus MM, is at least 130 times greater than the squared noise value (> 130 N^2). The positive fraction (PosFrac) is calculated for each probe set as the number of positive probe pairs divided by the total number of probe pairs in a probe set. An OTU is considered 'present' when its PosFrac for the corresponding probe set is > 0.92 (based on empirical data from clone library analyses). Replicate arrays can be used collectively in determining the presence of each OTU by requiring each to exceed a PosFrac threshold. Present calls are propagated upwards through the taxonomic hierarchy by considering any node (subfamily, family, order, etc.) as 'present' if at least one of its subordinate OTUs was present.

Hybridization intensity is the measure of OTU abundance and is calculated in arbitrary units for each probe set as the trimmed average (maximum and minimum values removed before averaging) of the PM minus MM intensity differences across the probe pairs in a given probe set. All intensities < 1 are shifted to 1 to avoid errors in subsequent logarithmic transformations.

Comparing PhyloChip and clone libraries

The breadth and accuracy of the PhyloChip in detecting diverse 16S rRNA gene sequence types compared to cloning-and-sequencing can be directly compared (DeSantis *et al.*, 2007). Using three environmental samples: urban aerosol, subsurface soil and subsurface water, 16S PCR products were classified using both methods. Approximately 8% of the clones could not be placed into a known subfamily and were considered novel. The microarray results confirmed the majority of clone-detected subfamilies and additionally demonstrated greater amplicon diversity (Fig. 5.2, left) extending into phyla not observed by the cloning method. Sequences within the phyla *Nitrospira*, *Planctomycetes*, and TM7, which were uniquely detected by the array, were verified with specific primers and subsequent amplicon sequencing. Communities dominated by few subfamilies are most likely to errantly appear well sampled (See accumulation curve for water, Fig 5.2, right) by the clone library. The PhyloChip enables observation of the rarer bacterial populations since the entire mass of PCR products (~10^{12} molecules) are sampled as opposed to a

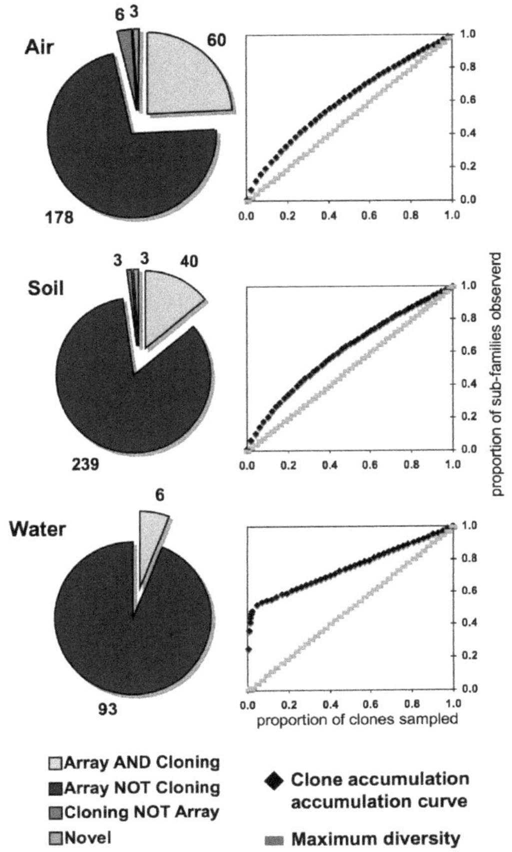

Figure 5.2 Comparison of cloning and PhyloChip (Array) analysis of air, soil and water samples. Left panel shows numbers of bacterial sub-families detected by array and cloning, array only, cloning only and novel sequences not detectable by the array. Right panel shows sub-family accumulation curves obtained by serial clone observations. The maximum diversity scenario where every clone sampled represents a new sub-family is indicated by grey line with constant slope = 1.

mere hundreds to thousands when cloning. In general, although the microarray is unreliable in identifying novel taxa, it reveals greater diversity in environmental samples than sequencing a typically sized (300–700) clone library. Deeper sequencing of up to several hundred thousand targets is possible with pyrosequencing methods. This has the potential to reduce the difference in diversity detected by both methods but at present such efforts are typically confined to fewer samples and not to fully replicated ecological studies.

Direct analysis of microbial community activity by microarray analysis of ribosomal RNA (rRNA).

While analysis of DNA by high-density phylogenetic microarrays provides information regarding relative changes in biomass within a microbial community, cell division in natural ecosystems may occur on the order of hours to decades or even millennia (Phelps *et al.*, 1994). Therefore the immediate environmental impacts on microorganisms may not be apparent through the analysis of DNA turnover rates. As in studies using stable-isotope labelled substrates such as ^{13}C (see Chapter 4), cellular RNA due to its higher synthesis rates, responds rapidly to environmental stimuli. For this reason, assaying the more labile RNA molecules permits immediate impacts to be monitored independent of replication.

Many standard approaches to sample preparation for RNA analysis require multiple enzymatic steps to replicate, amplify and label nucleic acid targets (e.g. reverse transcription, *in vitro* transcription or PCR). An advantage of these approaches is the ability to detect lower-abundance species but the quantitative representation of the original population may be altered during enzymatic amplification steps, such as PCR, or by the labelling procedure itself. Direct labelling of RNA on the other hand does not typically require amplification and also necessitates fewer manipulations, leading to a reduction in bias.

Direct labelling of RNA can be accomplished by various means such as base-incorporation with labelled nucleotides, or chemical modification. We have chosen to use a recently developed direct-labelling method using T4 RNA ligase to attach a biotinylated nucleotide donor molecule to the 3′ end of RNA targets according to the approach of Cole *et al* (Cole *et al.*, 2004). This approach results in uniform end-labelling of RNA fragments, avoiding sequence bias inherent in methods which require incorporation of labelled nucleotides during synthesis (e.g. biotin-dUTP). End-labelled RNA produced through this approach is predicted to have higher target–probe affinity because hybridization is unimpaired by label molecules, whereas methods that either label the nucleic acid strand internally, such as biotin–ULS (Vanbelkum *et al.*, 1994), or chemically modify a nucleotide (Kelly *et al.*, 2002) can result in decreased hybridization efficiency (Cook *et al.*, 1988). As the probes on the Affymetrix platform are tethered at the 3′ end, the 3′ position of the label will result in increased exposure of the biotin for more efficient binding by the streptavidin–fluorophore conjugate. Furthermore the donor molecule used to label targets with biotin contains biotin moieties tethered to the 3′-hydroxyl, rather than attached to the nucleobase. This permits multiple biotin molecules (three in this case) to be tethered sequentially to the donor nucleotide without any significant reduction in ligation efficiency and has the added potential of enhancing overall signal intensity to improve the detection of lower-abundance/activity taxa.

Other groups have also been successful with direct microarray-based rRNA analysis. Small *et al.* (2001) hybridized total RNA from a soil extract to a simple microarray and demonstrated that rRNA could be analysed without amplification, however they used biotin labelled detector probes rather than direct RNA target labelling. An important finding in this study was that fragmentation of the RNA produced greater hybridization specificity and sensitivity presumably through reduction of the complex secondary structure inherent in rRNA molecules. With this in mind, the choice of the most appropriate method for fragmentation must consider compatibility with the labelling procedure. Again, we have chosen to use the approach of Cole *et al* (2004) by using an enzymatic RNA fragmentation procedure. This procedure uses RNaseIII (a double strand-specific ribonuclease) and shrimp alkaline phosphatase to simultaneously perform controlled fragmentation and dephosphorylation of the target RNA, yielding fragments in the range of 20–200 nucleotides. The dephosphorylation step is necessary because of the requirement of T4 RNA ligase for a 3′-hydroxyl on the RNA target.

Example application of direct RNA labelling and PhyloChip analysis – monitoring bioremediation of uranium-contaminated sediments

A promising strategy for the containment of uranium within contaminated sites is *in situ* bioprecipitation. The valency state of uranium in contaminated groundwater and sediment,

U(VI), is soluble and thus capable of transport by advective and diffusive processes. The tetravalent form, U(IV) is highly insoluble and therefore has reduced mobility. Metal-reducing bacteria such as those that typically respire iron have been proposed as efficient catalysts for reductive precipitation of uranium and several laboratory and field studies have demonstrated the feasibility of this approach (e.g. Anderson *et al.*, 2003; Brodie *et al.*, 2006; Chang *et al.*, 2005; Peacock *et al.*, 2004; Reardon *et al.*, 2004; Wan *et al.*, 2005). One concern however, is the long-term stability of the bioreduced U(IV) and recent studies have demonstrated re-oxidation under anaerobic conditions (Ginder-Vogel *et al.*, 2006; Wan *et al.*, 2005).

In our studies we noted a carbonate-associated re-oxidation of U(IV) and further experiments were carried out to determine the influence of organic carbon (OC) electron donor concentration on the extent of re-oxidation (Wan *et al.*, 2005). In one set of columns, concentrations (32 mM OC) were retained at the concentration where re-oxidation was first noted after a period of successful reduction. In another set of columns, initially at 32mM OC, we increased the concentration to 100 mM OC while eliminating OC from a third set of columns (0mM OC). As hypothesized, uranium re-oxidation increased with increasing OC concentration and was correlated with increased carbonate concentrations. Uranium re-oxidation ceased in the columns without OC additions. To examine the impact of changing carbon concentrations on the microbial community structure we first employed PhyloChip analysis using DNA extracted from the column sediments. Following amplification of the 16S rRNA gene, array analysis between the three treatments (0, 32 and 100 mM OC) indicated no significant differences in hybridization intensity among the 271 subfamilies detected (of the 842 subfamilies represented on the array). Fig. 5.3A shows a comparison by DNA analysis between 0 and 100 mM OC treatments. At the point of sampling, columns had been running for almost two years with continuous OC addition, and most available electron acceptors (oxygen, nitrate, iron, manganese, sulphate) had been depleted or were of low bioavailability, therefore limitation on microbial replication may have been due to the limited electron acceptor and not the electron donor. Despite depletion of electron donors and the re-oxidation of uranium, several genera of metal-reducing bacteria including *Geobacter* and *Desulphovibrio* were still detected at similar intensities to those detected during the reducing phase. We hypothesized that cell-turnover was electron-acceptor limited but cellular activity was being maintained through interspecies electron transfer. To assess this, we extracted RNA from the sediment samples, gel-purified the 16S rRNA band, fragmented, labelled and hybridized the material directly to the PhyloChip without any primers or amplification steps.

Fig. 5.3B shows a comparative analysis of hybridization intensities between the 0 mM and 100 mM organic carbon (OC) treatments. Immediately apparent was the detection of

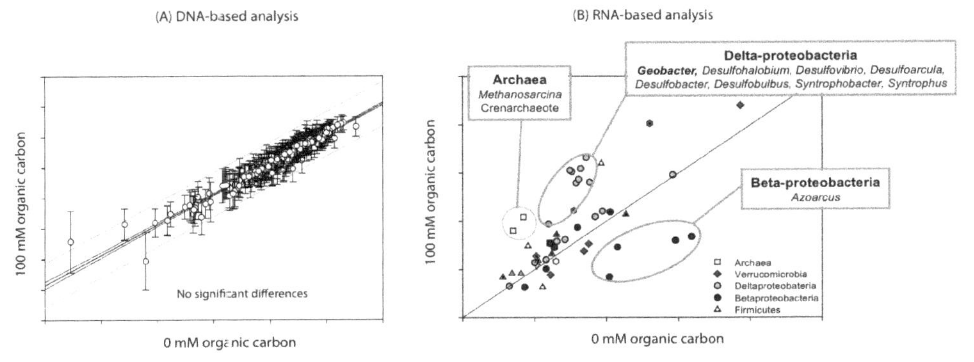

Figure 5.3 Comparison of microbial population dynamics under contrasting carbon amendment conditions using PhyloChip analysis of **A** DNA (PCR amplicons) and **B** RNA hybridization. X and Y axes display probe set intensities in arbitrary units under two carbon conditions (100 and 0 mM).

Geobacter 16S rRNA, indicating continued activity of this proven uranium reducing bacterial genus, despite observed re-oxidation. We also detected 16S rRNA of the methanogenic genus *Methanosarcina*. Various methanogens and sulphate reducers (also detected) are known to enter into syntrophic associations with the methanogens acting as biological electron acceptors in the absence of sulphate (Bryant *et al.*, 1977). Other active organisms detected such as the acetogens, *Syntrophobacter* and *Syntrophus*, are specialized in exocellular electron transfer and may also serve as a sink for electrons in the absence of sulphate or possibly iron. Intriguingly, *Geobacter* species may also enter into syntrophic electron transfer associations although to date their association with methanogens is unclear.

GeoChip, a functional gene array

Design and construction

Construction of FGAs faces many challenges in both microarray-based technologies (Zhou, 2003) and computational software development. First, a retrieval of sequences specific to a particular functional gene by key words alone is difficult because gene/protein names are not always specific, or they can be differently annotated in different organisms. On the other hand, using functionally characterized and known sequences to search databases usually leads to a large number of hits, and it really depends on the threshold used. Additional difficulties are that variants of a functional gene are often highly similar, and most of them are homologues; sequences from uncultured microbes or laboratory clones may not be complete; and the number of deposited sequences for each gene continuously increases. Thus, it is very difficult to select specific probes for all variants (sequences) of a functional gene, which leads to a low coverage for most functional genes. In addition, standardization of oligonucleotide probe design criteria and software development is still challenging. FGA oligonucleotide probes should be capable of specifically detecting their targets in a complex sample whose components are not known. In addition, for data normalization and analysis, FGAs are very different from the most commonly used gene expression arrays, and novel methods are needed. Finally, in the era of genomics and meta-genomics, microbial sequences are daily produced in a gigabyte or terabyte fashion. A comprehensive FGA must be periodically updated to reflect the latest information.

To address the above challenges, some strategies have been developed or are in development. First, functional gene sequences are first retrieved using key words, and then unrelated sequences are removed by the HMMER program (http://hmmer.wustl.edu/). Second, functional gene sequences are aligned using a multiple sequence alignment (MSA) program after verification by HMMER. Only the shared regions of the functional genes are used for probe design. Third, experimentally established oligonucleotide design criteria and a novel software tool specifically targeting highly similar sequences are used to select oligonucleotide probes. Fourth, to detect both divergent and closely related sequences, both gene- and group-specific probes can be designed.

Details for GeoChip 2.0 design and construction were described by He *et al.* (2007). Briefly, the major steps include:

1. Sequences of individual functional genes were retrieved from publicly available databases (e.g. NCBI GenBank).
2. 50mer oligonucleotides were designed after the removal of unrelated sequences using the oligonucleotide probe design software CommOligo (Li *et al.*, 2005) with a new feature for group-specific probe selection, and experimentally established oligonucleotide design criteria (He *et al.*, 2005b; Liebich *et al.*, 2006). Both gene-specific and group-specific probes were designed with the following criteria: (i) gene-specific probes with sequence identity ≤90%, stretch ≤20 bases, and free energy ≥−35 kcal/mol (Liebich *et al.*, 2006); and (ii) group-specific probes with sequence identity ≥96%, continuous stretch length ≥ 35 bases, and free energy ≤−60 kcal/mol (He *et al.*, 2005b).
3. Oligonucleotide probes were designed on the basis of all variants (sequences) of the same gene. To ensure the whole array specificity, all designed probes of different genes were verified using the same design criteria

against larger databases, such as NCBI and EMBL.

4 All verified probes were commercially synthesized, and spotted on glass slides (e.g. Corning UltraGAPS).

5 After printing, the slides were dried and then UV-cross-linked according to manufacturer's instructions. It should be kept in mind that since the number of sequences in databases increases rapidly, it is important to periodically update the FGA databases and chips to reflect the current status.

Specificity, sensitivity and quantitative capability

Specificity is one of most important parameters to ensure that high-quality microarray data can be obtained, and it is even more critical for FGA analysis of environmental samples. Microarray specificity can be controlled by probe design and hybridization conditions. Tiquia *et al.* (2004) showed that sequences could be differentiated with < 86% identity when hybridizations were at 50°C, and sequences with < 90% identity at 55°C using a 50mer FGA with 763 probes for nitrogen cycling (e.g. *nirS, nirK,* nifH) and sulphate reduction (e.g. *dsrA/B)* genes. With a 50mer FGA containing 1662 probes for genes involved in contaminant degradation, Rhee *et al.* (2004) showed that at hybridization conditions of 50°C and 50% formamide, the 50mer microarray was able to differentiate sequences with < 88% identities. In addition, Bozdech *et al.* (2003) showed that a significant cross-hybridization could occur if a 70mer probe had free energy < –35 kcal/mol. Therefore, based on sequence identity, continuous sequence stretches and free energy, we have experimentally established probe design criteria (He *et al.*, 2005b; Liebich *et al.*, 2006). Furthermore, those criteria have been implemented in a novel software tool, *CommOligo,* for microarrays probe design (Li *et al.*, 2005). Our GeoChip was designed using the newly developed software, and the evaluation results showed only a very small portion of false positives (0.002–0.004%) and no positive negatives (He *et al.*, 2007). Therefore, a well-designed FGA can achieve its desirable specificity.

Sensitivity is a critical parameter, particularly for environmental studies when biomass is low or for low-level molecules in individual cells. When cDNA-based (PCR-generated probes) functional gene arrays (FGAs) were used, the detection limit for *nirS* (nitrite reductase) genes was approximately 1 ng of pure gDNA and 25 ng of soil community DNA (Wu *et al.*, 2001). When oligonucleotide arrays with capture and detector probes were used, a previous study demonstrated that the detection sensitivity for *Geobacter chapellei* SSU rRNA gene sequences in soil extracts was approximately 500 ng of total RNA (Small *et al.*, 2001). Recently, studies with a 50mer FGA showed that the detection limit for some functional genes could be 5–10 ng of pure gDNA and 50–100 ng in a mixture of gDNA from different organisms (Rhee *et al.*, 2004; Tiquia *et al.*, 2004). By taking advantage of the WCGA approach, the 50mer FGA can detect subnanogram quantities of microbial community DNAs as low as 10 pg (Wu *et al.*, 2006). Therefore, the currently available technologies can generally obtain enough nucleic acids for FGA analysis, although challenges may still remain for those genes with low abundances in an environmental sample.

The quantitative capability of microarray-based technology is another central issue for environmental applications. Several previous studies showed that very good linear relationships were obtained between hybridization signal intensity and target DNA or RNA concentration from pure cultures, mixed cultures, and environmental samples (Rhee *et al.*, 2004; Wu *et al.*, 2006; Wu *et al.*, 2001). Recently we showed that reliable quantification could be obtained using a 50mer FGA with randomly amplified DNAs (Wu *et al.*, 2006), or randomly applied RNAs (Gao *et al.*, 2007), as targets. Thus, it is expected that FGAs can serve as a quantitative tool to analyse environmental samples.

Data normalization and analysis

Microarray data normalization is necessary to adjust microarray data for effects that arise from variations in the microarray technology rather than biological differences between samples, or probes on an array. Microarray technology variations may be due to dye bias, labelling efficiency, different scanning properties and settings, and use of different reagents, which can be systematically corrected. Normalization can be performed within a chip and among replicate chips. However, a normalization of microarray data needs to consider the following aspects: (i) what percentage of spots have positive signals on the array for each slide since a microarray normally contains a comprehensive set of probes and some microbial communities may be very simple, or complex; (ii) what spots are used as the control for normalization; (iii) the distribution of signals among positive spots or all spots on the array; (iv) normalization methods may be different based on different situations mentioned above. The following is a simple method for analysing digital array data output from image processing software (e.g. ImaGene, BioDiscovery Inc.). This method includes the following key steps: (i) poor-quality spots are removed, (ii) the signal intensity of each spot is normalized by calculating the mean intensity, (iii) spots with low signal intensities are removed based on the signal-to-noise ratio (SNR) (Wu *et al.*, 2006) and, normally, an SNR of 2.0 is generally used (Verdick *et al.*, 2002), and (iv) for outlier removal, if any of replicates (slides) has (signal – mean) more than three times the standard deviation, this replicate is removed. This process continues until no such replicates are identified.

Data analysis is the most difficult task for microarray analysis of microbial communities. First, a massive amount of data is generated by microarray hybridization. Second, although many methods and software have been developed for microarray data analysis, most of them are focused on the analysis of gene expression data, especially two-dye microarray hybridization, and they may not be suitable for microarray data analysis. Third, microarray data generally have large variations, and rigorous statistical analysis of such data is needed (He and Zhou, 2008). Statistical analysis of microarray data is complicated. Finally, our ultimate goal is to extract biological insights from such large datasets to understand microbial structure and function, so the results should be of biological relevance and statistical significance.

The following statistical approaches are commonly used to analyse microarray data:

1 *Scatterplot.* This is the simplest way to visualize microarray data. Scatterplots can display signal intensities for a single chip with two-dye hybridization, or for two chips with one-dye hybridization.
2 *Principal component analysis (PCA).* This is an exploratory multivariate statistical method for simplifying datasets that reduces the dimensionality of the variables by finding new variables, which are independent of each other. A few of the new variables, typically 2–3, are selected to explain the majority of variance in the original data. For microarray data analysis, genes or experiments can be considered as variables. The main advantage of PCA is that it identifies outliers in the data or genes that behave differently than most of the genes across a set of experiments.
3 *Cluster analysis.* One of the most commonly used methods is cluster analysis. Cluster analysis is used to identify groups of genes, or clusters that have similar profiles. Clusters and the genes within them can be subsequently examined for commonalities in functions and sequences for better understanding of how and why they behave similarly. Cluster analysis can help establish functionally related groups of genes to gain insights into structure and function of a given microbial community. A popular clustering method was developed by Eisen *et al.* (1998), and other algorithms were described by Heyer *et al.* (1999), Travazoie and Church (1998) and Zhou *et al.* (2000).
4 *Neural network analysis.* Since clustering methods have some serious drawbacks in dealing with data with a significant amount of noise, a fundamentally different network-based approach has been proposed for microarray data analysis (Herrero *et al.*, 2001; Tamayo *et al.*, 1999; Toronen *et al.*, 1999). Unsupervised neural networks, such as self-organizing maps (SOMs), are a more robust and accurate method for grouping large datasets. The main advantage of SOMs is that they are robust to noise, and SOMs are also reasonably fast and can be easily scaled up to large datasets. One disadvantage of SOMs is that they require pre-determined choices about geometry. In addition, it is very difficult to detect higher–order relationships between clusters of profiles due to the lack of a tree structure (Herrero *et al.*, 2001). To

overcome some of the limitations of SOMs, an unsupervised neural network, termed the self-organizing tree algorithm (SOTA), was proposed (Dopazo and Carazo, 1997). This new algorithm combines the advantages of hierarchical clustering (tree topology) and neural network (accuracy and robustness) and was used to analyse gene expression data (Herrero *et al.*, 2001).

There are many commercial and free software tools available for general microarray data analysis. Such tools are as simple as Excel (Microsoft), or as complicated as Matlab (The MathWorks), and those include R, GeneSpring (Agilent Technologies), Genesight (BisDiscovery), S-Plus (Insightful Corporation), SPSS (SPSS Inc.), SAS (SAS Institute Inc.), and SAM (Tusher *et al.*, 2001). It is noted that most currently available tools are focused on the analysis of gene expression data. Therefore, we need to carefully choose suitable tools for microarray data analysis. For example, to examine the correlations between the differences of uranium concentrations and those of various functional gene abundances, we used R to implement the Mantel test (He *et al.*, 2007).

Finally, such large datasets need to be simply presented and biologically interpreted. Richness of different gene categories in the community as a whole in the studied samples can be determined from the number of probes that detect their target(s). With probes by category (e.g. *nifH*) as indicators of individual taxa, Simpson's diversity index, Shannon diversity index, and Evenness based on Simpson's index can be calculated as described previously (Begon *et al.*, 1996). To compare different samples, some genes are specifically detected in one sample, and some in all samples tested. The numbers of those two types of genes can be calculated as unique and overlap genes as described previously (Wu *et al.*, 2006). Clustering is one of the most popular methods to analyse and visualize microarray data. Using a hierarchical clustering algorithm, the relationship between different samples taken at different times/sites and different clusters among those samples can be identified. Such analysis can be also applied to each gene/category with its variants. The software *Cluster* can be used for cluster analysis and *TreeView* for visualization (Eisen *et al.*, 1998). In addition, the network analysis of microarray data has received significant attention, and such a method may be used to present and interpret microarray results.

Example applications of functional gene arrays for microbial community analysis

FGAs have been extensively evaluated with artificial microbial communities (He *et al.*, 2007; Loy *et al.*, 2004; Rhee *et al.*, 2004; Steward *et al.*, 2004; Taroncher-Oldenburg *et al.*, 2003; Tiquia *et al.*, 2004; Wu *et al.*, 2001, 2006) and applied to investigate microbial communities in natural and contaminated environments (He *et al.*, 2007). Since all probes were designed using functional gene coding sequences, both DNA and RNA can be used as targets for measuring gene abundance and gene expression, respectively. Therefore, FGAs can be used in a variety of studies, including (but not limiting) (1) detection of functional genes and/or organisms in a particular environment; (2) linking microbial structures to function; and (3) estimation of gene abundance and activity. The GeoChip can be used for analysis of any environmental sample, including soil, water, sediments, oil fields, deep sea, animal guts, etc.

Here, an example study is presented for using an FGA (with 2006 50mer oligonucleotide probes) to analyse the structure and activity of microbial communities in groundwater (Wu *et al.*, 2006). The Environmental Remediation Science Program (ERSP) field research centre (FRC) sites are contaminated with nitrate, uranium, and technetium, as well as some residual organic compounds. Samples from five wells with different degrees of contamination were collected at such a site in Oak Ridge, TN. Microarray analyses showed more than 400 genes had positive hybridization signals (Table 5.2). As expected, the highest number of genes was detected for uncontaminated background samples (well FW300), while the lowest number of genes was detected for the highly contaminated sample (well FW010) (Table 5.2). Simpson's diversity indices showed that the diversities in the uncontaminated background well (FW300) and less contaminated well (FW003) were much higher than those in more heavily contaminated wells (FW021, FW024, and FW010), suggesting contaminants strongly

Table 5.2 The proportion of unique genes (grey shading in diagonal) in individual groundwater wells and a matrix representation of the overlapping number genes between wells

	FW300	FW003	FW021	FW010	FW024
FW300	61(20%)	189(36%)	174(35%)	80(21%)	111(23%)
FW003		25(11%)	144(35%)	61(17%)	84(20%)
FW021			10(5%)	64(20%)	90(24%)
FW010				6(5%)	118(37%)
FW024					30(16%)
Total genes detected	302	219	192	130	190
Genetic diversity, Simpson's (1/D)[a]	125.5	67.1	26.6	17.4	35.7

a Simpson's inverse (1/D) diversity index was calculated using the program Estimates S (http://viceroy.eeb.uconn.edu/estimates) with relative signal intensity ((measured/slide average)*100) used as a measure of abundance. Thus diversity in this case indicates the collective detected genetic diversity of each sample, rather than species diversity.

affected the microbial communities (Table 5.2). The proportion of overlapping genes in different samples was consistent with the contaminant level and geochemistry (Table 5.2).

Some important genes involved in denitrification (e.g. *nosZ* and *nirS*), degradation of organic contaminants (e.g. dienelactone hydrolase genes), and metal resistance (e.g. mercuric reductase gene) were observed in all samples (Fig. 5.4), suggesting that the microbial populations containing these genes are widespread. Dissimilatory sulphate-reducing bacteria are important in the reduction of uranium from soluble U(VI) to insoluble U(IV). In contrast to the results described above, while some dissimilatory sulphite-reducing organisms (*dsrAB*) were found in all of the samples (group D), the abundance and presence of most types (groups A, B, and C) seemed to vary with the origin of the sample (Fig. 5.4). The above results are consistent with 16S gene phylogenetic analysis, and recent microbial community sequencing data. Therefore, as a specific, sensitive and high-throughput tool, the FGA technology is capable of revealing biological processes of microbial communities in natural/contaminated environmental systems.

The GeoChip 2.0 has also been successfully used in our laboratory for tracking the dynamics of metal-reducing bacteria and associated communities for an *in situ* bioremediation study at the ERSP Oak Ridge FRC (Field Research Center) site, which is the first time to demonstrate that uranium can be bioremediated to the concentrations below the USA EPA maximum contaminant level (MCL) for drinking water (He *et al.*, 2007). In addition, a FGA has been recently used to characterize pure isolates, analyse soil microbial nitrogen and carbon cycles along a south polar latitudinal gradient (Yergeau *et al.*, 2007), to investigate microbial community structure during bioremediation of a hydrocarbon-contaminated aquifer, and to identify active members in a stable isotope experiment fed with labelled biphenyl. The GeoChip 2.0 has also been used to examine the gene–area relationship of microbial communities in soils, and the results suggest that a forest soil microbial community exhibited a relatively flat gene–area relationship, but the z (z is a measurement of the rate of species turnover across space in a power–law relationship: $S = cA^z$, where S is the number of species, A is the area, and c is the intercept in log–log space) values varied considerably across different functional and phylogenetic groups (Zhou *et al.*, 2008). All resulting data from these studies indicate that the functional gene array technology is a powerful tool for studying microbial communities in the environment.

Concluding remarks and future trends

We are entering a new era in environmental microbiology where we are beginning to understand some of the complex and interdependent

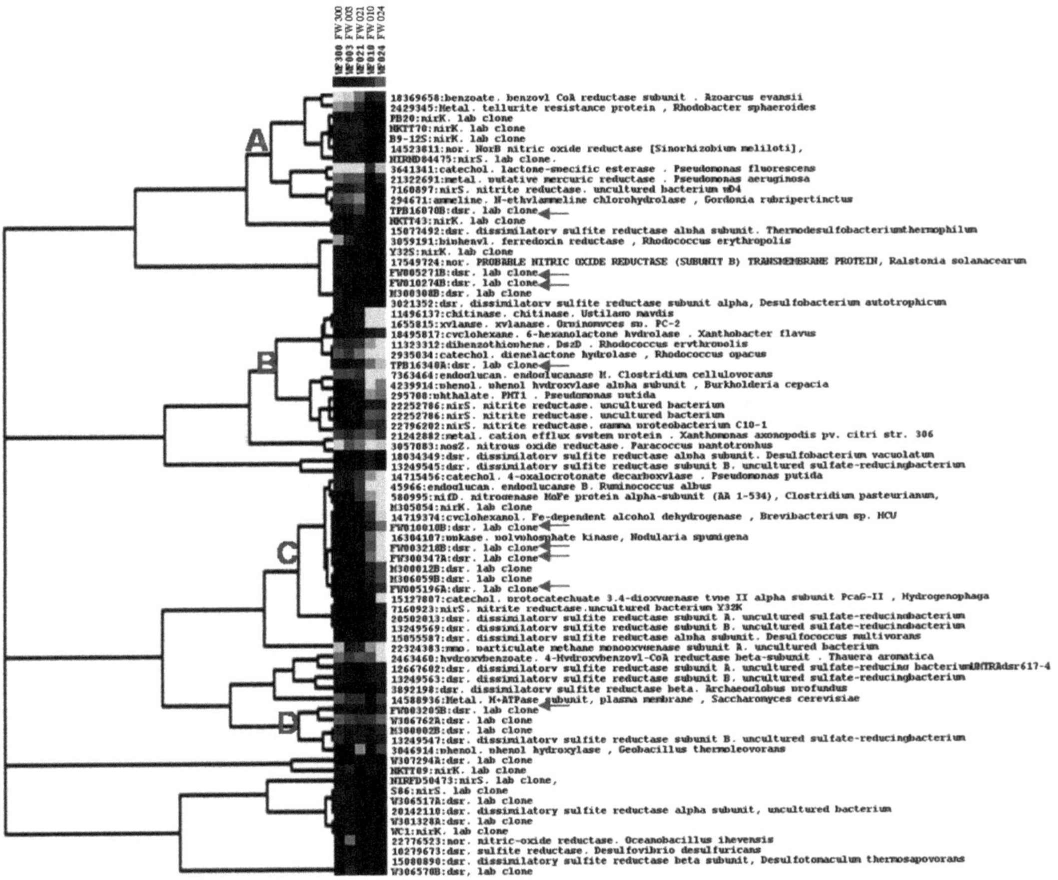

Figure 5.4 Hierarchical cluster analysis of gene relationships based on hybridization signal intensity. Representative genes from 5 groundwater samples collected from contaminated and uncontaminated sites within the FRC are shown. This figure was generated using CLUSTER and visualized with TREEVIEW. Black represents no detectable hybridization above background levels, while white represents positive hybridization signals. The intensity indicates the differences in hybridization signal intensity. Letters A, B, C, and D indicate similar detection patterns between samples. Genes within Clusters A, B and C showed especially high levels of co-occurence and similar intensity patterns between the highly contaminated samples (FW024 and FW010), or between the background and less contaminated samples (FW300 and FW003). Arrows indicated dsrA/B genes that originated from clone libraries developed from previous FRC groundwater studies.

relationships among prokaryotes in natural communities. Starting with an acid mine drainage (AMD) community (Tyson *et al.*, 2004) and working up in complexity to the Sargasso Sea (Venter *et al.*, 2004) and a Minnesota farm soil (Tringe *et al.*, 2005), shotgun microbial community sequencing (metagenomic sequencing) has implicated taxa and specific genes that contribute to the overall cellular respiration and other important functions of a microbial community (see Chapter 2). Phylogenetic arrays such as the PhyloChip that target the known diversity within bacteria and archaea are important for determining the composition of microbial communities in a number of different environments and conditions. Functional gene arrays such as the GeoChip are important for identifying physiological activities involved in certain biogeochemical processes.

The development and application of microarray technology for environmental studies has received a great deal of attention. Because of its high-density and high-throughput capacity, it is expected that such a technology will revolutionize the analyses of microbial community structure, function and dynamics. Current studies showed that microarray technology is able to provide specific, sensitive, and potentially quantitative analysis of microbial communities from a variety of natural environments. It is also useful for providing direct linkages of microbial genes/populations to ecosystem processes and functions. Unlike more costly and time consuming methods such as metagenomic, or 16S clone library sequencing, microarrays allow high-throughput sampling so that multiple replications and multiple treatments can be examined. Identifying the population dynamics of specific taxonomic groups in response to defined conditions is the first step towards understanding how they may contribute to the overall microbial community structure.

However, more rigorous and systematic assessment and development are needed to realize the full potential of microarrays for microbial ecology studies. Several key issues need to be addressed, including novel experimental designs and strategies for minimizing inherent high hybridization variations to improve FGA-based quantitative accuracy, novel approaches for increasing hybridization sensitivity to detect extremely low biomass in natural environments, novel computational tools for microarray data interpretation, and broad integration and application of microarray technologies with environmental studies to address ecological and environmental questions and hypotheses. Therefore, the future focuses will be not only on microarray technology and applications, but also on microarray-related data analysis, interpretation and modelling. First, novel strategies and approaches for experimental controls and design are needed to ensure that microarray hybridization data from different samples are comparable, interpretable and biologically significant because of the inherent variability in microarray hybridization signals. Second, automatic bioinformatic and computational tools are necessary to retrieve gene sequences, design oligonucleotide probes, construct databases, and update sequence, probe and array information. In addition, more advanced automatic mathematical and statistical approaches, such as multivariate analysis, time-series analysis, neural network, artificial intelligence, and differential equation-based modelling, should be extremely useful for rapid pattern recognition, visualization, data mining, cellular modelling, simulation and prediction.

References

Allison, D.B., Page, G.P., Beasley, T.M., and Edwards, J.W. (2007). A Review of: 'DNA Microarrays and Related Genomics Techniques: Design, Analysis, and Interpretation of Experiments'. J. Biopharm. Stat. *17*, 187–190.

Anderson, R.T., Vrionis, H.A., Ortiz-Bernad, I., Resch, C.T., Long, P.E., Dayvault, R., Karp, K., Marutzky, S., Metzler, D.R., Peacock, A., *et al.* (2003). Stimulating the *in situ* activity of Geobacter species to remove uranium from the groundwater of a uranium-contaminated aquifer. Appl. Environ. Microbiol. *69*, 5884–5891.

Ashelford, K.E., Chuzhanova, N.A., Fry, J.C., Jones, A.J., and Weightman, A.J. (2005). At least 1 in 20 16S rRNA sequence records currently held in public repositories is estimated to contain substantial anomalies. Appl. Environ. Microbiol. *71*, 7724–7736.

Ashelford, K.E., Chuzhanova, N.A., Fry, J.C., Jones, A.J., and Weightman, A.J. (2006). new screening software shows that most recent large 16S rRNA gene clone libraries contain chimeras. Appl. Environ. Microbiol. *72*, 5734–5741.

Begon, M., Harper, J.L., and Townsend, C.R. (1996). Ecology: individuals, populations, and communities, 3rd edn (Oxford: Blackwell Science).

Bozdech, Z., Zhu, J., Joachimiak, M.P., Cohen, F.E., Pulliam, B., and DeRisi, J.L. (2003). Expression profiling of the schizont and trophozoite stages of Plasmodium falciparum with a long-oligonucleotide microarray. Genome Biol. *4*, R9.

Braker, G., Zhou, J., Wu, L., Devol, A.H., and Tiedje, J.M. (2000). Nitrite reductase genes (nirK and nirS) as functional markers to investigate diversity of denitrifying bacteria in pacific northwest marine sediment communities. Appl. Environ. Microbiol. *66*, 2096–2104.

Brodie, E.L., DeSantis, T.Z., Joyner, D.C., Baek, S.M., Larsen, J.T., Andersen, G.L., Hazen, T.C., Richardson, P.M., Herman, D.J., Tokunaga, T.K., *et al.* (2006). Application of a high-density oligonucleotide microarray approach to study bacterial population dynamics during uranium reduction and reoxidation. Appl. Environ. Microbiol. *72*, 6288–6298.

Bryant, M.P., Campbell, L.L., Reddy, C.A., and Crabill, M.R. (1977). Growth of *Desulfovibrio* in lactate or ethanol media low in sulfate in association with H_2-utilizing methanogenic bacteria. Appl. Environ. Microbiol. *33*, 1162–1169.

Castiglioni, B., Rizzi, E., Frosini, A., Sivonen, K., Rajaniemi, P., Rantala, A., Mugnai, M.A., Ventura, S., Wilmotte, A., Boutte, C., *et al.* (2004). Development of a universal microarray based on the ligation detection reaction and 16S rrna gene polymorphism to target diversity of cyanobacteria. Appl. Environ. Microbiol. *70*, 7161–7172.

Chang, Y.J., Long, P.E., Geyer, R., Peacock, A.D., Resch, C.T., Sublette, K., Pfiffner, S.M., Smithgall, A., Anderson, R.T., Vrionis, H.A., *et al.* (2005). Microbial incorporation of C-13-labeled acetate at the field scale: Detection of microbes responsible for reduction of U(VI). Environmental Science & Technology *39*, 9039–9048.

Chee, M., Yang, R., Hubbell, E., Berno, A., Huang, X.C., Stern, D., Winkler, J., Lockhart, D.J., Morris, M.S., and Fodor, S.P. (1996). Accessing genetic information with high-density DNA arrays. Science *274*, 610–614.

Cole, K., Truong, V., Barone, D., and McGall, G. (2004). Direct labeling of RNA with multiple biotins allows sensitive expression profiling of acute leukemia class predictor genes. Nucleic Acids Res. *32*, e86.

Cole, J.R., Chai, B., Farris, R.J., Wang, Q., Kulam, S.A., McGarrell, D.M., Garrity, G.M., and Tiedje, J.M. (2005). The Ribosomal Database Project (RDP-II): sequences and tools for high-throughput rRNA analysis. Nucleic Acids R. *33*, D294–296.

Cook, A.F., Vuocolo, E., and Brakel, C.L. (1988). Synthesis and hybridization of a series of biotinylated oligonucleotides. Nucleic Acids Res. *16*, 4077–4095.

Curtis, T.P., and Sloan, W.T. (2005). Microbiology. Exploring microbial diversity--a vast below. Science *309*, 1331–1333.

DeSantis, T.Z., Hugenholtz, P., Keller, K., Brodie, E.L., Larsen, N., Piceno, Y.M., Phan, R., and Andersen, G.L. (2006a). NAST: a multiple sequence alignment server for comparative analysis of 16S rRNA genes. Nucleic Acids Res. *34*, W394–399.

DeSantis, T.Z., Hugenholtz, P., Larsen, N., Rojas, M., Brodie, E.L., Keller, K., Huber, T., Dalevi, D., Hu, P., and Andersen, G.L. (2006b). Greengenes, a Chimera-Checked 16S rRNA Gene Database and Workbench Compatible with ARB. Appl. Environ. Microbiol. *72*, 5069–5072.

DeSantis, T.Z., Brodie, E.L., Moberg, J.P., Zubieta, I.X., Piceno, Y.M., and Andersen, G.L. (2007). High-density universal 16S rRNA microarray analysis reveals broader diversity than typical clone library when sampling the environment. Microb. Ecol. *53*, 371–383.

Dopazo, J., and Carazo, J.M. (1997). Phylogenetic reconstruction using an unsupervised growing neural network that adopts the topology of a phylogenetic tree. J Mol Evol *44*, 226–233.

Edgar, R.C. (2004). MUSCLE: multiple sequence alignment with high accuracy and high throughput. Nucleic Acids Res. *32*, 1792–1797.

Eisen, M.B., Spellman, P.T., Brown, P.O., and Botstein, D. (1998). Cluster analysis and display of genome-wide expression patterns. Proc. Natl. Acad. Sci. U.S.A. *95*, 14863–14868.

Fields, M.W., Schryver, J.C., Brandt, C.C., Yan, T., Zhou, J.Z., and Palumbo, A.V. (2006). Confidence intervals of similarity values determined for cloned SSU rRNA genes from environmental samples. J. Microbiol. Methods *65*, 144–152.

Gans, J., Wolinsky, M., and Dunbar, J. (2005). Computational improvements reveal great bacterial diversity and high metal toxicity in soil. Science *309*, 1387–1390.

Gao, H., Yang, Z.K., Gentry, T.J., Wu, L., Schadt, C.W., and Zhou, J. (2007). Microarray-based analysis of microbial community RNAs by whole-community RNA amplification. Appl. Environ. Microbiol. *73*, 563–571.

Ginder-Vogel, M., Criddle, C.S., and Fendorf, S. (2006). Thermodynamic constraints on the oxidation of biogenic UO2 by Fe(III) (hydr) oxides. Environ. Sci. Technol. *40*, 3544–3550.

He, Z., Wu, L., Fields, M.W., and Zhou, J. (2005a). Comparison of microarray performance with different probe sizes for monitoring gene expression. Appl. Environ. Microbiol. *71*, 5154–5162.

He, Z., Wu, L., Li, X., Fields, M.W., and Zhou, J. (2005b). Empirical establishment of oligonucleotide probe design criteria. Appl. Environ. Microbiol. *71*, 3753–3760.

He, Z., Gentry, T.J., Schadt, C.W., Wu, L., Liebich, J., Chong, S.C., Huang, Z., Wu, W., Gu, B., Jardine, P., *et al.* (2007). GeoChip: A comprehensive microarray for investigating biogeochemical, ecological, and environmental processes. ISME J. *1*, 67–77.

He, Z., and Zhou, J. (2008). Empirical evaluation of a new method for calculating signal-to-noise ratio for microarray data analysis. Appl. Environ. Microbiol. *74*, 2957–2966.

Herrero, J., Valencia, A., and Dopazo, J. (2001). A hierarchical unsupervised growing neural network for clustering gene expression patterns. Bioinformatics *17*, 126–136.

Heyer, L.J., Kruglyak, S., and Yooseph, S. (1999). Exploring expression data: identification and analysis of coexpressed genes. Genome Res *9*, 1106–1115.

Huber, T., Faulkner, G., and Hugenholtz, P. (2004). Bellerophon: a program to detect chimeric sequences in multiple sequence alignments. Bioinformatics *20*, 2317–2319.

Hugenholtz, P. (2002). Exploring prokaryotic diversity in the genomic era. Genome Biol. *3*, 1–8.

Hurt, R.A., Qiu, X., Wu, L., Roh, Y., Palumbo, A.V., Tiedje, J.M., and Zhou, J. (2001). Simultaneous recovery of RNA and DNA from soils and sediments. Appl. Environ. Microbiol. *67*, 4495–4503.

Huse, S.M., Dethlefsen, L., Huber, J.A., Welch, D.M., Relman, D.A., and Sogin, M.L. (2008). Exploring microbial diversity and taxonomy using SSU rRNA hypervariable tag sequencing. PLoS Genet. *4*, e1000255.

Kelly, J.J., Chernov, B.K., Tovstanovsky, I., Mirzabekov, A.D., and Bavykin, S.G. (2002). Radical-generating coordination complexes as tools for rapid and effective fragmentation and fluorescent labeling of nucleic acids for microchip hybridization. Anal. Biochem. *311*, 103–118.

Kelly, J.J., Siripong, S., McCormack, J., Janus, L.R., Urakawa, H., El Fantroussi, S., Noble, P.A., Sappelsa, L., Rittmann, B.E., and Stahl, D.A. (2005). DNA microarray detection of nitrifying bacterial 16S rRNA in wastewater treatment plant samples. Water Res. *39*, 3229–3238.

Klerks, M.M., van Bruggen, A.H., Zijlstra, C., and Donnikov, M. (2006). Comparison of methods of extracting *Salmonella enterica* serovar Enteritidis DNA from environmental substrates and quantification of organisms by using a general internal procedural control. Appl. Environ. Microbiol. 72, 3879–3886.

Lehner, A., Loy, A., Behr, T., Gaenge, H., Ludwig, W., Wagner, M., and Schleifer, K.H. (2005). Oligonucleotide microarray for identification of Enterococcus species. FEMS Microbiol. Lett. *246*, 133–142.

Li, X., He, Z., and Zhou, J. (2005). Selection of optimal oligonucleotide probes for microarrays using multiple criteria, global alignment and parameter estimation. Nucleic Acids Res. *33*, 6114–6123.

Liebich, J., Schadt, C.W., Chong, S.C., He, Z., Rhee, S.K., and Zhou, J. (2006). Improvement of oligonucleotide probe design criteria for functional gene microarrays in environmental applications. Appl. Environ. Microbiol. *72*, 1688–1691.

Liu, W.T., Mirzabekov, A.D., and Stahl, D.A. (2001). Optimization of an oligonucleotide microchip for microbial identification studies: a non-equilibrium dissociation approach. Environ. Microbiol. *3*, 619–629.

Loring, J.F. (2006). Evolution of microarray analysis. Neurobiol Aging *27*, 1084–1086.

Loy, A., Kusel, K., Lehner, A., Drake, H.L., and Wagner, M. (2004). Microarray and functional gene analyses of sulfate-reducing prokaryotes in low-sulfate, acidic fens reveal cooccurrence of recognized genera and novel lineages. Appl. Environ. Microbiol. *70*, 6998–7009.

Ludwig, W., Strunk, O., Westram, R., Richter, L., Meier, H., Yadhukumar, Buchner, A., Lai, T., Steppi, S., Jobb, G., *et al.* (2004). ARB: a software environment for sequence data. Nucleic Acids Res. *32*, 1363–1371.

McGrath, K.C., Thomas-Hall, S.R., Cheng, C.T., Leo, L., Alexa, A., Schmidt, S., and Schenk, P.M. (2008). Isolation and analysis of mRNA from environmental microbial communities. J. Microbiol. Methods *75*, 172–176.

Pace, N.R. (1997). A molecular view of microbial diversity and the biosphere. Science *276*, 734–740.

Peacock, A.D., Chang, Y.J., Istok, J.D., Krumholz, L., Geyer, R., Kinsall, B., Watson, D., Sublette, K.L., and White, D.C. (2004). Utilization of microbial biofilms as monitors of bioremediation. Microb Ecol *47*, 284–292.

Phelps, T.J., Murphy, E.M., Pfiffner, S.M., and White, D.C. (1994). Comparison between geochemical and biological estimates of subsurface microbial activities. Microbial Ecol. *V28*, 335–349.

Reardon, C.L., Cummings, D.E., Petzke, L.M., Kinsall, B.L., Watson, D.B., Peyton, B.M., and Geesey, G.G. (2004). Composition and diversity of microbial communities recovered from surrogate minerals incubated in an acidic uranium-contaminated aquifer. Appl. Environ. Microbiol. *70*, 6037–6046.

Rhee, S.K., Liu, X., Wu, L., Chong, S.C., Wan, X., and Zhou, J. (2004). Detection of genes involved in biodegradation and biotransformation in microbial communities by using 50mer oligonucleotide microarrays. Appl. Environ. Microbiol. *70*, 4303–4317.

Rotthauwe, J.H., Witzel, K.P., and Liesack, W. (1997). The ammonia monooxygenase structural gene amoA as a functional marker: molecular fine-scale analysis of natural ammonia-oxidizing populations. Appl. Environ. Microbiol. *63*, 4704–4712.

Schadt, C.W., Liebich, J., Chong, S.C., Gentry, T.J., He, Z., Pan, H., and Zhou, J. (2004). Design and use of functional gene microarrays (FGAs) for the characterization of microbial communities, In Methods in Microbiology, Savidge, T., and Pothulakis, H., eds. (London: Academic Press), pp. 329–365.

Small, J., Call, D.R., Brockman, F.J., Straub, T.M., and Chandler, D.P. (2001). Direct detection of 16S rRNA in soil extracts by using oligonucleotide microarrays. Appl. Environ. Microbiol. *67*, 4708–4716.

Steward, G.F., Jenkins, B.D., Ward, B.B., and Zehr, J.P. (2004). Development and testing of a DNA macroarray to assess nitrogenase (nifH) gene diversity. Appl. Environ. Microbiol. *70*, 1455–1465.

Stoughton, R.B. (2005). Applications of DNA microarrays in biology. Annu Rev Biochem *74*, 53–82.

Tamayo, P., Slonim, D., Mesirov, J., Zhu, Q., Kitareewan, S., Dmitrovsky, E., Lander, E.S., and Golub, T.R. (1999). Interpreting patterns of gene expression with self-organizing maps: methods and application to hematopoietic differentiation. Proc. Natl. Acad. Sci. U.S.A. *96*, 2907–2912.

Taroncher-Oldenburg, G., Griner, E.M., Francis, C.A., and Ward, B.B. (2003). Oligonucleotide microarray for the study of functional gene diversity in the nitrogen cycle in the environment. Appl. Environ. Microbiol. *69*, 1159–1171.

Tavazoie, S., and Church, G.M. (1998). Quantitative whole-genome analysis of DNA–protein interactions by *in vivo* methylase protection in *E. coli*. Nat. Biotechnol *16*, 566–571.

Thompson, J.D., Higgins, D.G., and Gibson, T. (1994). CLUSTAL W: improving the sensitivity of progressive multiple sequence alignment through sequence weighting, position-specific gap penalties and weight matrix choice. Nucleic Acids Res *22*, 4673–4680.

Tiquia, S.M., Wu, L., Chong, S.C., Passovets, S., Xu, D., Xu, Y., and Zhou, J. (2004). Evaluation of 50mer oligonucleotide arrays for detecting microbial populations in environmental samples. Biotechniques *36*, 664–670, 672, 674–665.

Toronen, P., Kolehmainen, M., Wong, G., and Castren, E. (1999). Analysis of gene expression data using self-organizing maps. FEBS Lett *451*, 142–146.

Tringe, S.G., von Mering, C., Kobayashi, A., Salamov, A.A., Chen, K., Chang, H.W., Podar, M., Short, J.M., Mathur, E.J., Detter, J.C., *et al.* (2005). Comparative metagenomics of microbial communities. Science *308*, 554–557.

Tusher, V.G., Tibshirani, R., and Chu, G. (2001). Significance analysis of microarrays applied to the ionizing radiation response. Proc. Natl. Acad. Sci. U.S.A. *98*, 5116–5121.

Tyson, G.W., Chapman, J., Hugenholtz, P., Allen, E.E., Ram, R.J., Richardson, P.M., Solovyev, V.V., Rubin, E.M., Rokhsar, D.S., and Banfield, J.F. (2004). Community structure and metabolism through reconstruction of microbial genomes from the environment. Nature *428*, 37–43.

Urakawa, H., Noble, P.A., El Fantroussi, S., Kelly, J.J., and Stahl, D.A. (2002). Single-base-pair discrimination of terminal mismatches by using oligonucleotide microarrays and neural network analyses. Appl. Environ. Microbiol. *68*, 235–244.

Vanbelkum, A., Linkels, E., Jelsma, T., Vandenberg, F.M., and Quint, W. (1994). Nonisotopic labeling of DNA by newly developed hapten-containing platinum compounds. Biotechniques *16*, 148–153.

Venter, J.C., Remington, K., Heidelberg, J.F., Halpern, A.L., Rusch, D., Eisen, J.A., Wu, D., Paulsen, I., Nelson, K.E., Nelson, W., *et al.* (2004). Environmental genome shotgun sequencing of the Sargasso Sea. Science *304*, 66–74.

Verdick, D., Handran, S., and Pickett, S. (2002). Key considerations for accurate microarray scanning and image analysis, In DNA Array Image Analysis: Nuts and Bolts, Kamberova, G., ed. (Salem, MA: DNA Press), pp. 83–98.

Wan, J., Tokunaga, T.K., Brodie, E., Wang, Z., Zheng, Z., Herman, D., Hazen, T.C., Firestone, M.K., and Sutton, S.R. (2005). Reoxidation of bioreduced uranium under reducing conditions. Environ. Sci. Technol. *39*, 6162–6169.

Warsen, A.E., Krug, M.J., LaFrentz, S., Stanek, D.R., Loge, F.J., and Call, D.R. (2004). Simultaneous discrimination between 15 fish pathogens by using 16S ribosomal DNA PCR and DNA microarrays. Appl. Environ. Microbiol. *70*, 4216–4221.

Whitman, W.B., Coleman, D.C., and Wiebe, W.J. (1998). Prokaryotes: the unseen majority. Proc. Natl. Acad. Sci. U.S.A. *95*, 6578–6583.

Wilson, W.J., Strout, C.L., DeSantis, T.Z., Stilwell, J.L., Carrano, AV., and Andersen, G.L. (2002). Sequence-specific identification of 18 pathogenic microorganisms using microarray technology. Mol. Cell Probes *16*, 119–127.

Woese, C.R., Fox, G.E., Zablen, L., Uchida, T., Bonen, L., Pechman, K., Lewis, B.J., and Stahl, D. (1975). Conservation of primary structure in 16S ribosomal RNA. Nature *254*, 83–86.

Wu, L., Thompson, D.K., Li, G., Hurt, R.A., Tiedje, J.M., and Zhou, J. (2001). Development and evaluation of functional gene arrays for detection of selected genes in the environment. Appl. Environ. Microbiol. *67*, 5780–5790.

Wu, L., Liu, X., Schadt, C.W., and Zhou, J. (2006). Microarray-based analysis of subnanogram quantities of microbial community DNAs by using whole-community genome amplification. Appl. Environ. Microbiol. 72, 4931–4941.

Yergeau, E., Kang, S., He, Z., Zhou, J., and Kowalchuk, G.A. (2007). Functional microarray analysis of nitrogen and carbon cycling genes across an Antarctic latitudinal transect. ISME J. *1*, 163–179.

Zhang, Z., Schwartz, S., Wagner, L., and Miller, W. (2000). A greedy algorithm for aligning DNA sequences. J. Comput. Biol. 7, 203–214.

Zhou, J. (2003). Microarrays for bacterial detection and microbial community analysis. Curr. Opin. Microbiol. *6*, 288–294.

Zhou, J., Bruns, M.A., and Tiedje, J.M. (1996). DNA recovery from soils of diverse composition. Appl. Environ. Microbiol. *62*, 316–322.

Zhou, J., Kang, S., Schadt, C.W., Charles, T., and Garten, C.T.J. (2008). Spatial scaling of functional gene diversity across various microbial taxa. Proc Nat Acad Sci USA *105*, 7768–7773.

Zhou, Y., Kalocsai, P., Chen, J., and Shams, S. (2000). Information processing issues and solutions associated with microarray technology, In Microarray Biochip Technology, Schena, M., ed. (Natick, MA: Eaton Publishing), pp. 167–200.

The Soil Environment

6

Kornelia Smalla and Jan Dirk van Elsas

Abstract

Until fairly recently, the living soil has been considered as a functional black box that is intrinsically too difficult to be unravelled into its core components. However, this concept has changed with the advent of the modern methodologies. Here, we examine how our understanding of the intricacies of microbial life in soil has been impacted by the advanced, mainly molecularly based, approaches that have been unleashed on the soil habitat in recent years. Truly, although the soil 'microbial black box' is still far from being completely opened, the application of molecular and other advanced methods (allowing the so-called cultivation-independent analyses) has provided exciting new insights into microbial (in particular bacterial) life in soil.

Understandably, the extremely diverse and complex soil habitat – containing many microsites and gradients that define the niches for the soil inhabitants – supports an equally diverse microbial community. In this community we find the key ecosystem functions that determine important biogeochemical soil processes. Understanding how the living soil 'works', i.e. how the enormous microbial diversity relates to soil functioning and the stability thereof, as well as how the diverse microbes interact in the soil, can be considered as the great challenge in soil science. In addition, fostering our understanding of the factors that facilitate or limit the microbial colonization of soil microsites is another important goal.

Besides facilitating the unlocking of the rich untapped diversity of microbes in soil, the application of the novel methods is expected to also contribute to a better understanding of the ecology of soil microbes at the different scales in soil, i.e. from the macroscale (continent, landscape, field, plot) via the mesoscale to the microscale (cubic cm to μm of soil, including a dissection of soil into aggregate types or pore size classes). In this respect, soil sampling or processing is important. A special emphasis is thus placed on the particularities imposed by soil in respect of the required sampling strategies and the assessments of microbial community structure, diversity and function.

Introduction

Soil is an extremely diverse and complex habitat containing many microsites and gradients that form a range of different biogeochemical interfaces. Depending on the proportion of sand, silt and clay, the surface area in soil can vary from 11 cm^2 up to 8 million cm^2 per gram of soil (Standing and Killham, 2007). The aggregates formed by minerals, soil organic matter, fungal hyphae, roots and plant debris offer a range of potential niches for microorganisms with different lifestyles. The architecture of the soil pore network essentially defines the habitat colonized by the microorganisms and the pore space strongly influences the nature and extent of the interactions between the organisms inhabiting the soil (Young and Ritz, 2000; Ranjard and Richaume, 2001). The heterogeneous physical structure of soil affects the spatial distribution of water, oxygen and nutrients, which in turn influences the composition and activity of the microbial communities themselves (Nunan *et al.*, 2002, 2003; Zhou *et al.*, 2002). As an example, the spatial distribution of bacteria in topsoil and subsoil was found to be different (Nunan *et al.*, 2002), but lateral variations in spatial distributions are also likely to occur. In terms of their occurrence in microsites, bacteria can be found in soil as single cells but most often they occur as microcolonies, i.e. small agglomerates of cells that can be regarded as primitive soil biofilms (Nunan *et al.*, 2001; Van Elsas *et al.*, 2007).

Microorganisms are the major drivers of geochemical and biotransformation processes in soil. In concert with the soil's inorganic and organic constituents, microbes are influential in actively shaping the architecture of the soil matrix by the formation and restructuring of soil aggregates (Preston *et al.*, 2001). In addition, the diversity of microbial communities is extremely high in most soils (Fierer *et al.*, 2007b). There are only a few quantitative estimates of the numbers of microbial taxa that can co-exist in just a single gram of soil, but an advanced analysis of nucleic acid-based analyses, based on re-association kinetics, has suggested that prokaryotic diversity can reach 1 million species genomes per gram, which by far exceeds the common estimates of bacterial richness in soil obtained from cultivation-based studies (Gans *et al.*, 2005). A major driving force that spurs the microbial diversity of soil is the enormous heterogeneity of the soil habitat, allowing the formation of numerous niches. Different factors, such as the presence or absence of water, soil pH, temperature, redox potential and the soil organic matter content do not only influence the types of microbes colonizing the respective microniche but also their activity. All these factors can vary greatly between the different microhabitats and, thus, not only the composition but also the activity and interactions of the microbiota will largely vary due to the spatial and temporal heterogeneity between as well as within the microsites.

A key determinant of microbial fitness in soil is the ability of microbial cells to fine-tune their cellular metabolism to the abiotic and biotic conditions that prevail locally. In addition, the rate of adaptation of microorganisms to changing environmental conditions might be enhanced by horizontal gene transfer processes (Smets and Barkay, 2005; Smalla and Heuer, 2006; Heuer *et al.*, 2008). Undoubtedly, the most important prerequisite for microbial life in soil is the availability of water. Next to being indispensable for microbial life, the water in soil carries dissolved gases, ions and nutrients to microorganisms, and, in cases of saturation, may quickly establish anaerobic conditions. For instance, an increase of the moisture content of soil can greatly influence the microbial communities that are locally present, in particular by connecting pore spaces in and among aggregates that were unconnected without water, thus increasing the aggregate connectivity. Predation by protozoa or *Bdellovibrio* species will therefore be particularly enhanced in relatively wet soils (often at > 60% of the soil's water holding capacity).

In this chapter we will focus on the intricacies of microbial, in particular bacterial, life in bulk soil, i.e. the portion of soil that is not affected by plant roots. Until fairly recently, microbial life in soil has largely remained a 'black box', mainly because the tools required to study the structural and functional diversity, and the actual ecosystem roles, of soil microbes have long been missing. Although the 'soil black box' is far from being completely opened, the application of molecular and other approaches allowing cultivation-independent methods has provided exciting new insights into microbial life in soil (Tiedje *et al.*, 1999) that will be briefly summarized in this chapter. Because microbial communities in soils are assumed to be very diverse and complex, and can vary at small scales, any attempt to link soil microbial community structure with function can be considered as a 'Decadal Grand Challenge' (Stres and Tiedje, 2006). Still very little is actually known of the factors that facilitate or limit the microbial colonization of, and activity in, soils. Besides exploring the vast untapped diversity of microbes in soil by means of novel molecular tools, the application of these methods is expected to also contribute to a better understanding of the ecology of soil microbes which need to be studied at different scales, i.e. from the macroscale (continent, landscape, field, plot) to the microscale (cubic cm to μm of soil, including a dissection of soil into aggregate types or pore size classes). We will place special emphasis on the particularities imposed by soil on the required sampling strategies and the assessments of microbial community structure, diversity and function. Bacterial life in the plant rhizosphere will be addressed separately in Chapter 7.

How to sample soil?

In any soil study, the questions that are posed ultimately determine the sampling design that is appropriate (Van Elsas *et al.*, 2002). In fact, the sampling strategy can be often identified as the weakest point in any study that applies molecular tools to study the soil microbiota. The reason for this is that molecular tools most often deal with numbers and sizes of samples for which their representation for the whole system is a question of concern. Hence, before we highlight findings and observations made on the basis of molecular methods employed to uncover microbial life in soil, some considerations about soil sampling strategies and the microscale-versus-macroscale discrepancy should be made.

On the one hand, it is clear that to study how microorganisms perceive their immediate environment, the microscale level needs to be studied. Such studies are required to determine to what extent microbes in soil indeed live in spatial isolation from others, with locally differently adapted microbial assemblages, or whether there is sufficient transport and connectivity to attenuate this biogeochemical and evolutionary separation. Microscale-level studies are also required to explore how adaptation and evolution of microbial communities occurs under conditions of spatial isolation or at least in highly structured environments (Nielsen *et al.*, 2007).

In contrast, field-scale studies are generally required if effects of soil treatments, such as, for instance, different agricultural management practices or land use or even bioremediation treatments, are to be investigated. Obviously, at this scale the phenomena – that are actually determined at the microscale level – are integrated into an overall, higher-level, type of response. It is primordial that our thinking about soil microbial processes in relation to the soil microbial communities residing in their microhabitats always takes this microscale-to-macroscale discrepancy into account.

The need to use a proper sampling design

Sampling of soil requires the careful design of a strategy in which the estimated heterogeneity of the soil system and the expected variability of the data are considered (Van Elsas *et al.*, 2002). This pre-sampling consideration strongly determines the number of replicate plots to be sampled and/or of samples to be taken. The adequate sampling design depends on the hypotheses to be tested and often pre-experiments might assist in determining the numbers of samples to be analysed. In addition, sample sizes for molecular analyses are necessarily small, and there is often a lack of information about how representative such small soil samples can be even for the composite sample. Thus, in the majority of studies to date, composite samples of soil taken from a plot or a field have been analysed, often in a number of replicates. In multiple-plot experimental designs, compositing several samples over each plot allows to obtain an averaged value over the plot, whereas (plot) replication will indicate the variation over the field, allowing cross-comparisons between e.g. treatments (plots) to be made. The main argument here is that the samples analysed should, in a representative manner, report on the characteristics of the field soil under a particular treatment and on the variation among the average value of that characteristic.

Considering what the statistical pre-assessment of variability tells us, the use of a minimal number of composite samples per treatment taken from independent plots (often four) is recommended to assess the effect of a treatment under field conditions (Van Elsas *et al.*, 2002). Each composite sample may consist of several soil cores taken in a randomized fashion from a plot. Mixing of the soil is usually achieved by sieving the material of the various soil cores through a 2- or 4-mm mesh size sieve. A subsample from each composite sample is used for each of the analyses, which include the molecular ones. However, the investigation of composite samples may obscure any meaningful biological variation over the plot. Thus, in order to study plot or field heterogeneity, characterization of independent single samples would be required, which allows preservation of the picture of environmental variability and may provide insights into the abundance of the microbes present in the sample (Sliwinski and Goodman, 2004; Becker *et al.*, 2006).

Also, sample size has been seen to matter. Nicol *et al.* (2003) found that archaeal fingerprint profiles for replicate 10 g soil samples from the same plot were quite similar, while those for 1 g samples and 0.1 g samples were more variable. However, when subsamples of homogenized composite samples are analysed, highly similar bacterial communities are commonly revealed even if the total community DNA is extracted from 0.5 g soil (see Fig. 6.1: Cuban soils with replicates next to each other from 0.5 g samples).

Previous work, in which top soils of several agricultural fields and other soils were sampled across location and time (season), has shown that, as determined by PCR with denaturing gradient gel electrophoresis (DGGE), the community composition of the dominant bacterial populations was fairly similar in a similar soil type, indicating commonality of the respective bacterial communities per soil type across time and space (Gelsomino *et al.*, 1999). How this observation impinges upon the microscale level is, however, still unknown.

The cost of molecular analysis limits the number of samples processed

A main thrust to the enhancement of our insight in the intricacies of the soil microbiota has been provided by the novel molecular methods that have emerged (Kowalchuk *et al.*, 2007; Chapters 2, 4 and 5). However, the costs as well as laboriousness of the molecular tools that are currently being employed, for instance in cloning and sequencing or gene array hybridization to determine the microbial diversity of soil, often do not permit the analysis of a sufficiently high number of (composite) samples per treatment to allow the characterization of variation over a field or even over a plot. Thus, many studies on the microbial diversity and community composition of soil have been performed on the basis of just a single sample, which was – in the best of cases – a composite sample. The question as to how representative the collective sequence information based on the ribotypes or functional genes analysed actually was for a given soil has thus often remained unclear. Some studies have exclusively focussed on the question whether a sufficiently high number of 16S rRNA gene sequences had been determined for just a single soil sample (Schloss and Handelsman, 2006), and hence relevant questions as to field heterogeneity have been completely ignored. In a groundbreaking paper, 139,000 sequences were analysed across four soils (three agricultural and one forest), but again replicates were not included (Fulthorpe *et al.*, 2008). However, when direct sequencing, cloning and sequencing or array hybridization approaches are

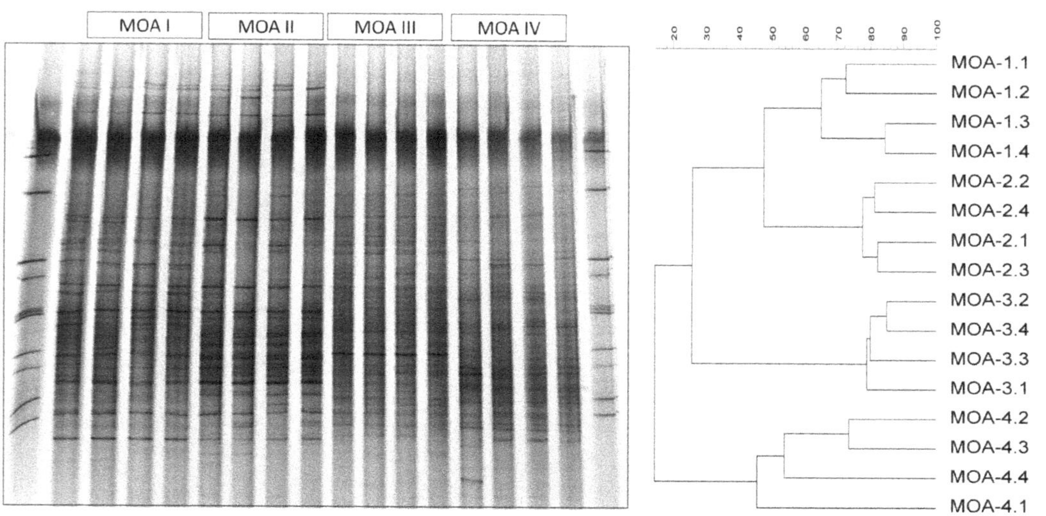

Figure 6.1 Bacterial community fingerprints of four composite soil samples, each taken from four different sites in Moa, Cuba.

combined with cheaper molecular fingerprinting techniques such as DGGE or terminal restriction fragment length polymorphism (T-RFLP), which allows the extension of the number of samples analysed, as a range of samples can often be run in parallel, the representation of a sample used for cloning and sequencing can easily be shown.

The microscale level – insights and caveats

In the light of the known microheterogeneity of soil and the importance of the microscale in the interaction of bacteria with their surroundings (Dechesne *et al.*, 2003), several researchers have voiced the feeling that it is important to sample and analyse soil at the microscale level (i.e. cubic mm to cm; Grundmann and Gourbière, 1999; Sessitsch *et al.*, 2001; Poll *et al.*, 2003; Izquierdo and Nüsslein, 2006). Data on the structure and function of the soil microbiota at this level have been achieved by different strategies. Soil fractionation techniques – such as wet sieving – provide insights into the distribution of microorganisms over the different particle size fractions that make up soil. This approach has been used in many studies (e.g. Sessitsch *et al.*, 2001; Poll *et al.*, 2003). Mummey *et al.* (2006) used soil fractionation based on pressing liquid-immersed soil through a 2000-μm sieve first, followed by passage through 250 and 53 μm sieves. Then, a high-energy treatment was applied to break up and isolate the inner-microaggregate fractions. Macroaggregates, defined by diameters > 250 μm, are formed by the temporary association of microaggregates, minerals and particulate organic matter. Water-stable microaggregates, of diameters < 250 μm, typically form within macroaggregates by microbial action. They are largely dependent on persistent organic binding agents for structural activity. Soil fractionation techniques do not provide data on the distribution of microorganisms at the microscale level, i.e. the localization of communities along a microscale transect through the soil. Soil architecture and spatial organization determine the proximity of bacterial cells and, thus, the likelihood that microbial processes occur that depend on interactions, including horizonal gene transfer (Grundmann, 2004).

One strategy to analyse microbes at the microscale level is based on a soil microsampling procedure (Grundmann and Gourbière, 1999). Extracting nucleic acids, in parallel, from minute amounts of soil has been employed in order to explore the microscale diversity of microorganisms in soil and to characterize the three-dimensional spatial distribution of bacterial cells in soil by analysis of a large number of microsamples in a defined volume (Grundmann and Gourbière, 1999; Grundmann *et al.*, 2001). The number of microsamples required to achieve representation will obviously differ between well-mixed (e.g. arable) or unmixed (stratified) soils. This approach enables an increased inventory of microbial diversity in soil and the study of rare microbial populations, which are difficult to discern in overall soil microbial analyses.

It is clear that the potentials and biases of the different soil sampling approaches and how these should be implemented in the investigation of soil architecture, pollutant distribution (not discussed in this chapter) and microbial community structure and function need further systematic investigation.

Application of molecular tools to study the soil microbiota: insights and challenges

The great plate count anomaly and tools to tackle it

Staley and Konopka (1985) coined the term 'great plate count anomaly' when reviewing data on scientists' ability to cultivate microbes from environmental samples. The observation that the vast majority of microbial cells which can be observed in a microscope cannot form colonies on common nutrient agar media was then recognized as a severe limitation in studying and understanding microbial communities in natural habitats (Amann *et al.*, 1995). Thus, it was estimated that only 0.1–1% of the bacteria in soil can form colonies under standard cultivation conditions. In the last decade, soil microbiology has witnessed great progress in techniques that allow the cultivation-independent analysis of the microbial communities in many soils (Kowalchuk *et al.*, 2007). The tool set available to unlock the 'black box' of microbial life in soil has become quite impressive (see Fig. 6.2). It ranges from microscopy-based methods such as fluorescence *in situ* hybridization (FISH), microautoradiography-FISH (MAR-FISH; see Chapter 11 for a description of the use

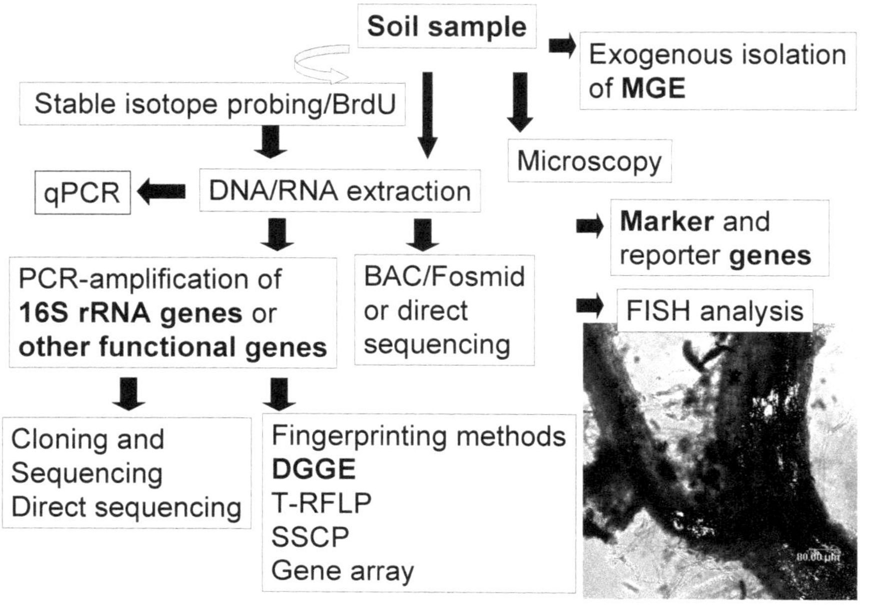

Figure 6.2 Molecular methods of studying microbial diversity in soil. MGE, mobile genetic elements.

of this technique for studying biofilms) and methods based on reporter genes such as the *gfp* gene, encoding the green fluorescent protein (Wagner *et al.*, 2003; Sørensen and Nybroe, 2006; Rogers *et al.*, 2007) to a range of DNA- or RNA-based methods that aim to study the structural and functional diversity of microbial communities and targeted cell isolation and sequencing (Podar *et al.*, 2007; Chapter 4). A difficulty for the application of FISH to soils is the low signal-to-noise ratio due to the presence of the solid matrix in soil, in combination with the low *in situ* metabolic activity of most soil microorganisms that tends to suppress ribosome numbers.

Over the last decade, DNA- or RNA-based analyses of the soil microbiota have developed into routine methods in many labs (Van Elsas *et al.*, 2000; Kowalchuk *et al.*, 2007). The basis for these advances in the analyses of soil microbial communities has clearly been the successful extraction of nucleic acids from soils (Torsvik *et al.*, 1990). Whereas previous methods were research laboratory based, commercially available soil DNA or RNA extraction kits are now overwhelmingly used, which, after harsh cell lysis, greatly facilitate the extraction and purification of nucleic acids of sufficient purity for molecular analysis. However, it is a matter of truth that any nucleic acid based assessment still provides a more or less biased view of the microbiota of soil. First of all, differential lysis of soil bacterial cells is an important issue. Cells from different species and sometimes even different physiological stages of the same population often differ in their susceptibility to the lysis conditions used. When applying harsh lysis methods, the DNA released from easy-to-lyse bacterial cells might be more sheared and fragmented than DNA from more recalcitrant cells, which might result in the loss of probe or primer annealing sites. Also, the genomic DNA from bacteria that are difficult to lyse might be underrepresented in metagenomic clone libraries, as the generation of high molecular weight DNA requires gentle lysis techniques to avoid shearing. Secondly, extracellular DNA or DNA originating from dead cells is also obtained, which blurs assessments in cases when the metabolically active, or viable, microbiota is to be studied. Thirdly, as a result of the extreme complexity of soil microbial communities, only the tip-of-the-iceberg of soil microbial diversity is commonly seen unless pre-fractionation techniques of different kinds are applied. For instance, the average G+C content of the DNA can be used as the basis for a pre-separation, allowing the enrichment of particular organisms or genes of interest (Nüsslein and Tiedje, 1998). However, it is inherently impossible to assess 'all' microorganisms that abound in soil. It is foreseen that, now that many labs have mastered the basic technology to molecularly describe soil microbial communities, considerable investment in improvement of nucleic acid extraction technologies from soil will still be needed. Recently, several methodological approaches were developed in order to extract nucleic acids from metabolically active populations or populations involved in particular degradation processes. Parallel extraction of total RNA and DNA from soil has become a well established approach (Griffith *et al.*, 2000; Hurt *et al.*, 2001; Gomes *et al.*, 2005). The use of directly extracted ribosomal RNA is assumed to provide taxonomic information on the metabolically active bacteria. However, the observation that the ribosome content of oligotrophic bacteria is consistently low irrespective of growth rate, but also that resting stages and spores of e.g. bacilli contain ribosomal RNA amounts not very much lower than those in cells needs to be taken into consideration. Often, the molecular fingerprints generated from either DNA or RNA based cDNA obtained from soil do not show pronounced differences (Griffith *et al.*, 2003; Gomes *et al.*, 2005), which indicates a slow growth rate or low metabolic activity of the dominant bacteria in soil (Fig. 6.3).

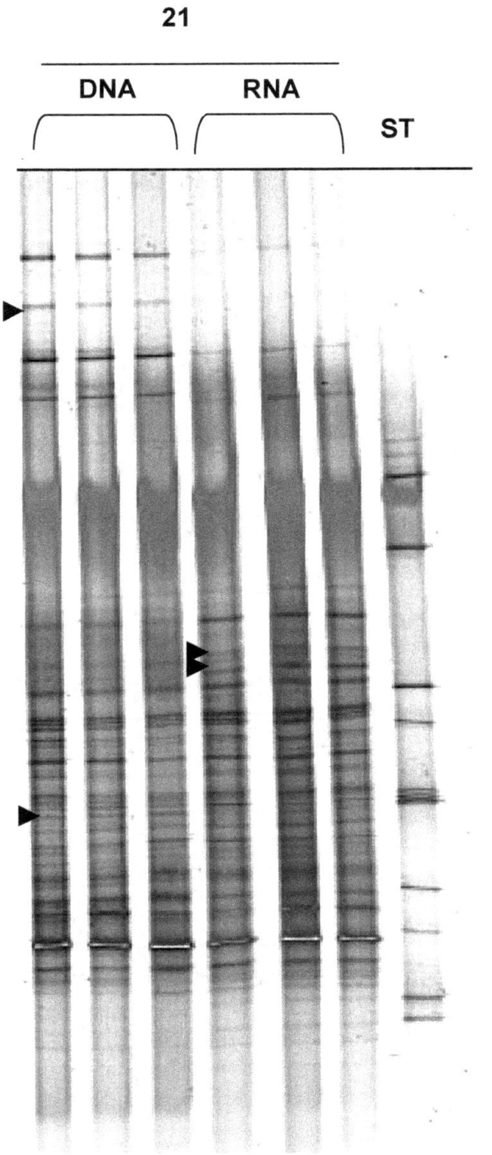

Figure 6.3 Comparison of bacterial DGGE fingerprints generated from 16S rRNA gene fragments amplified from DNA or RNA extracted from soil samples 21 days after artificial contamination with naphthalene (Gomes *et al.*, 2005).

What do molecular approaches tell us about soil microbial life?

Linking DNA:DNA reassociation kinetics and diversity

Measuring the microbial diversity in complex environmental samples remains problematic for several reasons. On the one hand, the bacterial species concept is still under debate and hence a robust criterion for defining bacterial species is lacking (Stackebrandt *et al.*, 2002; Chapter 1). Another major problem for censing the number of species present, e.g. in a soil sample, is the high number of rare species relative to the low number of dominant ones – another major dilemma termed 'the tragedy of the uncommon' by Bent and Forney (2008). It is difficult to experimentally determine the abundance of rare species and therefore estimates of these are based on species abundance distributions or abundances based on assumed species numbers (Curtis and Sloan, 2005). Therefore, Haegeman *et al.* (2008) argued that diversity indices based on richness values (number of species) – which take all species into account without regarding their relative abundance – are not useful while other diversity indices, such as the Shannon and Simpson or Hill ones, are more appropriate because they take relative abundances into account. Already in 1990, reassociation kinetics of single-stranded soil DNA was used by Torsvik *et al.* (1990) to estimate the bacterial richness of a forest soil. The estimates of bacterial richness in soil based on the so-called C_0t (in jargon: cot) curves obtained at the time were recently reassessed (Gans *et al.*, 2005) and species numbers above 10^6 per gram of soil predicted. However, estimates from other recent studies indicate that this number is most likely an overestimation (Schloss and Handelsman, 2006; Roesch *et al.*, 2007). While DNA reassociation kinetics provides an estimate of the species richness in a soil, information on the taxonomic affiliation of the different types can be obtained either by cloning and sequencing of rRNA gene fragments amplified from total DNA or cDNA (Janssen, 2006) or by (pyro)sequencing of 16S rRNA gene amplicons amplified from DNA (Roesch *et al.*, 2007; Fulthorpe *et al.*, 2008) or direct pyrosequencing of cDNA (Urich *et al.*, 2008).

Clone libraries – information on dominant phyla

The first publication describing a 16S ribosomal RNA gene based clone library obtained from soil was published only 17 years ago by Liesack and Stackebrandt (1992). Since then, such clone libraries have been generated from a wide range of different soils and the method has become almost routine, allowing us to obtain a greater insight into the nature and extent of the microbial diversity that lies just below our feet. In a meta-analysis, the available information on the contribution of different genera to 16S rRNA gene clone libraries from soils was recently assembled (Janssen, 2006), based on a total of 3,240 sequences from 32 libraries assigned to genus level by the RDP classifier program (Cole *et al.*, 2005). Although 16S rRNA gene sequences from soil bacteria could be affiliated to 32 phyla, the dominant phyla in the libraries belonged to nine phyla only (Janssen, 2006), i.e. the *Proteobacteria* (average 39%, range 10–77%), *Acidobacteria* (average 20%, range 5–46%), *Actinobacteria* (average 13%, range 0–34%); *Verrucomicrobia* (average 7%, range 0–21%), *Bacteroidetes* (average 5%, range 0–18%), *Chloroflexi* (average 3%, range 0–16%), *Planctomycetes* (average 2.5%, range 0–8%) and Gemmatimonadetes (average 2%, range 0–4%). Interestingly, the phylum *Firmicutes,* including the classes *Bacilli* and *Clostridia* that are considered to be typical soil bacteria, contributed up to only 2% of the sequences in the clone libraries (range 0–8%). However, the composition of the clone libraries, e.g. the relatively low abundance of the *Firmicutes,* might have been strongly affected by DNA extraction biases as mentioned above.

A remarkable proportion of the sequences from the dominant phyla did not show high similarity to known isolates, indicating that a high proportion of soil bacteria still awaits isolation in order to study their physiologies and their potential roles in the soil habitat. However, cultivation-independent analysis of microbial communities has already helped to isolate as-yet-uncultured organisms. Recent advances in this area suggest that a larger fraction of bacterial species in soil might be culturable than previously thought, and that we often just ignore the conditions under which they grow. Recently, isolates were obtained from phyla for which only very few or no isolates were available just a few years ago (Sait *et al.*, 2006; Eichhorst *et al.*, 2007; Nichols, 2007). Advancements in cultivating bacteria were achieved by using growth conditions which simulate *in situ* conditions with low concentrations of nutrients (mirroring the oligotrophic conditions in soil), extended incubation times, addition of cell-to-cell signalling compounds, or applying dilution-to-extinction culturing. Members of subdivision 1 of the acidobacteria favour slightly to moderately acidic growth conditions which correlates with the finding that these organisms are most abundant in libraries from soils of $pH < 6$ but rare in soils of $pH > 6.5$ (Sait *et al.*, 2006).

In a comprehensive survey, Fierer *et al.* (2007b) studied the richness and evenness of bacteria, archaea, fungi and viruses in transcontinental prairie, desert and rainforest soils using a metagenomic and small-subunit rRNA-based sequence analysis approach. In each of the soils, the numbers of unique archaeal and fungal operational taxonomic units (OTU) equalled or exceeded the numbers of unique bacterial OTU.

While the estimated richness of bacterial OTU (approx. 10^4) was quite similar across the different soils, the highest number of archaeal OTU was found in desert soil and the highest number of fungal OTU in prairie soil. Although the size of the clone libraries analysed in this study was larger than that in other studies, rarefaction curves – which indicate the completeness of sampling – still indicated that for all microbial groups, only a fraction of the extant richness was assessed.

Quantifying the genetic richness of the soil microbiota is a crucial step and a prerequisite for developing an ecological framework on soil functioning. Schloss and Handelsman (2006) developed a statistical model that makes richness estimation statistically accessible using an evaluation of the characteristics of samples of simulated communities with parametric community distributions. Based on the distribution of 16S rRNA gene sequences in two soils (Alaskan and Minnesotan soils), simulated communities were constructed with richness of 5,000 and 2,000 OTUs (16S rRNA gene sequences with less than 3% sequence distance). As none of the methods presently used captures more than a small proportion of the total richness of the microbial community and the majority of rare species are difficult to access, the true richness needs to be estimated. The methods commonly used for this estimation are based on nonparametric tools such as provided by the Chao1 and ACE estimators of richness or parametric estimation based on model fitting (Curtis *et al.*, 2006). The merits of the different diversity indices for microbial ecology are still under debate. In one instance, an influence of soil management on soil bacterial communities could not be detected by means of diversity indices but was evidenced by molecular fingerprints (Hartmann and Widmer, 2006). This study showed that changes in the soil community structure do not necessarily result in altered diversity.

Several recent reviews further discuss the particular problems connected with the use of a conceptual framework that was basically developed in macroecology, in microbial ecology (Curtis and Sloan, 2005; Bent and Forney, 2008).

Pyrosequencing of the living soil

Recently, studies have been published which employ mass pyrosequencing for assessing soil microbial diversity. Although based on relatively short sequences (in the studies discussed here, around 100 bp; whereas in later-generation platforms up to 400 bp are possible), these studies provided the largest datasets on rRNA gene sequences obtained for soil so far. In the study by Roesch *et al.* (2007), 16S rRNA gene regions were amplified from total community DNA in 96 independent PCR reactions using 20 cycles – followed by three cycles to add the adapter – and a total of 3 μg DNA was then subjected to 454 pyrosequencing. The OTUs of defined sequence similarity, that ranged from 0 to 20% dissimilarity, were used to calculate rarefaction curves and the richness estimators Chao1 and ACE. Over 139,000 16S rRNA gene fragment sequences from four different soils (three agricultural, one forest) from North and South America were analysed. At the highest phylogenetic resolution (0% dissimilarity) the maximum OTU number was about 52,000 (Roesch *et al.*, 2007). The data indicated that the number of species per g of soil estimated by Gans *et al.* (2005) was an overestimate. More importantly, sequencing of soil microbial communities was shown to be no longer impracticable as suggested by Gans *et al.* (2005). The *Proteobacteria* represented the dominant phylum in each soil, followed by the *Bacteriodetes,* with 40% and 15–25% of the bacterial sequences affiliated to these phyla, respectively. In the agricultural soils, a significant proportion of the sequences had high similarity to the Crenarchaeota, with one-third belonging to the ammonia-oxidizing Archaea. The dataset obtained for each soil (Roesch *et al.*, 2007) was analysed after removing shorter sequences and the sequences were further grouped into phyla by Fulthorpe *et al.* (2008). At 100% similarity, only few genera were detected in all four soils. *Chitinophaga* (Bacteroidetes) and *Acidobacterium* (Acidobacteria) were the only genera detected in all four soils. When using 100% similarity to delineate OTUs, only 1.5% of all OTUs were found in all soils, and at 97% only 5%, indicating that distantly sampled soils carry only few species in common. In contrast to the PCR-based 16S rRNA gene pyrosequencing datasets obtained by Roesch *et al.* (2007) and Fulthorpe *et al.* (2008), an RNA-based metatranscriptomic approach as recently used by Urich *et al.* (2008) omitted the use of a PCR amplification step prior to analysis. Total community RNA directly extracted from soil after harsh cell lysis in a 'fast prep' machine was used as the template for hexamer-primed reverse transcription and the products were subjected to pyrosequencing. More than 250,000 high quality RNA sequences were obtained, the majority representing so-called ribotags (74.8%) with a similar proportion of SSU and LSU ribotags. The ribotags originated from Eukarya (10.3% SSU; 13.3% LSU), Bacteria (87.2% SSU; 83.8% LSU) and Archaea (1.5% SSU; 1.4% LSU) and thus resulted in a dataset which enabled a broad and holistic picture of microbial communities. The untargeted approach had the potential of simultaneously identifying taxa belonging to all three domains of life and also those that will not be amplified with the primer sets that are presently available. Bacteria were – with more than 165,000 ribotags – by far the most abundant and diverse taxon in the soil sample analysed. Nineteen out of the 24 validly described phyla and 20 candidate divisions were identified. In the soil analysed, the most frequently detected phyla were *Actinobacteria, Proteobacteria, Firmicutes, Planctomyces* and *Acidobacteria*. Within the Proteobacteria, by far the majority of OTUs was assigned to the *Alphaproteobacteria* (around 60%). This remarkable dominance of the *Alphaproteobacteria* (Urich *et al.*, 2008) contrasts with the findings by pyrosequencing of PCR amplified ribosomal DNA. Based on the ribotags obtained for fungi, the phylum Ascomycota accounted for two-thirds of the sequences, followed by the phyla Glomeromycota and Basidiomycota. SSU- and LSU-based ribotags showed that the Crenarchaeal candidate division Group I.1b was most abundant in the soil analysed. Aligning the sequences which did not give a hit with the rRNA database (totalling 65,192), following *in silico* translation, with the protein database provided evidence for the *in situ* transcription of the underlying genes. A comparison of these metatranscriptomic data with metagenomic data from the same soil and a different farm soil revealed a remarkable level of congruence, indicating a pool of functions present and expressed in soil microbial communities (Urich *et al.*, 2008). However, the categories RNA and protein metabolism were significantly overrepresented in the metatranscriptome compared to the metagenome data. Presently, a major problem of identifying protein-encoding genes is the bias due to the underrepresentation of full genome sequences available for abundant phyla such as the Planctomycetes and Acidobacteria. Both phyla are represented in the database with fewer than 10 sequences, while several hundreds of full genome sequences are available for different *Proteobacteria* and *Firmicutes*.

Molecular fingerprints

To study biogeographical aspects of the occurrence of particular groups in soil or elucidate how spatial and/or temporal dynamics of microbial communities occur in response to environmental gradients or experimental factors, investigations must rely on a rapid comparative analysis of multiple samples (Muyzer and Smalla, 1998; Forney *et al.*, 2004; Kowalchuk *et al.*, 2006). The molecular fingerprinting techniques based on 16S or 18S rRNA gene fragments or on alternative marker genes amplified from total community DNA still represent the best compromise between the number of processed samples and the level of information that is obtained (see Chapter 4 for a description of fingerprinting techniques). Another advantage of microbial community fingerprint analysis is the quick overview of the composition of dominant sequences, which reflects the relative abundance of bacterial populations or a specific taxonomic or functional group. When combined with the analysis of sequences of differentiating bands, information on the phylogenetic placement of these dominant community members can be rapidly gathered. The question of whether the analysis of 16S rRNA gene fragments amplified from soil DNA by DGGE, single-strand conformation polymorphism (SSCP) or T-RFLP would, for the same sample, yield comparable results between methods, was recently addressed (Smalla *et al.*, 2007). DNA extracted directly from replicate composite soil samples originating from four long-term observation field sites (16 samples in total) was used for PCR amplification of 16S rRNA genes, and subsequent analysis by DGGE, SCCP and T-RFLP. Although the target regions varied, DGGE, T-RFLP and SSCP analyses all led to similar clustering. The clustering of the fingerprints correlated with soil physicochemical properties and little variability between the four replicates of the same soil was observed. The fingerprints of the different soil types revealed significant differences between them in a permutation test (Kropf *et al.*, 2004). In a continental-scale study, including 98 soil samples from North and South America, T-RFLPs of 16S rRNA gene fragments amplified from total community DNA were used to compare bacterial communities across sites (Fierer and Jackson, 2006). The study showed that the composition of soil bacterial communities differed between ecosystems and that most of the differences could be explained by soil pH.

T-RFLP and cloning and sequencing of 16S rRNA gene fragments were also used to study the bacterial community structures in different soil particle size fractions (Sessitsch *et al.*, 2001). The composition and relative abundance of bacteria was clearly affected by the particle size fraction. While *Alphaproteobacteria* dominated in the coarse (sand) fraction, bacteria assigned to the *Holophaga/Acidobacterium* division were found more frequently in the smaller-sized soil particle fraction. Interestingly, 16S rRNA gene sequence analyses revealed the presence of aerobic as well as anaerobic bacteria in the fraction of particles smaller than 2 μm, indicating the presence of both aerobic and anaerobic microsites in this fraction. As dispersion of soil into water by low-energy sonication and subsequent wet sieving was performed to obtain the different particle size fractions, it might well be that the particle size differences found are also a result of the different strengths of interaction with mineral surfaces.

The resolving power of current PCR-based community fingerprinting methods can be considerably increased by using taxon-specific primers. During the PCR reaction with universal primers, a particular target sequence should be present at least at about 0.1% to 1% of the total target sequences to be amplified and appear as a visible band in a fingerprint. To analyse the less common ribotypes and also reduce the complexity of the patterns, nested or semi-nested PCR approaches with taxon-specific primers have been quite successful (Heuer *et al.*, 1997; Garbeva *et al.*, 2003; Gomes *et al.*, 2001; Boon *et al.*, 2002; Salles *et al.*, 2002; Gomes *et al.*, 2005; Freitag *et al.*, 2005; Costa *et al.*, 2006a,b). Such methods have been implemented in many laboratories, allowing the study of phenomena such as shifts of bacterial communities in response to pollution or agricultural management regimes (Nakatsu *et al.*, 2000; Garbeva *et al.*, 2004). A specific analysis of the *Bacillus* and *Pseudomonas* communities in soil allowed Garbeva *et al.* (2003, 2004, 2006) to assess the effects of agricultural management practices on the diversity of these groups, which encompass several organisms with high antagonistic activity against phytopathogens. A major initial finding was the shifted and higher diversity of bacilli found in permanent grassland soils as compared to that in arable soils (Garbeva *et al.*, 2003), and this finding was later extended in another study to the pseudomonads (Garbeva *et al.*, 2004). Moreover, in another study a high relative abundance of non-thermophilic *Crenarchaeota* within the DGGE patterns of archaeal communities, as well as a pronounced influence of grassland management practices on the soil archaeal community structure was found (Nicol *et al.*, 2003). The latter study differentiated between the presence and presumed activity status of different ribotypes by comparing DGGE patterns based on16S rRNA amplified from DNA or cDNA (following RNA extraction and a reverse transcription step). Rather surprising was the observation made by different researchers that DNA- and RNA-based fingerprints of soil bacterial communities are often rather similar (see, for example, Fig. 6.3; Gomes *et al.*, 2005). However, the use of community profiling on the basis of rRNA gene amplicons can bring particular biases, such as those caused by the presence, in different species, of different numbers of rRNA operons per genome (from 1 to up to 15) and inherent within-cell sequence heterogeneities. Moreover, preferential amplification due to mismatches in the target region to the primer used for amplification or the presence of more than one population in one band have also been found (von Wintzingerode *et al.*, 1997). To overcome particular problems incurred with the 16S rRNA gene, marker genes alternative to this, such as *rpoB, cpn60* and *gacA*, have been proposed (Dahllöf *et al.*, 2000; Case *et al.*, 2007; Costa *et al.*, 2007; Chapter 1). These alternative targets offer particular advantages. Given the fact that *rpoB* or *gacA* are, as far as is known, single-copy genes, their use would bypass the disadvantage of potential sequence heterogeneity of the 16S rRNA gene contained in different operons within a single genome (Dahllöf *et al.*, 2000). However, the numbers of *rpoB* or *gacA* sequences present in public databases are still low, and they dwindle in the light of the enormous dataset of 16S rRNA gene sequences.

Due to the fact that we still lack knowledge on whether key functional genes of interest occur scattered or phylogeny-consistent in a community, a major challenge in the use of molecular fingerprinting methods remains to link microbial community structure with function. Thus, the 16S rRNA gene marker may, in specific cases, truly report on (potential) function. This is the case of the ammonia oxidizers, in which the 16S rRNA gene appears as a good marker (Kowalchuk *et al.*, 1998; Prosser *et al.*, 2002; Webster *et al.*, 2002). DGGE profiling of the ammonia oxidizers revealed that the composition of these groups in soils with different agricultural histories was quite similar although significant changes in potential rates of nitrification and in the sizes of ammonia oxidizer populations were observed (Philips *et al.*, 2000). In particular, the relative dominance of *Nitrosospira* cluster 3 in all soils was shown. As an alternative to the 16S rRNA gene-based molecular fingerprinting techniques, methods that are directly based on functional genes have been used to assess the diversity of these genes in soils and to ascertain how their abundances and diversities are affected by treatment (Liesack and Dunfield, 2002; Costa *et al.*, 2007; Webster *et al.*, 2002). Thus, the gene encoding nitrogenase (the key enzyme in nitrogen fixation), i.e. *nifH*, has been used to assess the diversity of organisms involved in this process, and so has the gene involved in the first step in ammonia oxidation, *amoA*. The resulting fingerprints allow an assessment of the particular functional types that are or become dominant under particular conditions in the soil. However, a major task ahead of us still is to link the information on functional gene diversity to functioning of the system, as well as to address the issue of stability of function in the face of stress imposed on the soil.

Soil metagenomics and shotgun sequencing

The holy grail in soil microbiology is considered to be the genetic and functional mapping of the complete microbiota in samples of interest. Whereas, on the one hand, the 16S rRNA gene-based studies have been useful in assessing the relative abundance and diversity of different taxa in soil and have led to striking discoveries, such as the high abundance of *Alphaproteobacteria, Acidobacteria* and/or *Crenarchaeota* in soils, metagenomics-based approaches, in which libraries are generated by direct cloning of large genomic fragments into bacterial artificial chromosome (BAC) or fosmid libraries, enable us to analyse large genomic fragments of as-yet-uncultured organisms. In addition to clone-based analyses, metagenomics (see Chapter 2) can also be based on the direct (454 or pyro)sequencing of soil-derived nucleic acids. Two parallel strategies can be used in the analyses (1) untargeted sequence-based analysis and (2) gene/function-targeted approaches. To obtain sequence information of particular uncultured taxa which are assumed to be abundant in the soil sample, in a first step specific primers are used to identify clones carrying the respective genes, either based on 16S rRNA or even on functional genes. By this approach, genomic fragments of uncultured *Acidobacteria* or Archaea could be obtained and subjected to sequence-based analyses (Ochsenreiter *et al.*, 2003; Treusch *et al.*, 2005; Van Elsas *et al.*, 2008a,b). Sequencing of a large fosmid library obtained from a grassland soil revealed that a fragment with an rRNA gene, which showed the highest similarity to group 1.1b of the crenarchaeota, carried genes encoding copper nitrite reductase and two subunits of ammonia monooxygenase or particulate methane monooxygenase (Treusch *et al.*, 2005). Furthermore, these authors could demonstrate the expression of an *amo*A homologue in soil and increased expression levels in the presence of elevated ammonia concentrations. The discovery that ammonia oxidizing Archaea are often as abundant in soil (Leininger *et al.*, 2006) as their bacterial counterparts showed the potential of soil metagenomics studies to elucidate the role of uncultured organisms in, for example, soil nutrient cycles. Considering that the majority of soil bacteria occur at relatively low abundance but might have an important ecological role, approaches such as targeted access to low-abundance bacteria are of considerable interest. Recently, Podar *et al.* (2007) combined taxon-specific separation by flow cytometry and whole-genome amplification to obtain partial characterization of the genome sequence of an uncultivated soil bacterium belonging to the candidate division TM7. Although TM7 bacteria are usually low in abundance and have not yet been cultivated, they have been reported to occur in many soils. Besides information on major metabolic routes, the partial genome sequence revealed many interesting biological features, such as protection mechanisms against toxins or foreign DNA, genes linked to environmental stress responses and type IV pili which may be important for colonizing interstitial space in soil or biofilm formation at surfaces (see Chapter 11).

Gene microarrays – the next generation of fingerprinting techniques

To increase the throughput of the detection of microorganisms in complex samples such as soil, high-density 16S rRNA gene probe arrays have been developed. These so-called PhyloChips allow the comprehensive detection of bacteria and archaea in complex environmental samples (Chapter 5). Recently, the applicability of the PhyloChip was explored for soils. Using this approach, 16S rRNA gene regions are amplified from total community DNA, after which the amplicons are fragmented by DNase treatment and biotin end-labelled. The labelled fragments are hybridized to oligonucleotide probes of which 500,000 can be held on a glass slide (see Chapter 5 for a detailed description of this technique). In a very complex sample, such as soil, distinguishing potential sequence diversity and cross-hybridization is the greatest problem. To overcome the problem (with potential sequence diversity) a minimum of 11 or more short oligonucleotides were designed by Brodie *et al.* (2007) for each taxonomic group. Cross-hybridization becomes a key issue, in particular, when highly abundant 16S rRNA gene fragments share sequence similarity to non-target probes resulting in weak false-positive signals. The use of the perfect match (PM)-mismatch (MM) pair approach has been shown to substantially increase the chance that positive hybridization is due to hybridization to a target sequence and not by cross-hybridization to a non-target probe. Using a phyloarray with the potential to detect 8741 OTUs, over 2000 OTUs could be detected in DNA from agricultural land and cactus forest soils from two sites in Baja California, Mexico (Smalla *et al.*, in preparation). The approach was suitable to identify OTUs which contributed to the differences in the soils under different land use. De Santis *et al.* (2007) detected a higher degree of diversity by this approach than that observed by 'classical' 16S rRNA gene based PCR followed by cloning and sequencing. Moreover, the approach was useful to overcome the problem of dominance, i.e. the overshadowing of members of the rare soil biosphere by the dominant ones. The phyloarrays that are presently available can thus complement the 16S rRNA gene-based cloning and sequencing as well as community fingerprinting by DGGE, SSCP or T-RFLP.

In order to achieve a comprehensive investigation of functional genes, so-called GeoChip (microarrays containing over 24,000 probes covering more than 10,000 genes distributed among more than 150 functional groups involved in nitrogen, carbon, sulphur and phosphorus cycling) were recently also employed for soil studies (Schadt and Zhou, 2006; Wu *et al.*, 2006; Gentry *et al.*, 2006; He *et al.*, 2007; Chapter 5). The direct hybridizations were, however, not very sensitive, and to increase sensitivity of the approach, a preamplification step (e.g. using rolling circle amplification) is now included. Although numerous challenges need to be solved, the GeoChip will be a powerful tool that allows linking functional microbial dynamics to particular ecosystem processes.

Probe development, hybridization quality and data evaluation are crucial steps for an appropriate use of DNA microarrays to study the soil microbiota. Recently, the GeoChip was applied to study N- and C-cycle genes in DNA extracted from soil taken from different sites in the Antarctic (Yergeau *et al.*, 2007). The N- and C-cycle genes detected differed significantly across sampling locations and vegetation types. Given the novelty of the functional gene array, it was felt that substantiation of the data by another technique was required and hence quantitative

PCR and/or enzymatic assays were used to confirm the hybridization results. In most cases a good correlation was found between the microarray signals and the abundance of particular genes as detected by q-PCR (Yergeau *et al.*, 2007). However, high costs are still a limiting factor in the broad use of this approach. For instance, to compare the effects of treatments, the variability of technical as well as field replicates needs to be considered. Statistical evaluation of the array data is yet another bottleneck. Furthermore, DNA microarrays cannot generate information on new sequence types and thus only the breadth of known functions can be assessed (Yergeau *et al.*, 2007).

Despite such remaining challenges, the all-at-once view of (potential) soil functioning offered by DNA microarrays is attractive. The data obtained can be placed in the context of (local) soil conditions to obtain gene level – habitat correlations. More specifically, quantitative methods, such as q-PCR, might be employed at the microscale level to work out to what extent local soil conditions affect gene and gene expression levels within the microbial communities present. Hence, even in the face of the daunting heterogeneity of soil and the consequent high diversity of microorganisms and functions, it will be possible to map microbial function to soil space and come closer to the '*who does what, where, with whom, when and why*' question.

Approaches for linking microbial community structure and function

In the 1980s the ^{14}C radioactive isotope (a weak alpha emitter) was extensively used as a tracer to indicate the fate of radioactively-marked C (for example from plant material) into members of the soil microbial community. Hence, the concept of using molecular tags to follow carbon transformations in the soil system was established relatively early.

In a modern variant of this work, and in combination with DNA fractionation approaches, a method coined 'stable isotope probing' (SIP) was recently developed (see Chapter 4). In a number of studies, the use of stable isotopes like ^{13}C (Radajewski *et al.*, 2000; Manefield *et al.*, 2002, 2006), ^{15}N, or of the nucleotide homologue, bromodeoxyuridine (BrdU) (Artursson and Jansson, 2003; Hjort *et al.*, 2007), have allowed the identification of the microbes that are involved in key processes, such as carbon and nitrogen cycling and/or the degradation of organic chemicals (Dumont and Murrell, 2005). A combination of stable isotope probing (SIP) with biomarker fingerprints enabled identification of soil microorganisms involved in the decomposition of labelled cellulose (Haichar *et al.*, 2007), plants (Bernard *et al.*, 2007), and tracking of photosynthates as they are translocated below ground (Vandenkoornhuyse *et al.*, 2007). These studies have enabled us to better understand the carbon flow in soil. Expanding SIP to other applications certainly has great potential, not only to unravel primary responders but also to better understand multitrophic food webs (Lueders *et al.*, 2006). In the latter study it was demonstrated – by means of rRNA-SIP – that as-yet-uncultivated relatives of gliding bacterial micropredators were involved in carbon sequestration and mineralization of added ^{13}C-labelled *E. coli* under *in situ* soil conditions. Free-living diazotrophs in soil were recently identified by employing $^{15}N_2$-DNA-SIP (Buckley *et al.*, 2007). The *nifH* diversity detected provided evidence for three microbial groups belonging to the rhizobiales and two deeply branching lineages of uncultured *Betaproteobacteria* and *Actinobacteria*.

How can pending scientific challenges in soil microbiology be addressed by the emerging molecular tools?

As indicated in the foregoing, the decadal challenge of modern soil microbiology is to link microbial community structure to function. This improved understanding of who does what ideally allows insights – and this should include the microscale level – in the network of key processes in soil and the organisms that play roles therein. This rather simple statement has daunting implications if we consider the enormous microbial diversity of soil, which by its sheer magnitude seems prohibitive for any meaningful and comprehensive type of analysis that leads to enhanced understanding. For instance, any molecular fingerprinting method that is executed on the basis of directly extracted soil DNA or RNA will only provide insight into the most dominant organisms present in the system unless the extraction is deliberately biased or taxon-specific primers are used. Many soil microbiologists have voiced the contention that microbial function rather than community structure is key to soil systems. In other words, what matters would be the functioning of the system, directed by the presence of a sufficient amount of genes or operons that are functional, and not so much the precise nature (affiliation) of the microorganisms that are the carriers of these functions. When it comes to the preservation of key functions in the soil environment, conservation of microbial function might then be the crucial asset rather than conservation of community structure. Extending this argument a bit, one would expect the *redundancy* of the function investigated to be a key element for study, that is, the level to which a particular functional capability is shared across members of a given community. Molecular methods are then helpful to ascertain robust soil functioning when and if they are able to provide information about the functional redundancy levels; functions can be characterized as either broadly shared (high functional redundancy) or poorly shared (low functional redundancy). Examples of the former are any function linked to the central cellular metabolism, whereas the latter may be exemplified by e.g. the capacity to utilize ammonia by oxidation (shared by just a narrow subgroup of bacteria and archaea).

It should be borne in mind, though, that the actual functioning of soil is determined by the levels of gene expression at any point in time, and that functionality therefore should be tackled by assessing the relevant functional genes involved in a given process in terms of their expression. Gene expression may be very instantaneous, i.e. it may vary over even minute time spans. Furthermore, given the fact that microorganisms perceive the conditions in their microenvironment very acutely, gene expression may vary even over small distances in the soil. Thus, in spite of the obvious need to address *in situ* gene expression, one can argue about what the finer-scale data would then tell us (in an overall fashion) about soil function and stability, if every microsite may be functionally different at different points in time.

Conclusions

In the light of the soil's heterogeneity and the multitude of chemical, physical and biological gradients in soil, it is a daunting and probably impracticable task to even try to come to grips with all microbial processes and interactions taking place at the microscale level in soil. Scientific progress in this area will very much depend on how successful we will be in applying the promising novel molecular tools to the microscale level. In particular, the enormous analytical power of high-throughput (454) pyrosequencing should be harnessed to encompass soil metagenomes at a level that we can comprehensively analyse. There should be a thorough investigation of the real needs of performing microscale analyses prior to any massive sequencing effort. Moreover, the emerging fields of soil transcriptomics (Urich *et al.*, 2008) and soil proteomics (Chapter 3) should also be taken the same way. In fact, a reduction of data complexity on the basis of educated guesses about the desired level at which the system should be analysed, space- and time-wise, will form a major asset in many studies. Major challenges in these areas will thus be the use of intelligent bioinformatics tools that will guide us, in our scientific questions, towards the types of data that are most likely to provide answers to the relevant questions.

Future trends

Future mining and exploration of the soil metagenome is greatly facilitated by the ever-expanding high-throughput sequencing facilities, and direct deep sequencing of the soil microbiota has become a realistic scenario. What one will gather from this endeavour is, first of all, an inventory of all genetic/metabolic potential that is present in the system. Only in a second stage, following meticulous work that is strongly dedicated to the unravelling of particular metabolic routes (e.g. the ammonia oxidation pathway, its diversity and functioning), will it be possible to truly link the metagenomic information to soil functioning. In addition, the development of vectors that facilitate gene expression in hosts alternative to *Escherichia coli*, or even vectors that allow shuttling between *E. coli* and such alternative hosts will spur a wider range of expression screenings in different hosts, which is important to understand aspects of gene expression and functioning in such soil-derived hosts (Van Elsas *et al.*, 2008b). However, untargeted soil metagenomics approaches will allow us to address gene expression and genomes of the most abundant organisms only (Van Elsas *et al.*, 2008b). Hence, to facilitate focussing on target organisms, genes or operons, intelligent approaches based on, for instance, pre-fractionation of cells (using the flow cytometer) or genes/DNA (based on, for instance, the G+C%), will be required to allow targeted metagenomics-based analyses of less abundant organisms (Van Elsas *et al.*, 2008b). Given the enormous complexity of the soil microbiota, such targeted approaches are highly needed to enhance our understanding of organisms or genes involved in particular functions. In a general sense, there are also concerns regarding the representation of the soil microbiota in many directly prepared libraries (e.g. van Elsas *et al.*, 2008a,b) and specific statistical tools are required to allow assessments of the degree of representation.

Acknowledgements

We thank Ms Ilse-Marie Jungkurth for critically assessing the typographical quality of this manuscript. Furthermore, the DFG funded SPP1315 is acknowledged for multidisciplinary discussions of the challenging research field 'soil'. JDVE was supported by NATO Collaborative Research grant 981785.

References

Amann, R.I., Ludwig, W., and Schleifer, K.-H. (1995). Phylogenetic identification and *in situ* detection of individual microbial cells without cultivation. Microbiol. Rev. *59*, 143–169.

Artursson, V., and Jansson, J.K. (2003). Use of bromodeoxyuridine immunocapture to identify active bacteria associated with arbuscular mycorrhizal hyphae. Appl. Environ. Microbiol. *69*, 6208–6215.

Becker, J.M., Parkin, T., Nakatsu, C.H., Wilbur, J.D., and Konopka, A. (2006). Bacterial activity, community structure, and centimeter-scale spatial heterogeneity in contaminated soil. Microbial Ecol. *51*, 220–231.

Bent, S.J., and Forney, L.J. (2008). The tragedy of the uncommon: understanding limitations in the analysis of microbial diversity. ISME *2*, 689–695.

Bernard, L., Mougel, C., Maron, P.-A., Nowak, V., Lévêque, J., Henault, C., el Zahar Haichar, F., Berge, O., Marol, C., Balesdent, J., Gibiat, F.P., and Ranjard, L. (2007). Dynamics and identification of soil microbial populations actively assimilating carbon from ^{13}C-labelled wheat residue as estimated by DNA- and RNA-SIP techniques. Environ. Microbiol. *9*, 752–764.

Boon, N., de Windt, W., Verstraete, W., and Top, E.M. (2002). Evaluation of nested PCR-DGGE (denaturing gradient gel electrophoresis) with group-specific 16S rRNA primers for the analysis of bacterial communities from different wastewater treatment plants. FEMS Microbiol. Ecol. *39*, 101–112.

Brodie, E.L., DeSantis, T.Z., Moberg Parker J.P., Zubietta, I.X., Piceno, Y.M., and Andersen G.L. (2007). Urban aerosols harbor diverse and dynamic bacterial populations. Proc. Natl. Acad. Sci. U.S.A. *104*, 299–304.

Buckley, D.H., Huangyutitham, V., Hsu, S.-F., and Nelson, T.A. (2007). Stable isotope probing with $^{15}N_2$ reveals novel noncultivated diazotrophs in soil. Appl. Environ. Microbiol. *73*, 3196–3204.

Case, R.J., Boucher, Y., Dahllöf, I., Holmström, C., Doolittle, W.F., and Kjelleberg, S. (2007). Use of 16S rRNA and *rpoB* genes as molecular markers for microbial ecology studies. Appl. Environ. Microbiol. *73*, 278–288.

Cole, J.R., Chai, B., Farris, R.J., Wang, Q., Kulam, S.A., McGarrell, D.M., Garrity, G.M., and Tiedje, J.M. (2005). The ribosomal database project (RDP-II): sequences and tools for high-throughput rRNA analysis. Nucl. Acids Res. *33*, D294-D296.

Costa, R., Salles, J.F., Berg, G., and Smalla, K. (2006a). Cultivation-independent analysis of *Pseudomonas* species in soil and in the rhizosphere of field-grown *Verticillium dahliae* host plants. Environ. Microbiol. *8*, 2136–2149.

Costa, R., Götz, M., Mrotzek, N., Lottmann, J., Berg, G., and Smalla, K. (2006b). Effect of site and plant species on rhizosphere community structure as revealed by molecular analysis of different microbial guilds. FEMS Microbiol. Ecol. *56*, 236–249.

Costa, R., Gomes, N.C.M., Krögerrecklenfort, E., Opelt, K., Berg, G., and Smalla, K. (2007). *Pseudomonas* community structure and antagonistic potential in the rhizosphere: insights gained by combining phylogenetic and functional gene-based analyses. Environ. Microbiol. *9*, 2260–2273.

Curtis, T.P., and Sloan, W.T. (2005). Microbiology: Exploring microbial diversity – a vast below. Science *309*, 1331–1333.

Curtis, T.P., Head, I.M., Lunn, M., Woodstock, S., Schloss, P.D., and Sloan, W.T. (2006). What is the extent of prokaryotic diversity? Phil. Trans. R. Soc. Lond. B Biol. Sci. *361*, 2023–2037.

Dechesne, A., Pallud, C., Debouzie, D., Flandrois, J.P., Vogel, T.M., Gaudet, J.P., and Grundmann, G.L. (2003). A novel method for characterizing the microscale 3D spatial distribution of bacteria in soil. Soil Biol. Biochem. *35*, 1537–1546.

DeSantis, T.Z., Brodie, E.L., Moberg, J.P., Zubieta, I.X., Piceno, Y.M., and Andersen, G.L. (2007). High-density universal 16S rRNA microarray analysis reveals broader diversity than typical clone library when sampling the environment. Microb. Ecol. *53*, 371–383.

Dumont, M.C., and Murrell, J.C. (2005). Stable isotope probing – linking microbial identity to function. Nature *3*, 499–504.

Dunbar, J., Ticknor, L.O., and Kuske, C.R. (2000). Assessment of microbial diversity in four Southwestern United States soils by 16S rRNA gene terminal restriction fragment analysis. Appl. Environ. Microbiol. *66*, 2943–2950.

Eichhorst, S.A., Breznak, J.A., and Schmidt, T.M. (2007). Isolation and characterization of soil bacteria that define Terriglobus gen. nov. in the phylum *Acidobacteria*. Appl. Environ. Microbiol. *73*, 2708–2717.

Fierer, N., Jackson, J.A., Vilgalys, R., and Jackson, R.B. (2005). Assessment of soil microbial community structure by use of taxon-specific quantitative PCR assays. Appl. Environ. Microbiol. *71*, 4117–4120.

Fierer, N., and Jackson, R.B. (2006). The diversity and biogeography of soil bacterial communities. Proc. Natl. Acad. Sci. U.S.A. *103*, 626–631.

Fierer, N., Bradford, M.A., and Jackson, R.B. (2007a). Toward an ecological classification of soil bacteria. Ecology *88*, 1354–1364.

Fierer, N., Breitbart, M., Nulton, J., Salamon, P., Lozupone, C., Jones, R., Robeson, M., Edwards, R.A., Felts, B., Rayhawk, S., Knight, R., Rohwer, F., and Jackson, R.B. (2007b). Metagenomic and small subunit rRNA analyses reveal the genetic diversity of Bacteria, Archaea, Fungi and Viruses in soil. Appl. Environ. Microbiol. *73*, 7059–7066.

Forney, L.J., Zhou, X., and Brown, C.J. (2004). Molecular microbial ecology: land of the one-eyed king. Curr. Opin. Microbiol. *7*, 210–220.

Freitag, T.E., Chang, L., Clegg, C.D., and Prosser, J.I. (2005). Influence of inorganic nitrogen management regime on the diversity of nitrite-oxidizing bacteria in agricultural grassland soils. Appl. Environ. Microbiol. *71*, 8323–8334.

Fulthorpe, R.R., Roesch, L.F.W., Riva, A., and Triplett, E.W. (2008). Distantly sampled soils carry few species in common. The ISME J, *2*, 901–910.

Gans, J., Wolinsky, M., and Dunbar, J. (2005). Computational improvements reveal great bacterial diversity and high metal toxicity in soil. Science *309*, 1387–1390.

Garbeva, P. (2004). Assessment of the diversity and antagonism towards *Rhizoctonia solani* AG3 of *Pseudomonas* species in soil from different agricultural regimes. FEMS Microbiol. Ecol. *47*, 51–64.

Garbeva, P., Van Veen, J.A., and Van Elsas, J.D. (2003). Predominant *Bacillus* spp. in agricultural soil under different management regimes detected via PCR-DGGE. Microb. Ecol. *45*, 302–316.

Garbeva, P., Postma, J., Van Veen, J.A., and Van Elsas, J.D. (2006). Effect of aboveground plant species on soil microbial community structure and its impact on suppression of *Rhizoctonia solani* AG3. Environ. Microbiol. *8*, 233–246.

Gelsomino, A., Keijzer-Wolters A.C., and Van Elsas, J.D. (1999). Assessment of bacterial community structure in soil by polymerase chain reaction and denaturing gradient gel electrophoresis. J. Microbiol. Meth. *38*, 1–15.

Gentry, T.J., Wickham, G.S., Schadt, C.W., He, Z., and Zhou, J. (2006). Microarray applications in microbial ecology research. Microb. Ecol. *52*, 159–175.

Gomes, N.C.M., Heuer, H., Schönfeld, J., Costa, R., Hagler-Mendonça, L., and Smalla, K. (2001). Bacterial diversity of the rhizosphere of maize (*Zea mays*) grown in tropical soil studied by temperature gradient gel electrophoresis. Plant and Soil *232*, 167–180.

Gomes, N.C.M., Kosheleva, I.A., Abraham, W.-R., and Smalla, K. (2005). Effects of the inoculant strain *Pseudomonas putida* KT2442 (pNF142) and of naphthalene contamination on the soil bacterial community. FEMS Microbiol. Ecol. *54*, 21–33.

Griffith, R.I., Whiteley, A.S., O'Donell, A.G., and Bailey, M.J. (2000). Rapid method for coextraction of DNA and RNA from natural environments for analysis of ribosomal DNA- and rRNA-based microbial community composition. Appl. Environ. Microbiol. *66*, 5488–5491.

Griffith, R.I., Whiteley, A.S., O'Donell, A.G., and Bailey, M.J. (2003). Physiological and community responses of established grassland bacterial populations to water stress. Appl. Environ. Microbiol. *69*, 6961–6968.

Grundmann, G.L. (2004). Spatial scales of soil bacterial diversity – the size of a clone. FEMS Microbiol. Ecol. *48*, 119–127.

Grundmann, G.L., and Gourbière, F. (1999). A microsampling approach to improve the inventory of bacterial diversity in soil. Appl. Soil Ecol. *13*, 123–126.

Grundmann, G.L., Dechesne, A., Bartoli, F., Flandrois, J.P., Chassé, J.L., and Kizungu, R. (2001). Spatial modeling of nitrifier microhabitats in soil. Soil Sci. Soc. Am. J. *65*, 1709–1716.

Hackl, E., Zechmeister-Boltenstern, S., Bodrossy, L., and Sessitsch, A. (2004). Comparison of diversities and compositions of bacterial populations inhabiting natural forest soils. Appl. Environ. Microbiol. *70*, 5057–5065.

Haegeman, B., Vanpeteghem, D., Godon, J.-J., and Hamel, J. 2008. DNA reassociation kinetics and diversity indices: richness is not rich enough. Oikos *117*, 177–181.

Haichar, F.Z., Achouak, W., Christen, R., Heulin, T., Marol, C., Marais, M.-F., Mougel, C., Ranjard, L., Balesdent, J., and Berg, O. (2007). Identification of cellulytic bacteria in soil by stable isotope probing. (2007). Environ. Microbiol. *9*, 625–634.

Hartmann, M., and Widmer, F. (2006). Community structure analyses are more sensitive to differences in soil bacterial communities than anonymous diversity indices. Appl. Environ. Microbiol. *72*, 7804–7812.

He, Z., Gentry, T.J., Schadt, C.W., Wu, L., Liebich, J., Chong, S.C., Huang, Z., Wu, W., Gu, B., Jardine, P., Criddle, C., and Zhou, J. (2007). GeoChip: a comprehensive microarray for investigating biogeochemical, ecological and environmental processes. ISME *1*, 67–77.

Heuer, H., Krsek, M., Baker, P., Smalla, K., and Wellington, E.M.H. (1997). Analysis of actinomycete communities by specific amplification of genes encoding 16S rRNA and gel-electrophoretic separation in denaturing gradients. Appl. Environ. Microbiol. *63*, 3233–3241.

Heuer, H., Abdo, Z., and Smalla, K. (2008). Patchy distribution of flexible genetic elements in bacterial populations mediates robustness to environmental uncertainty. FEMS Microbiol. Ecol. *65*, 361–371.

Hjort, K., Lembke, A., Speksnijder, A., Smalla, K., and Jansson, J.K. (2007). Community structure of actively growing bacterial populations in plant pathogen suppressive soil. Microbial Ecol. *53*, 399–413.

Hurt, R.A., Qiu, X.Y., Wu, L.Y., Roh, Y., Palumbo, V., Tiedje, J.M., and Zhou, J. (2001). Simultaneous recovery of RNA and DNA from soils and sediments. Appl. Environ. Microbiol. *67*, 4495–4503

Izquierdo, J.A., and Nüsslein, K. (2006). Distribution of extensive *nifH* gene diversity across physical soil microenvironments. Microb. Ecol. *51*, 441–452.

Janssen, P.H. (2006). Identifying the dominant soil bacterial taxa in libraries of 16S rRNA and 16S rRNA genes. Appl. Environ. Microbiol. *72*, 1719–1728.

Kowalchuk, G.A., Bodelier, P.L.E., Heilig, G.H.J., Stephen, J.R., and Laanbroek, H.J. (1998). Community analysis of ammonia-oxidizing bacteria, in relation to oxygen availability in soils and root-oxygenated sediments, using PCR, DGGE and oligonucleotide probe hybridisation. FEMS Microbiol. Ecol. *27*, 339–350.

Kowalchuk, G.A., Drigo, B., Yergeau, E., and Van Veen, J.A. (2006). Assessing bacterial and fungal community structure in soil using ribosomal RNA and other structural gene markers. In Nucleic Acids and Proteins in Soil, Nannipieri, P. and Smalla, K., eds (Berlin Heidelberg, Germany: Springer-Verlag), pp. 159–188.

Kowalchuk, G.A., Speksnijder A.G.C.L., Zhang, K., Goodman, R.M., and van Veen, J.A. (2007). Finding the Needles in the Metagenome Haystack. Microb. Ecol. *53*, 475–485.

Kropf, S., Heuer, H., Grüning, M., and Smalla, K. (2004). Significance test for comparing complex microbial community fingerprints using pairwise similarity measures. J. Microbiol. Methods *57*, 187–195.

Leininger, S., Urich, T., Schloter, M., Schwark, L., Qui, J., Nicol, G.W., Prosser, J.I., Schuster, S.C., and Schleper, C. (2006). Archaea predominate among ammonia-oxidizing prokaryotes in soils. Nature *442*, 806–809.

Liesack, W., and Stackebrandt, E. (1992). Occurrence of novel groups of the domain Bacteria as revealed by analysis of genetic material isolated from an Australian terrestrial environment. J. Bacteriol. *174*, 5072–5078.

Liesack, W., and Dunfield, P.F. (2002). Biodiversity in soils: use of molecular methods for its characterization. In Encyclopedia of Environmental Microbiology, Bitton, G., ed (New York: Wiley & Sons), pp. 528–544.

Lueders, T., Kindler, R., Miltner, A., Friedrich, M.W., and Kaestner, M. (2006). Identification of bacterial micropredators distinctively active in a soil microbial food web. Appl. Environ. Microbiol. *72*, 5342–5348.

Manefield, M., Whiteley, A.S., Griffiths, R.I., and Bailey. M.J. (2002). RNA stable isotope probing, a novel means of linking microbial community function to phylogeny. Appl. Environ. Microbiol. *68*, 5367–5373.

Manefield, M., Griffiths, R.I., Bailey, M.J., and Whiteley, A.S. (2006). Stable Isotope Probing: a critique of its role in linking phylogeny and function, In Nucleic Acids and Proteins in Soil, Nannipieri, P. and Smalla, K., eds (Berlin: Springer-Verlag), pp. 205–216.

Mummey, D., Holben, W., Six, J., and Stahl, P. (2006). Spatial stratification of soil bacterial populations in aggregates of diverse soils. Microb. Ecol. *51*, 404–411.

Muyzer, G., and Smalla, K. 1998. Application of denaturing gradient gel electrophoresis (DGGE) and temperature gradient gel electrophoresis (TGGE) in microbial ecology. Antonie van Leeuwenhoek *73*, 127–141.

Nakatsu, C.H., Torsvik, V., and Øvreås, L. (2000). Soil community analysis using DGGE of 16S rDNA polymerase chain reaction products. Soil Sci. Soc. Am. J. *64*,1382–1388

Nichols, D. (2007). Cultivation gives context to the microbial ecologist. FEMS Microbiol. Ecol. *60*, 51–57.

Nicol, G.W., Glover, L.A., and Prosser, J.I. (2003). Spatial analysis of archaeal community structure in grassland soil. Appl. Environ. Microbiol. *69*, 7420–7429.

Nielsen, K.M., Johnsen, P., and van Elsas, J.D. (2007). Horizontal gene transfer and microevolution in soil. In: Modern Soil Microbiology II, van Elsas, J.D. Jansson, J.K., and Trevors, J.T., eds (Boca Raton: CRC Press), pp. 55–82.

Nunan, N., Ritz, K., Crabb, D., Harris, K., Wu, K., Crawford, J.W., and Young, I.M. (2001). Quantification of the *in situ* distribution of soil bacteria by large-scale imaging of thin sections of undisturbed soil. FEMS Microbiol. Ecol. *36*, 67–77.

Nunan, N., Wu, K., Young, I.M., and Crawford, J.W. (2002). *In situ* spatial patterns of soil bacterial populations, mapped at multiple scales, in an arable soil. Microb. Ecol. *44*, 296–305.

Nunan, N., Wu, K., Young, I.M., Crawford, J.W., and Ritz, K. (2003). Spatial distribution of bacterial communities and their relationships with the micro-architecture of soil. FEMS Microbiol. Ecol. *44*, 203–215.

Nüsslein K., and Tiedje, J.M. (1998). Characterization of the dominant and rare members of a young Hawaiian soil bacterial community with small-subunit ribosomal DNA amplified from DNA fractionated on the basis of its guanine and cytosine composition. Appl. Environ. Microbiol. *64*, 1283–1289.

Ochsenreiter, T., Selezi, D., Quaiser, A., Bonch-Osmolovskaya, L., and Schleper, C. (2003). Diversity and abundance of Crenarchaeota in terrestrial habitats studied by 16S RNA surveys and real time PCR. Environ. Microbiol. *5*, 787–797.

Phillips, C.J., Harris, D., Dollhopf, S.L., Gross, K.L., Prosser, J.I., and Paul, E.A. (2000). Effects of agronomic treatments on structure and function of ammonia-oxidizing communities. Appl. Environ. Microbiol. *66*, 5410–5418.

Podar, M., Abulencia, C.B., Walcher, M., Hutchison, D., Zengler, K., Garcia, J.A., Holland, T., Cotton, D., Hauser, L., and Keller, M. (2007). Targeted access to genomes of low abundance organisms in complex microbial communities. Appl. Environ. Microbiol. 73, 3205–3214.

Poll, C., Thiede, A., Wermbter, N., Sessitsch, A., and Kandeler, E. (2003). Micro-scale distribution of microorganisms and microbial enzyme activities in a soil with long-term organic amendment. European J. Soil Science *56*, 715–724.

Preston, S., Wirth, S., Ritz, K., Griffiths, B.S., and Young, I.M. (2001). The role played by microorganisms in the biogenesis of soil cracks: importance of substrate quantity and quality. Soil Biol. & Biochem. *33*, 1851–1858.

Prosser, J.I. (2002). Molecular and functional diversity in soil micro-organisms. Plant and Soil *244*, 9–17.

Quaiser, A., Ochsenreiter, T., Lanz, C., Schuster, S.C., Treusch, A.H., Eck, J., and Schleper, C. (2003). Acidobacteria form a coherent but highly diverse group within the bacterial domain: evidence from environmental genomics. Mol. Microbiol. *50*, 563–575.

Radajewski, S., Ineson, P., Parekh, N.R., and Murrell, J.C. (2000). Stable isotope probing as a tool in microbial ecology. Nature *403*, 646–649.

Ramette, A., and Tiedje, J.M. (2007). Multiple responses of microbial life to spatial distance and environmental heterogeneity in a patchy ecosystem. Proc. Natl. Acad. Sci. U.S.A. *104*, 2761–2766.

Ranjard, L., and Richaume, A. (2001). Quantitative and qualitative microscale distribution of bacteria in soil. Res. Microbiol. *152*, 707–716.

Roesch, L.F.W., Fulthorpe, R.R., Riva, A., Casella, G., Hadwin, A.K.M., Kent, A.D., Daroub, S.H., Camargo, F.A.O., Farmerie, W.G., and Triplett, E.W. (2007). Pyrosequencing enumerates and contrasts soil microbial diversity. ISME *1*, 283–290.

Rogers, S.W., Moorman, T.B., and Ong, S.K. (2007). Fluorescent *in situ* hybridization and microautoradiography applied to ecophysiology in soil. Soil Sci. Soc. Am. J. *71*, 620–631.

Sait, M., Davis, K.E.R., and Janssen, P.H. (2006). Effect of pH on isolation and distribution of members of subdivision 1 of the phylum *Acidobacteria* occurring in soil. Appl. Environ. Microbiol. *72*, 1852–1857.

Salles, J.F., de Souza, F.A., and van Elsas, J.D. (2002). Molecular method to assess the diversity of *Burkholderia* species in environmental samples. Appl. Environ. Microbiol. *68*, 1595–1603.

Schadt, C.W., and Zhou, J. (2006). Advances in microarray-based technologies for soil microbial community analyses, In Nucleic Acids and Proteins in Soil, Nannipieri, P. and Smalla, K., eds (Berlin Heidelberg, Germany: Springer-Verlag), pp. 189–203.

Schloss, P.D., and Handelsman, J. (2006). Toward a census of bacteria in soil. PLOS Computational Biology 2, 0786–0793.

Sessitsch, A., Weilharter, A., Gerzabek, M.H., Kirchmann, H., and Kandeler, E. (2001). Microbial population structures in soil particle size fractions of a long-term fertilizer field experiment. Appl. Environ. Microbiol. *67*, 4215–4224.

Sliwinski, M.K., and Goodman, R.M. (2004). Spatial heterogeneity of Crenarchaeal assemblages within mesophilic soil ecosystems as revealed by PCR-Single-Stranded Conformation Polymorphism profiling. Appl. Environ. Microbiol. *70*, 1811–1820.

Smalla, K., and Heuer, H. (2006). How to assess the abundance and diversity of mobile genetic elements (MGE) in soil bacterial communities? In Nucleic Acids and Proteins in Soil, Nannipieri, P. and Smalla, K., eds (Berlin Heidelberg, Germany: Springer-Verlag), pp. 313–330.

Smalla, K., Oros-Sichler, M., Milling, A., Heuer, H., Baumgarte, S., Becker, R., Neuber, G., Kropf, S., Ulrich, A., and Tebbe, C.C. (2007). Bacterial diversity of soils assessed by DGGE, T-RFLP and SSCP fingerprints of PCR-amplified 16S rRNA gene fragments: do the different methods provide similar results? J. Microbiol. Meth. *69*, 470–479.

Smets, B.F., and Barkay, T. (2005). Horizontal gene transfer: Perspectives at a crossroads of scientific disciplines. Nature Rev. Microbiol. *3*, 675–678.

Sørensen, J., and Nybroe, O. (2006). Reporter genes in bacterial inoculants can monitor life conditions and functions in soil, In Nucleic Acids and Proteins in Soil, Nannipieri, P. and Smalla, K., eds (Berlin Heidelberg, Germany: Springer-Verlag), pp. 375–395.

Stackebrandt, E., Frederiksen, W., Garrity, G.M., Grimont, P.A.D., Kämpfer, P., Maiden, M.C.J., Nesme, X., Rosselló-Mora, R., Swings, J., Trüper, H.G., Vauterin, L., Ward, A.C., and Whitman, W.B. (2002). Report of the ad hoc committee for the re-evaluation of the species definition in bacteriology. Int. J. Syst. Evol. Microbiol. *52*, 1043–1047.

Staley, J.T., and Konopka A. (1985). Measurement of *in situ* activities of nonphotosynthetic microorganisms in aquatic and terrestrial habitats. Annu Rev Microbiol *39*, 321–346.

Standing, D., and Killham, K. (2007). The soil environment. In: Modern Soil Microbiology II, JD van Elsas, JK Jansson and JT Trevors, eds (Boca Raton: CRC Press), pp. 1–22.

Stres, B., and Tiedje, J.M. (2006). New frontiers in soil microbiology: How to link structure and function of microbial communities. In Nucleic Acids and Proteins in Soil, Nannipieri, P. and Smalla, K., eds (Berlin Heidelberg, Germany: Springer-Verlag), pp. 1–22.

Tiedje, J.M., Asuming-Brempong, S., Nüsslein, K., Marsh, T.L., and Flynn, S.J. (1999). Opening the black box of soil microbial diversity. Appl. Soil Ecol. *13*, 109–122.

Torsvik, V., Goksøyr, J., and Daae, F.L. (1990). High diversity in DNA of soil bacteria. Appl. Environ. Microbiol. *56*, 782–787.

Treusch, A.H., Leininger, S., Kletzin, A., Schuster, S.C., Klenk, H.-P., and Schleper, C. (2005). Novel genes for nitrite reductase and amoA-related proteins indicate a role of uncultivated mesophilic crenarchaeota in nitrogen cycling. Environ. Microbiol. *7*, 1985–1995.

Urich, T., Lanzén, A., Ji Qi, Huson, D.H., Schleper, C., and Schuster, S.C. (2008). Simultaneous assessment of soil microbial community structure and function through analysis of the Meta-transcriptome. PLoS One *3*, 1–13.

Vandenkoornhuyse, P., Mahé, S. Ineson, P., Staddon, P., Ostle, N., Cliquet, J.-B., Francez A.-J., Fitter, A.H., and Young, J.P.W. (2007). Active root-inhabiting microbes identified by rapid incorporation of plant-derived carbon into RNA. Proc. Natl. Acad. Sci. U.S.A. *104*, 16970–16975.

Van Elsas, J.D., Smalla, K., and Tebbe, C.C. (2000). Extraction and analysis of microbial community nucleic acids from environmental matrices. In: Tracking genetically-engineered microorganisms, Jansson, J.K., van Elsas, J.D., and Bailey, M.J., eds (Austin, Texas: Eurekah), pp 29–51.

Van Elsas, J.D., Smalla, K., Lilley, A.K., and Bailey, M.J. (2002). Methods for sampling soil microbes. In Manual of Environmental Microbiology, 2nd edition, Hurst, C.J., Crawford, R.L., Knudsen, G.R., McInerney, M.J., and Stetzenbach, L.D., eds (Washington, USA: ASM Press), pp. 505–526.

Van Elsas, J.D., Tam, L., Finlay, R.D., Killham, K., and Trevors, J.T. (2007). Microbial interactions in soil. In: Modern Soil Microbiology II, Modern Soil Microbiology II, van Elsas, J.D. Jansson, J.K., and Trevors, J.T., eds (Boca Raton: CRC Press), pp. 177–210.

Van Elsas, J.D., Speksnijder, A.J., and van Overbeek, L.S. (2008a). A procedure for the metagenomics exploration of disease-suppressive soils. J. Microbiol. Meth. *75*, 515–522.

Van Elsas, J.D., Costa, R., Jansson, J.K., Sjoling, S., Bailey, M.J., Nalin R., Vogel, T.M., and van Overbeek, L.S. (2008b). The metagenomics of disease-suppressive soils – Experiences from the METACONTROL project. Trends Biotechnol. *26*, 591–601.

Von Wintzingerode, F.V., Göbel, U.B., Stackebrandt, E. (1997). Determination of microbial diversity in environmental samples: pitfalls of PCR-based rRNA analysis. FEMS Microbiol. Rev. *21*, 213–229.

Wagner, M., Horn, M., and Daims, H. (2003). Fluorescence *in situ* hybridisation for the identification and characterisation of prokaryotes. Curr. Opin. Microbiol. *6*, 302–309.

Webster, G., Embley, T.M., and Prosser, J.I. (2002). Grassland management regimens reduce small-scale heterogeneity and species diversity of ß-Proteobacterial ammonia oxidizer populations. Appl. Environ. Microbiol. *68*, 20–30.

Wu, L., Liu, X., Schadt, C.W., and Zhou, J. (2006). Microarray-based analysis of subnanogram quantities of microbial community DNAs by using whole-community genome amplification. Appl. Environ. Microbiol. *72*, 4931–4941.

Yergeau, E., Kang, S., He, Z., Zhou, J., and Kowalchuk, G. (2007). Functional microarray analysis of nitrogen and carbon cycling genes across an Antarctic latitudinal transect. ISME *1*, 163–179.

Young, I.M., and Ritz, K. (2000). Tillage, habitat space and function of soil microbes. Soil Tillage Res. *53*, 201–213.

Zhou, J. (2003). Microarrays for bacterial detection and microbial community analysis. Curr. Opin. Microbiol. *6*, 288–294.

Zhou, J., Xia, B., Treves, D.S., Wu, L.-Y., Marsh, T.L., O'Neill, R.V., Palumbo, A.V., and Tiedje, J.M. (2002). Spatial and resource factors influencing high microbial diversity soil. Appl. Environ. Microbiol. *68*, 326–334.

Plant-associated Microbial Communities

7

George A. Kowalchuk, Etienne Yergeau, Johan H.J. Leveau, Angela Sessitsch and Mark Bailey

Abstract

Plants, both above and below ground, offer diverse habitats for microbial colonization and growth. Plant–microbe interactions lie at the heart of plant performance and ecology. Plants provide various growth substrates and physical habitats for microbes on both sides of the air–soil interface, and numerous plant-associated niches have been exploited by specific microbial species, either by specializing on the distinct environmental conditions available or by entering into commensal, mutualistic or parasitic interactions with plants. This chapter seeks to examine the state of the art with respect to our ability to characterize the structure, function and interactions of plant-associated microbial communities, with a particular focus on the role of molecular biological methods and environmental genomics strategies in promoting this field. We will pay particular attention to bacterial and fungal colonization of above- and below-ground plant surfaces (phyllosphere and rhizosphere, respectively), as well as *in planta* (endosphere) interactions of endophytic, parasitic and symbiotic microorganisms. Of particular importance to advancing this research field are emerging methodologies, including novel '-omics' approaches, that seek to link microbial identity to *in situ* functioning, and holistic approaches that capture the complexities involved in multiple plant–microbe interactions.

Introduction

The habitats

The 200,000 or more species of vascular plants on record provide an exceptionally diverse range of habitats for microorganisms. Even within an individual plant, several highly disparate habitats can be identified, broadly falling into three distinct compartments: the phyllosphere, the endosphere and the rhizosphere (Box 7.1 and Fig. 7.1). Although very different in their environmental conditions, these compartments are interconnected, and microorganisms in different compartments may exert influence over distant populations and communities indirectly by stimulating systemic plant responses. These plant-associated habitats provide resident microorganisms with nutrients and energy (across a wide range of availabilities), as well as physical structures for attachment and protection. However, local conditions of water availability, temperature and solar radiation, as well as specific plant defence mechanisms, and a range of other induced plant stresses directly affect the growth and activities of microorganisms in these habitats. The sum of these positive and negative mechanisms provides the selective pressures that shape the microbial communities associated with plants.

Plants are the dominant net primary producers in the vast majority of terrestrial ecosystems. Consequently, plants are the main energy and

Box 7.1 Definition of the different plant environments

The phyllosphere

The leaf surface, or the phyllosphere, represents the largest biosphere–atmosphere interface on Earth (some 10^9 km^2) and with an estimated total load of 10^{26} bacteria, it harbours a significant reservoir of the world's total bacterial population. At a small scale, leaf surfaces are highly heterogeneous in terms of topography, protection against stresses, and nutrient availability. Most relevant niche-specific stresses are desiccation, UV radiation, and low nutrient availability.

The rhizosphere

The rhizosphere is a term that was first coined by Hiltner (1904) and is defining the root surface (the rhizoplane) and the zone around roots where plants exert their influence. It was shown to be a hotspot of plant–microbe interaction. Plant exudates are the major source of carbon in this environment, but this is not a free meal. Plants are known to exude other compounds that are selecting for some microorganisms, and normally diversity in the rhizosphere is lower than diversity in nearby bulk soil (Kowalchuk *et al.*, 2002). Most relevant stresses are plant-secreted inhibitory substances and high competition for nutrients.

The endosphere

The endosphere designates everything inside the plant. This habitat is relatively nutrient- and water-rich. However, osmotic stress is high and oxygen availability is relatively low. The endosphere represent a huge variety of habitats, with each plant tissue offering different conditions.

nutrient source for heterotrophic terrestrial microbes. Indeed, plants transfer the energy fixed by photosynthesis to associated microbes in numerous ways. Microbes in the rhizosphere benefit from exudation of organic compounds from roots and from dead root material; microbes living on aboveground plant surfaces can benefit from nutrients leaching from plant tissues, while microbes living inside plants can access nutrients directly. The availability and the diversity of these nutrients and energy sources are likely to apply a selection pressure on resident microorganisms. For instance, organic carbon is mainly available as simple sugars and organic acids in the rhizosphere, while more complex organic molecules are normally present in the endosphere. A variety of nutrients have been identified that are utilized by microbial colonizers of the phyllosphere (Tukey, 1970), including sugars (Mercier and Lindow, 2000) and C_1 compounds such as methanol (Holland and Polacco, 1994). Nutrient-specific bioreporters have revealed that the availability of nutrient resources to individual bacteria on leaves can vary greatly between different locations on the same leaf (Leveau and Lindow, 2002).

In addition to providing nutrients and energy to their associated microorganisms, plant environments also act as a physical support and offer an environment that is relatively protected from the elements. The protection that the plant provides is highly environment-specific: for instance, microbes on leaf surfaces experience more variable conditions (i.e. wind, rainfall, temperature, UV, etc.) than those in the endosphere, which are to a much greater extent protected from environmental fluctuations. Most soil-borne microbes are attached to soil particles or roots, and microorganisms in the rhizosphere would be expected to have a slightly more protected position as compared to bulk soil microbes. Plants alter their physiology in response to changing environmental conditions, and it has been demonstrated that associated rhizosphere, and particularly endophytic, microbial communities can respond rapidly to these plant changes (Bailey, 2004; Rasche *et al.*, 2006; Sessitsch *et al.*, 2002). The solid surfaces for attachment provided by plants also function to promote the formation of microbial biofilms and micro-aggregates, which represent the most common forms of organization in terrestrial microbial communities. The biofilm way of life is especially important for plant-associated microorganisms, as reviewed by Morris and Monier (2003) and Ramey *et al.*

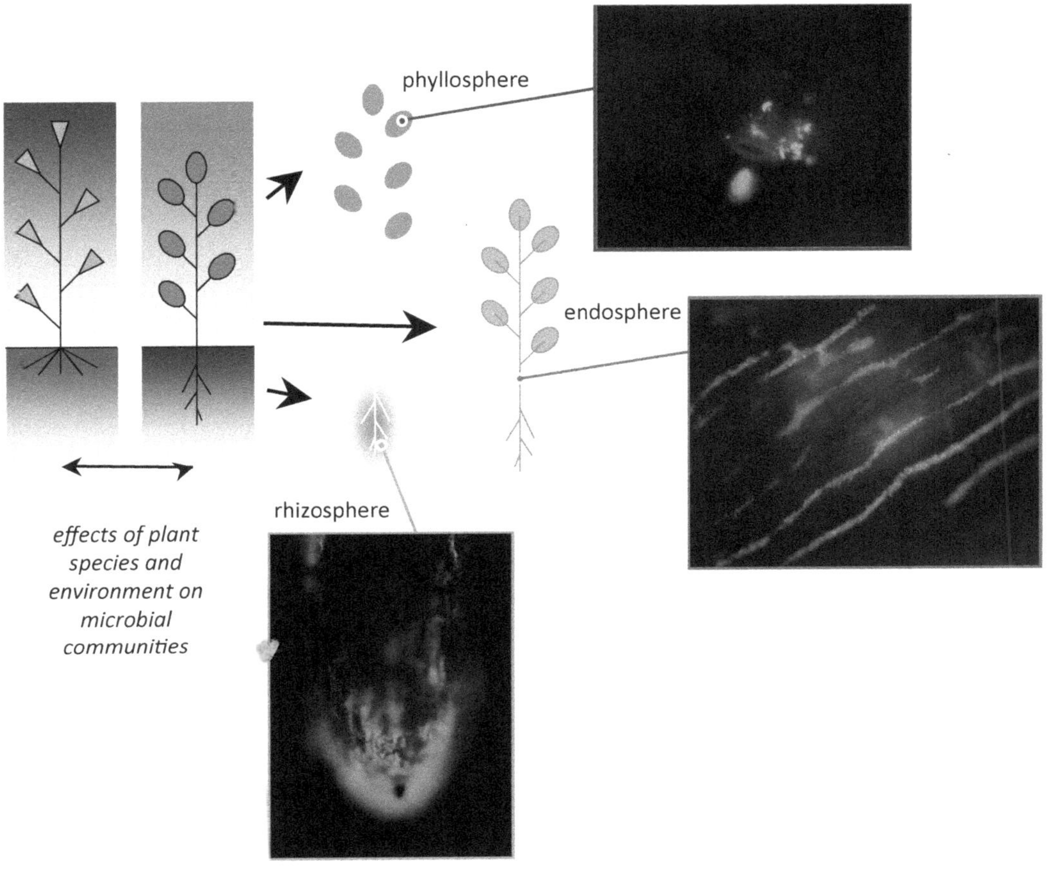

Figure 7.1 Schematic of plant-associated environments colonized by bacteria, highlighting the phyllosphere, endosphere and rhizosphere. Images depict bacterial colonization by GFP-labelled cells of *Burkholderia phytofirmans* strain PsJN, from top to bottom, on a leaf surface (surrounding a stomate), between epidermal cells, and at the root tip. Images courtesy of Stephane Compant. A colour version of this figure is located in the plate section at the back of the book.

(2004). For instance, patterns of aggregation are quite common among bacteria on leaf surfaces (Monier and Lindow, 2004), and grouped bacteria in these biofilms are thought to better withstand environmental stresses than individual cells, thereby increasing survival chances (Monier and Lindow, 2003; Monier and Lindow, 2005a).

The inhabitants

The diversity of plant-associated microorganisms varies significantly with the compartment sampled and with the plant species and environment under study. Table 7.1 lists some diversity estimates derived from both cultivation-dependent and -independent studies of microbial communities associated with several different plant species and terrestrial environments. These numbers are ultimately dependent on the methods used, the depth of sampling within the study and the cut-off used to define microbial taxa or operational taxonomic units (OTUs), complicating the cross-study comparisons. Despite our limited ability to compare studies of different environments, it is apparent that different plant species offer very different suites of microbial niches. Plant-specific habitats are created not only due to differential patterns of plant architecture above and below ground and inside the plants, but also differ in their niches due to differences in the chemical composition.

Table 7.1 Illustrative examples of the diversity of microorganisms associated with plants

Species (part)	Diversity	Methods	References
Endosphere			
Oryza sativa (roots)	52 bacterial OTUs	16S rRNA genes library	Sun *et al.*, 2008
Agrostis capillaries and *Trifolium repens* (roots)	17 AMF OTUs	18S rRNA genes library	Vandenkoornhuyse *et al.*, 2007
Agrostis stolonifera (roots)	17 bacterial OTUs	16S rRNA genes library	Vandenkoornhuyse *et al.*, 2007
Zea mays (roots)	27 (isolation) or 74 (cloning) bacterial OTUs	ARDRA and isolation	Chelius and Triplett, 2001
Phyllosphere			
Eucalyptus globulus	282 fungal species	Isolation	Sankaran *et al.*, 1995
Zea mays	387 fungal species	Isolation	Hawksworth and Rossman, 1997
Lantana camara (pernicious woody shrub)	55 fungal species	Isolation	Barreto *et al.*, 1995
Tree leaf canopy of a tropical Atlantic forest	95–671 bacterial species	16S rRNA genes library	Lambais *et al.*, 2006
Snap bean plants	78 bacterial species, representing 37 genera	Isolation	Beattie and Lindow, 1999
Halophila stipulacea	51 bacterial OTUs	ARDRA and 16S rRNA genes library	Weidner *et al.*, 2000
Rhizosphere			
Picea mariana (black spruce)	84 bacterial OTUs and 31 fungal OTUs	ARDRA	Filion *et al.*, 2004
Picea mariana (black spruce)	45 fungal taxa	Isolation	Vujanovic *et al.*, 2007
Several different native grassland plants	198 OTUs, estimated richness of 866 OTUs	16S rRNA genes clone libraries	Kielak *et al.*, 2008
Several different native grassland plants	239 bacterial OTUs	ARDRA	Kowalchuk *et al.*, 2002

OTU, operational taxonomic unit; ARDRA, amplified ribosomal DNA restriction analysis.

Specific plant environments clearly have a strong selective effect on the associated microbial communities. For instance, *Alphaproteobacteria* were reported to dominate over *Acidobacteria* in the rhizosphere, while the opposite trend has been observed in soils where no plants were growing (Griffiths *et al.*, 2006). The overall soil chemistry was similar, and the authors therefore concluded that the local plant inputs in the form of rhizodeposits were responsible for driving changes in the relative abundance of dominant taxa. Similarly, Kielak *et al* (2008) observed a greater proportion of *Acidobacteria* in bulk soil samples, as compared to the rhizosphere, and an opposite trend for *Verrucomicrobia*. However, these authors did not see strong evidence for a plant species-specific effect in shaping rhizosphere bacterial community structure in the former arable field studied. This latter result was in contrast to other studies that have observed clear plant species-specific microbial communities in the rhizosphere (Smalla *et al.*, 2001; Kowalchuk *et al.*, 2002). A recent study using stable isotope probing (SIP) reported that, in addition to arbuscular mycorrhizal fungi, *Betaproteobacteria* (more specifically *Burkholderiales*) were among the dominant active consumers of newly formed photosynthetates in mixed turf and peatland vegetation communities (Vandenkoornhuyse *et al.*, 2007). These authors hypothesized that these latter dominant assimilators of plant-derived carbon were novel potential plant symbionts.

Numerous different bacterial species and genera have been isolated from the endosphere (Hallman *et al.*, 1997). A number of detailed studies have recently highlighted the abundance and importance of endophytes, especially members of the *Proteobacteria* and *Actinomycetes*,, many of which have been suggested to promote the growth and health of various plant species (Berg *et al.*, 2005a; Castillo *et al.*, 2007; Chi *et al.*, 2005; Conn and Franco, 2004b). These groups have also frequently been detected by cultivation-independent analysis of endophytic bacterial communities (Conn and Franco, 2004a; Rasche *et al.*, 2006; Reiter and Sessitsch, 2006).

Bacterial species that can be readily isolated and cultured from leaves most commonly belong to the genera *Pseudomonas, Erwinia, Xanthomonas* and *Bacillus* (Thompson *et al.*, 1993; Lindow and Brandl, 2003). The few culture-independent studies of leaf surface communities have revealed much greater diversity than previously assumed based upon cultivation alone (Weidner *et al.*, 2000; Yang *et al.*, 2001; Whipps *et al.*, 2008). The bacterial community structure of several plants was found to be more similar between leaves of the same plant species than between leaves from different plant species (Yang *et al.*, 2001), suggesting that the plant may indeed select its associated phyllosphere community.

Interactions between plants and microorganisms cover the whole span of biotic interactions: from beneficial to parasitic. One of the most important and widespread beneficial interactions with plants is that of arbuscular mycorrhizal fungi (AMF), which interact with plant cells inside roots and assist in uptake of the plant's mineral nutrition in exchange for assimilated C (Smith and Read, 2008). Parasitic interactions include all the plant pathogenic bacteria and fungi, but can also include endophytes that are not necessarily causing diseases but utilising plant resources. These interactions are shaped by environmental conditions. For instance, mycorrhiza are typically beneficial for plants in conditions of low nutrient availability, but this effect is diminished or negated under higher nutrient conditions, and can even become parasitic in the case of certain fungus–plant combinations and environmental conditions (Smith and Read, 2008). Similarly, several pathogens, like *Fusarium*, are ubiquitous in the endosphere, but often only cause diseases in combination with specific environmental conditions (Yergeau *et al.*, 2006a,b). Most of these interactions have been observed in the endosphere, where the microorganisms and the plants interact more intimately, but direct plant–microbe interactions have also been identified in the rhizosphere and the phyllosphere. For instance, plant growth promoting rhizobacteria (PGPR – or also plant growth promoting bacteria (PGPB)) interact with the plant in the rhizosphere thereby improving plant growth and suppressing plant diseases. Their position in the rhizosphere in turn makes them well poised to receive significant amounts of energy from the plant via root exudates (Bais *et al.*, 2006) that regulate PGPR gene expression (Rainey 1991; Zhang *et al.*, 2004). Many endophytic bacteria have been reported to confer plant beneficial effects by utilizing strategies similar to those described for rhizosphere bacteria (Feng *et al.*, 2006; Sessitsch *et al.*, 2004; Sturz *et al.*, 2000). Phyllosphere bacteria are also thought to play a major role in antagonizing pathogens via a variety of mechanisms such as competition (Marois and Coleman, 1995), antibiosis (Lindow, 1988) or by providing the resident microbial community with nutrients through nitrogen fixation (Fürnkranz *et al.*, 2008).

Ecological niches and specific adaptation

The characteristics of the three plant-associated environments addressed in this chapter create highly distinct suites of niches that require specific microbial adaptations. One of the few adaptations that seems to be shared by organisms of all the plant-associated environments is the capacity to form aggregates or biofilms (Rudrappa *et al.*, 2008).

The phyllosphere

As already alluded to above, the phyllosphere represents a relatively harsh environment with reduced access to nutrients, high fluctuations in water availability, and exposure to UV radiation. Two types of strategies have evolved that allow microorganisms to survive, and sometimes thrive, on leaf surfaces: tolerance and avoidance (Beattie and Lindow, 1999). Pigmentation (Jacobs *et al.*, 2005) and DNA repair mechanisms (Kim and Sundin, 2000) are the classical examples of tolerance mechanisms that aid in mediating the potentially damaging effects of UV light. The ability of bacteria to form biofilms is a typical avoidance strategy, protecting bacteria against desiccation and at the same time trapping nutrients. The production and/or degradation of phytohormones, such as indole 3-acetic acid (IAA), is a common trait among diverse plant-associated bacteria (Libbert and Risch, 1969; Leveau and Lindow, 2005; Spaepen *et al.*, 2007; Leveau and Gerards, 2008), which is thought to alter plant physiology so as to benefit the leaf-colonizing bacteria. Microbes may also change the properties of the cuticle to increase the flow of nutrients from the leaf interior (Knoll and Schreiber, 2000; Schreiber *et al.*, 2005). Another type of avoidance strategy that some microorganisms on leaf surfaces may opt for is to enter the leaf interior (Manceau and Kasempour, 2002), but this niche requires its own set of specific adaptations (see below).

The rhizosphere

In order for microorganisms to be able to proliferate in the rhizosphere, they first have to be able to colonize this soil compartment, and several microbial traits have been identified that facilitate rhizosphere colonization (Lugtenberg *et al.*, 2001), including quorum sensing mechanisms (Espinosa-Urgel and Ramos, 2004). Indeed, it is possible to steer the evolution of bacteria toward the development of high rhizo-competence (de Weert *et al.*, 2004).

Actively growing roots release organic compounds into the rhizosphere that modify the nutritional and chemical status of the surrounding soil. These compounds support the growth of specific microbial community members in the rhizosphere and this normally results in an increased population density and a community structure distinct from that of bulk soil (Smalla *et al.*, 2001; Kowalchuk *et al.*, 2002). The rhizosphere is thereby selectively enriched in microorganisms that are adapted to the utilization of specific plant organic compounds (Berg *et al.*, 2002; 2005b; Gomes *et al.*, 2003; el Zahar Haichar *et al.*, 2008).

Roots also affect the microbes in their vicinity through competition for nutrients. Being able to gather and efficiently compete for resources is a key adaptive trait in the rhizosphere. Indeed, in this compartment, microorganisms not only compete for nutrients with other soil organisms, but also with plants and their associated symbionts. However, it should be noted that competition for nutrients may be reduced in agricultural settings and in naturally nutrient-rich soils. Other adaptive strategies include the ability to utilize novel and recalcitrant compounds, as well as the ability to detoxify or resist toxic antibiotic compounds produced by plants and other organisms in the rhizosphere.

The endosphere

The rhizosphere and the phyllosphere are likely sources of endophytic microorganisms, which can penetrate inside plants via various entry points including stomata and radicles (Sturz *et al.*, 2000). Furthermore, endophytic microorganisms can be transmitted vertically or via seed contamination. Once inside the plant, bacteria have to respond to different environmental conditions than those encountered in the rhizosphere and on leaf surfaces; these include reduced oxygen availability, higher osmotic pressure and altered availability of different carbon sources. Many endophytes are able to resist systemic plant defence mechanisms (Iniguez *et al.*, 2005; Miché *et al.*, 2006), however, little is known about the molecular interactions that occur between non-pathogenic endophytes and their plant hosts. Efficient xylem colonization, the secretion of cell wall-degrading enzymes as well as being outcompeted by other microorganisms will influence the colonization success of endophytes.

Recent advances in plant-associated microbial ecology

Until recently, the study of plant-associated microorganisms has been hampered by a number of factors. Their microscopic size makes them difficult to study *in situ*, and, as we have yet to devise appropriate methods for the propagation of pure (or even mixed) cultures of many Bacteria, Archaea or fungi, the vast majority remains poorly described. Indeed, in view of the particular nature of the plant environment, several microorganisms cannot be cultivated on synthetic media while several others are obligate biotrophs and cannot live without the host plant. For instance, a limited range of AMF can be routinely cultivated together with roots (St-Arnaud *et al.*, 1996), but such systems are not available for most obligate plant-associated microorganisms. Furthermore, the complexity of the environment in which plant-associated microorganisms reside makes it difficult to extract them without significant bias. Fortunately, several recent breakthroughs have been made possible by the introduction of molecular methods that overcome several of the limitations mentioned above. Modern molecular methods now permit an assessment of the genetic diversity of microorganisms present in a sample. Such methods are generally considered to generate a clearer picture of the identities of dominant organisms compared to traditional culture-based methods (Hugenholtz *et al.*, 1998). The application of 'omics'-based approaches to study plant-associated microorganisms is very promising and might yield novel and more comprehensive information about the microorganisms inhabiting this unique environment.

The phyllosphere

The study of phyllosphere microbial communities has exploited several recent methodological advances that have boosted our understanding of the structure and function of microbial communities associated with plant leaf surfaces. Given the open character of the leaf surface, it is an ideal environment for approaches that rely on fluorescent microscopy, e.g. fluorescence *in situ* hybridization (FISH), GFP-based bioreporters, and confocal microscopy. Such approaches have revealed the micrometer variation that exists on leaves in terms of local bacterial densities (Monier and Lindow, 2004), survival or interactions (Monier and Lindow, 2005a,b), as well as in the exposure of bacteria to nutrients (Leveau and Lindow, 2002), water (Axtell and Beattie, 2002), and signalling compounds (Dulla and Lindow, 2008).

The traditional reliance on culture-dependent approaches for the study of phyllosphere biodiversity is being re-balanced by studies based on PCR–denaturing gradient gel electrophoresis (DGGE) (Yang *et al.*, 2001), quantitative PCR (Heuser and Zimmer, 2002, 2003), 16S rRNA gene clone libraries (Lambais *et al.*, 2006), and ribosomal intergenic spacer analysis (RISA) (Knief *et al.*, 2008). Proteomics has also made its way into the toolbox of phyllosphere microbiologists. For instance, in *Methylobacterium extorquens* AM1, 45 proteins were found to be more abundantly expressed on plant surfaces than on synthetic medium, including nine proteins that were more abundant during colonization on leaves compared to roots (Gourion *et al.*, 2006). Interestingly, one of these proteins was identified as PhyR, a gene known as a master regulator of stress responses (Gourion *et al.*, 2008). This gene is an absolute requirement for the plant colonization phenotype of *M. extorquens*, as demonstrated by deletion mutants (Gourion *et al.*, 2006). The availability of an increasing number of complete genome sequences of plant- and leaf-associated bacteria (Feil *et al.*, 2005) has opened up new possibilities for microarray approaches to study gene expression in plant-associated bacteria (Okinaka *et al.*, 2002). Directed mutagenesis (Darsonval *et al.*, 2008) and *in vivo* expression technologies (Marco *et al.*, 2003, 2005) have also been used successfully to identify genes with an essential or accessory contribution to phyllosphere colonization, growth and persistence. Recent studies have also begun to address the changes in gene expression in the leaves of the plant hosts as they are being colonized by bacteria (Kürkcüoglu *et al.*, 2007).

The rhizosphere

Rhizosphere-inhabiting microbial communities have been studied using a wide range of advanced methods. For instance, stable-isotope probing has been used with success to track microorganisms associated with the rhizosphere (See Chapter 4 for a detailed discussion of this approach). One interesting way to follow active microorganisms in the rhizosphere is to label plants using $^{13}CO_2$ and then follow the incorporation of ^{13}C in rhizosphere microbial DNA, RNA or phospholipid fatty acids (PLFA) (Radajewski *et al.*, 2000;

Manfield *et al.*, 2002; Ostle *et al.*, 2003). The organisms rapidly incorporating labelled C should be active and feeding directly on plant exudates. However, this is true only for a short period of time, since other rhizosphere inhabitants might incorporate labelled C through secondary carbon flow from the exudates-utilizing bacteria. Still, this method was recently used in combination with PLFA analyses to follow microbial communities in the rice rhizosphere over time (Lu *et al.*, 2007) and in combination with RNA-based DGGE to observe the changes in rhizosphere microbial communities following soil liming (Rangel-Castro *et al.*, 2005) or plant exposure to elevated atmospheric CO_2 (Drigo *et al.*, unpublished). Another application of SIP in the rhizosphere was through direct labelling by feeding ^{13}C-labelled biphenyls to the rhizosphere of pine growing in polychlorinated biphenyls (PCBs) soils. Following labelling, the authors then used functional gene microarrays (the GeoChip, He *et al.*, 2007; see Chapter 5) to identify the functional genes associated with the active rhizosphere microbial community. 16S rRNA gene microarrays were also used to follow specific bacterial communities in the rhizosphere (Sanguin *et al.*, 2006a,b). Large-insert metagenomic strategies have also been developed for examination of soil-borne metagenomic information (Rondon *et al.*, 2000), and these are generally applicable to rhizosphere soil, although modifications are often necessary to account for smaller soil volumes. Large-insert metagenomic studies in the rhizosphere have allowed the identification and characterisation of metal resistance genes in the rhizosphere of plants adapted to acid mine drainage (Mirete *et al.*, 2007) and antifungal genes in the rhizosphere of pine trees (Chung *et al.*, 2008) and crop plants (Van Elsas *et al.*, 2008). Although such studies have provided insight into the great metabolic diversity within rhizosphere communities (Daniel, 2005), the development of efficient screening methods to interrogate metagenomic libraries for activities of interest remains a significant challenge (Kowalchuk *et al.*, 2007).

The endosphere

For a long time, endophytes were only addressed by cultivation-dependent analysis. However, in the last several years, 16S rRNA gene-based analyses have revealed an unexpected diversity of bacterial endophytes, even in plants growing in extreme environments (Chelius and Triplett, 2001; Idris *et al.*, 2004). By targeting rRNA in addition to DNA it has been shown that endophytes are not only diverse but also highly active (Reiter *et al.*, 2003a). Using plants labelled with $^{13}CO_2$, stable isotope probing revealed that endophytes derive a large part of their carbon from plant photosynthates (Rasche *et al.*, 2008; Vandenkoornhuyse *et al.*, 2007) and that the transfer of newly fixed carbon to endophytes is extremely rapid. In addition, endophyte functions that are important for plant performance, such as nitrogen fixation, have been addressed by targeting *nifH* genes in RNA and DNA isolated directly from plants (Knauth *et al.*, 2005; Reiter *et al.*, 2003; Tan *et al.*, 2003). Although genomic analysis of endophytes revealed how a single endophyte might interact with plants (Krause *et al.*, 2006), this remains to be elucidated at the community level. Ongoing metagenomic studies will undoubtedly yield additional functional information on yet uncultivated endophytes.

Fluorescent microscopy techniques are ideal to address the localization of endophytic bacteria in planta. For instance, Pirttilä *et al.* (2000) applied *in situ* hybridization to track bacterial cells in pine bud cells, while Chi *et al.* (2005) followed GFP-tagged rhizobia in rice from the colonization of the root surface all the way to migration through stems and colonisation of leaves. Similarly, systemic colonization of grapevine plants by a GFP-tagged plant growth-promoting *Burkholderia* strain was demonstrated (Compant *et al.*, 2008; Compant *et al.*, 2005). This strain was chronologically detected on root surfaces, in the endorhiza, grape inflorescence stalks and inside berries. In contrast, a GFP-tagged *Paenibacillus polymyxa* strain, formed biofilms and invaded *Arabidopsis* roots, but did not colonize above-ground plant tissues (Timmusk *et al.*, 2005).

Future trends

Utilization of emerging techniques

The utilization of emerging techniques for the study of plant-associated microorganisms should yield large amounts of novel, high-resolution information. The most promising emerging technologies include the following: Raman microspectroscopy, flow cytometry, stable isotope probing, microarrays, new sequencing techniques, metagenomics, metatranscriptomics, metaproteomics, metabolomics and new data analysis methods. Furthermore, novel combinations of two or more of these or other techniques should also be seen as a way to retrieve new and more robust information about plant-associated microorganisms.

Raman microspectroscopy (Huang *et al.*, 2004) could be useful to follow cells in plant-associated environments and has already shown some promise in the analysis of cells associated with plants (Huang *et al.*, 2007a). Raman microspectroscopy could be particularly powerful for the analysis of the metabolic histories of cells in plant environments, providing resolution to discriminate species from one another based upon metabolic signatures, as well as examine metabolic status within a species. Of particular interest are emerging methods that allow the coupling of microbial identification and metabolic activities, such as Raman-FISH (Huang *et al.*, 2007b; Neufeld *et al.*, 2007).

A wide range of physical and chemical cell characteristics can be used to sort cells with a flow cytometer. Adding GFP constructs or fluorescently labelled antibodies targeting specific characteristics or taxonomic groups could sort cells having properties of interest from the remaining cells. Once the cells are sorted, other downstream applications can be used on the cells having the required characteristics. However, for plant-associated environments, especially in the endosphere, it could be difficult to extract microbial cells from their environment without damaging them or changing their physical/chemical properties.

SIP is one method particularly amenable to combination with others (Radajewski *et al.*, 2000; Manfield *et al.*, 2002; Vandenkoornhuyse *et al.*, 2007; Neufeld *et al.*, 2007). Indeed, this approach has already been applied to study plant environments (mainly the rhizosphere), and was recently used in combination with functional microarray analysis (Leigh *et al.*, 2007). Basically, once the labelled nucleic acids are retrieved, it is possible to use almost any molecular technique available. The wide range of substrates that can be used to label microorganisms, or their plant hosts, is also making this method more versatile and highly appropriate for microbial studies in the plant environment.

Recent expansions in genetic databases (Cole *et al.*, 2005; DeSantis *et al.*, 2006) have allowed the design of comprehensive microarray platforms encompassing a very large number of functional genes or taxa. Recent examples include a 16S rRNA gene microarray, the PhyloChip (DeSantis *et al.*, 2007), and a functional gene microarray, the GeoChip (He *et al.*, 2007); see Chapter 5 for a more detailed description of these approaches. The adaptation and application of such existing comprehensive platforms for specific plant environments could yield highly valuable data. For instance, functional gene microarrays could be adapted to follow specifically N-fixation functions in the rhizosphere (Zhang *et al.*, 2007), or C_1 compound utilization in the phyllosphere. As it has been demonstrated recently, taxa-specific microarray platforms could be designed to specifically follow microorganisms that have important functions in plant environments. Combining functional and phylogenetic platforms in a single study shows promise in revealing relevant community responses to specific environmental conditions, although bioinformatic interpretation of such results is far from a trivial task (Yergeau *et al.*, 2009). One major challenge, especially of functional microarrays, is to adapt the existing platforms to carry out analyses at the RNA level, allowing thereby access to the actively expressed genes.

New sequencing technologies, such as massively parallel pyrosequencing (Margulies *et al.*, 2005) now allow the very rapid sequencing of huge amount of DNA. Indeed, the present version (Titanium) of the Genome Sequencer FLX of Roche can generate over 400 Mbp in a single run (~10 h). This technique was recently used to conduct the largest soil 16S rRNA gene diversity study undertaken to date (Roesch *et al.*, 2007); see Chapters 2 and 4 for more detailed information about this approach. Such an approach could be used to sequence 16S rRNA genes in plant environments and could help evaluate with a high degree of precision the diversity in the different plant compartments. It could also help to evaluate the community structure and the potential differences between the microbial communities inhabiting the different compartments of a single plant, potentially replacing less quantitative community profiling methods. Furthermore, this new sequencing technology should be used to rapidly sequence the genomes of microorganisms having particularly interesting functions in plant-associated environments. Expanding full genome comparisons are already providing novel insights into key microbial properties involved in plant–microbe interactions, including quorum sensing, siderophore production, and type three secretion systems (van Sluys *et al.*, 2002; Pühler *et al.*, 2004).

Metagenomics studies have been carried out in several different environments in recent years (for reviews see Handelsman, 2004; Keller and Zengler, 2004; Lorenz and Eck, 2005; Streit and Schmitz, 2004; and Chapter 2). Some metagenomic studies have already been reported for the rhizosphere (Chung *et al.*, 2008; Mirete *et al.*, 2007; van Elsas *et al.*, 2008), the endosphere and the phyllosphere (Lambais *et al.*, 2006) as well as soil (Rondon *et al.*, 2000). In the plant environment, metagenomic approaches would be particularly interesting to combine with SIP or flow cytometry. In the relatively simpler plant environments (phyllosphere, endosphere), shotgun analysis of community genomes could yield manageable size metagenomes that would provide insight into functions carried out in these environments, as has been done previously for simple environments (Tyson *et al.*, 2004). In contrast, soil environments were shown to be difficult to sequence completely or in a way that would allow production of contigs (Tringe *et al.*, 2005), and shotgun analysis of metagenomes could be similarly difficult in the rhizosphere. However, functional or taxonomic screening of metagenomic libraries could yield highly interesting information. Large-insert metagenomic libraries were already shown to yield interesting information regarding adaptations to life in the plant environment (Chung *et al.*, 2008; Mirete *et al.*, 2007).

Recent studies have directly assessed the diversity of transcripts in the environment, both for Bacteria (Poretsky *et al.*, 2005) and Eukarya (Bailly *et al.*, 2007). Combining meta-transcriptomic approaches with massive parallel sequencing methods has been demonstrated to be a very powerful approach in the study of marine microbial assemblages (Frias-Lopez *et al.*, 2008), but such approaches are more problematic in their application to plant-soil environments. Several important hurdles still need to be addressed, especially for Bacterial and Archaeal communities, for meaningful application of meta-transciptomics to plant-microbe systems. Indeed, Bacteria and Archaea lack the poly-A tail, which makes it harder to separate the mRNA from the overwhelming amount of rRNA. Furthermore, the identification of the transcripts remains a problem, with high amounts of sequences returning hypothetical or unknown functions in database searches. Using new sequencing technologies to sequence the plant-associated microbial meta-transcriptome would yield a high number of environmental transcripts probably related to interesting plant-associated functions. But, then again, a large proportion of this information would be hardly usable, since it could not be readily identifiable in databases. It appears therefore that for this technique to work satisfactorily in environmental settings, genetic databases need to be complemented by a large number of identified sequences other than 16S rRNA genes. Transcriptomic approaches that are dedicated to individual organisms or in model systems have proven to be more tractable and have identified numerous traits important to plant–microbe interactions (Mark *et al.*, 2005).

Proteomics could be an excellent tool to study interactions between individual strains, or microbial communities and plants. Metaproteomics analysis (Wilmes and Bond, 2004; 2006) of the proteins associated with a particular plant environment or growth stage could yield information about the functions that are important for that environment or that stage. The identification of proteins from the environment is greatly facilitated when focussing on fully sequenced organisms and relatively simple systems (Gourion *et al.*, 2006). Although this approach has yet to be applied to more complex ecosystems, it was recently demonstrated for the relatively complex human gut microbiota (Verberkmoes *et al.*, 2009; in press) and could be adapted to study soil and rhizosphere microbial communities in the future (see Chapters 3 and 9).

Meta-metabolomics is one step further down the regulatory pathway than meta-proteomics. This approach focuses on the products of the microbial community metabolism and could highlight the functions and the related products of microbial communities in the plant environment. Similarly, it would also be of interest in the context of this chapter to study plant metabolomics, for instance the metabolic reactions of plants to colonisation by microbial communities. Metabolomic approaches are of particular interest in the rhizosphere, where plant root exudates are playing a strong shaping role. Such approaches have already been implemented for the selection of enhanced plant–microbe interactions associated with the removal of polychlorinated biphenyls (Narasimhan *et al.*, 2003). Although this field is vastly unexplored, putative metabolomic data can already be derived from large-scale metagenomics, meta-transcriptomics and meta-proteomics. The combination of this type of data with meta–metabolomic data is clearly the next step and will probably bring us closer to an understanding of the functioning of microbial communities in the environment (Turnbaugh and Gordon, 2008).

With all these high-throughput techniques generating large amounts of data, a crucial bottleneck is data analysis and interpretation. Bioinformatics tools are used in the first part of the analysis, and bring raw data to a format that can be understood and manipulated. Sometimes this step is enough, especially when focusing on a small number of samples or a single characteristic of interest. However, when a large number of samples are to be analysed for several different characteristics, advanced statistical tools are necessary to extract the major trends in the datasets, relate different datasets with one another and bring data in a format amenable to publication. Statistical tools, especially multivariate analyses, have been available to ecologists for several decades (see Legendre and Legendre, 1998), but have only recently been applied systematically to microbial ecology studies (Fromin *et al.*, 2002; Morris *et al.*, 2002; Ramette, 2007). In the context of plant-associated microbial ecology, the utilization of bioinformatics and statistical tools will mainly depend on the techniques used and on the aims of the study. Studies with clearly defined hypotheses and a strong conceptual approach are clearly calling for the use of state-of-the-art statistical tools.

Application of ecological theories: towards a conceptual approach

In recent years, there has been considerable interest in applying to microorganisms the ecological theories devised for macroorganisms. These theoretical advances, if applied to various plant-associated environments, could help understand the patterns and interactions observed between plants and microorganisms. A strong theoretical background might also help in identifying interesting trends that might otherwise go unnoticed. In our opinion, two theories (among others) would be, or have been, interesting to apply and test in plant environments: biogeography (Andrews and Harris, 2000) and succession. Bacteria were recently reported to show spatial patterns (Horner-Devine *et al.*, 2004; Yergeau *et al.*, 2007) and some of the patterns observed could be applicable to the distribution of microbes inside a plant. Spatial distribution of organisms is indeed thought to be related to two different factors: environmental conditions and historical contingencies (Martiny *et al.*, 2006). Similarly, the local availability of nutrients in the plant environment together with the history of the colonisation of the plant during plant developmental stages could help explain the distribution of microorganisms in plant environments. Historical events can, however, only be maintained if there is some degree of isolation between the communities and it is very uncertain if this is the case for plant-associated environments, which have been shown to exhibit a significant degree of linkage to one another (Compant *et al.*, 2008).

Soil microorganisms were also reported to show clear succession patterns with ecosystem ageing (Tarlera *et al.*, 2008; van der Wal *et al.*, 2006). Ecological theory of succession predicts that colonisation will be dominated by r-strategists that can grow rapidly, are generalists and in the case of microbes, prefer easily metabolizable nutrients. In later stages, K-strategists become the dominant members of the community and are particularly adapted to conditions of high competition, grow slowly, are normally specialists and can degrade more complex organic molecules. There might be such a succession pattern in plant environments, either through the life cycle of the plant or through the seasons. Indeed, plant compounds are changing with plant age and physiological stage and are probably shaping the associated microbial community. For instance, senescing leaves are a completely different environment than young leaves. In general, bacteria are the pioneers on young leaves, while yeasts and fungi dominate as the leaf becomes older and senesces (Ercolani, 1991). Some studies already highlighted succession in plant environment, with a succession of AMF species associated with asparagus roots aging (Wacker *et al.*, 1990; Yergeau *et al.*, 2006b). It is however uncertain if it would be the case for other organisms and in other plant parts.

From a conservation perspective, it is highly important to understand the distribution of microbial diversity in plant environments. Indeed, some plants might be harbouring highly adapted microorganisms found exclusively on that plant. Similar to the concept of endemic species, such exclusive relationships might bolster efforts for plant diversity conservation. In addition to the fundamental interest for biodiversity conservation, microorganisms are an almost endless source of enzymes and products that might have highly interesting uses for biotechnology.

Scaling up and scaling down

One of the fundamental problems in microbial ecology is that the scales at which the studies are carried out are seldom appropriate for microorganisms. Indeed, microbial interactions and spheres of influence are at the micrometer scale, yet the collective impacts of microbial processes are only manifest at scales that are multiple orders of magnitude larger. It is probably not relevant to work with average values over a leaf or a gram of rhizosphere soil, since the conditions at the micrometer scale can vary widely. In complex plant-associated environments, the concept of an average bacterium in a population is a misnomer.

There are basically two ways around this problem: scaling up or down. Scaling-down can zoom in to the study of microbial aggregates or even single cells in the context of variation in environmental conditions at the micrometer scale, and methodologies for such studies are rapidly emerging (Brehm-Stecher and Johnson, 2004). Several interesting studies (see above), especially in the phyllosphere, have been able to follow microorganisms at the micrometer scale, revealing a staggering degree of heterogeneity within, for instance, environments on a single leaf surface. It is interesting to note that, although microbial densities can be quite high in plant-associated environments, their clumped distribution means that most of the actual volume is devoid of microbial cells. Given this spatial separation of organisms, it may be that microbial aggregates are similar to islands that are separated by a considerable distance, as least relative to the organisms themselves, from other microbial populations. Emerging methods of small-scale sampling and molecular analyses should provide novel insights into the connectivity of spatially separated microbial aggregates in these environments to examine if micro-scale island biogeography may help to explain the diversity and spatial structure of plant-associated microbial communities.

The other way to proceed would be to scale up the study. Indeed, at a larger scale, most of the complicated small-scale details would average out, leaving a simplified picture. This would require large sampling schemes that would allow meaningful statistical comparisons. As already discussed, this could be made possible by the application of multivariate statistical analyses to plant-environment studies. However, several of the state-of-the-art molecular methods listed above were not tested thoroughly for large-scale ecological studies, and the cost involved in assessing large amount of samples using most of these methods are still prohibitive.

Conclusions

There is a large diversity of plant environments available for microorganisms, and in terrestrial environments, these environments harbour a large proportion of the total microbial diversity. The environmental characteristics of specific plant habitats impose strong selective pressures on resident microbes, resulting in large structural and functional differences between communities in the different parts of a single plant, with interactions between microbes and plants range from symbiotic to pathogenic. However, relatively little information is as yet available concerning the organisms inhabiting these environments.

Emerging technologies, especially those within the growing '-omics' toolbox, hold considerable promise for increasing our knowledge of these unique environments. In the longer-term, a detailed knowledge of the organisms, relationships and driving forces involved might be applied to devise new strategies to manipulate the plant-associated microflora in agriculture and forestry, thereby ultimately providing better control plant diseases and more efficient food production.

References

Adamczyk, J., Hesselsoe, M., Iversen, N., Horn, M., Lehner, A., Nielsen, P.H., Schloter, M., Roslev, P., and Wagner, M. (2003). The isotope array, a new tool that employs substrate-mediated labeling of rRNA for determination of microbial community structure and function. Appl. Environ. Microbiol. *69*, 6875–6887.

Andrews, J.H., and Harris, R.F. (2000). The ecology and biogeography of microorganisms on plant surfaces. Annual Rev. Phytopath. *38*, 145–180.

Axtell, C.A., and Beattie, G.A. (2002). Construction and characterization of a proU-gfp transcriptional fusion that measures water availability in a microbial habitat. Appl. Environ. Microbiol. *68*, 4604–4612.

Bailey, M.J. (2004). Life in the phyllosphere. In Pseudomonas, genomics, life style and molecular architecture., J..L. Ramos, ed. (New York: Kluwer Academic/ Plenum Publishers), pp. 431–456.

Bailly, J., Fraissinet-Tachet, L., Verner, M.C., Debaud, J.C., Lemaire, M., Wesolowski-Louvel, M., and Marmeisse, R. (2007). Soil eukaryotic functional diversity, a metatranscriptomic approach. ISME J. *1*, 632–642.

Bais, H.P., Weir, T.I., Perry, L.G., Gilroy, S., and Vivanco, J.M. (2006). The role of root exudates in rhizosphere interactions with plants and other organisms. Ann. Rev. Plant Biol. 5, 233–266 .

Barreto, R.W., Evans, H.C., and Ellison, C.A. (1995). The mycobiota of the weed *Lantana camara* in Brazil, with particular reference to biological control. Mycol. Res. *99*, 769–782.

Beattie, G.A., and Lindow, S.E. (1999). Bacterial colonization of leaves: A spectrum of strategies. Phytopathology *89*, 353–359.

Benndorf, D., Balcke, G.U., Harms, H., and Von Bergen, M. (2007). Functional metaproteome analysis of protein extracts from contaminated soil and groundwater. ISME J., 1, 224–234.

Berg, G., Krechel, A., Ditz, M., Sikora, R.A., Ulrich, A., and Hallman, J. (2005a). Endophytic and ectophytic potato-associated bacterial communities differ in structure and antagonistic function against plant pathogenic fungi. FEMS Microbiol. Ecol. *51*, 215–229.

Berg, G., Roskot, N., Steidle, A., Eberl, L., Zock, A., and Smalla, K. (2002). Plant-dependent genotypic and phenotypic diversity of antagonistic rhizobacteria isolated from different *Verticillium* host plants. Appl. Environ. Microbiol. *68*, 3328–3338.

Berg, G., Zachow, C., Lottmann, J., Gotz, M., Costa, R., and Smalla, K. (2005b). Impact of plant species and site on rhizosphere-associated fungi antagonistic to *Verticillium dahliae* Kleb. Appl. Environ. Microbiol. *71*, 4203–4213.

Brehm-Stecher, B.F., and Johnson, E.A. (2004). Single-cell microbiology: Tools, technologies, and applications. Microbiol. Mol. Biol. Rev. *68*, 538–559.

Castillo, U.F., Browne, L., Strobel, G., Hess, W.M., Ezra, S., Pachego, G., and Ezra, D. (2007). Biologically active endophytic streptomycetes from *Nothofagus* spp. and other plants in Patagonia. Microb. Ecol. *53*, 12–19.

Chelius, M.K., and Triplett, E.W. (2001). The diversity of archaea and bacteria in association with the roots of *Zea mays* L. Microb. Ecol. *41*, 252–263.

Chi, F., Shen, S.-H., Cheng, H.-P., Jing, Y.-X., Yanni, Y.G., and Dazzo, F. (2005). Ascending migration of endophytic rhizobia, from roots to leaves, inside rice plants and assessment of benefits to rice growth physiology. Appl. Environ. Microbiol. *71*, 7271–7278.

Chung, E.J., Lim, H.K., Kim, J.C., Choi, G.J., Park, E.J., Lee, M.H., Chung, Y.R., and Lee, S.W. (2008). Forest soil metagenome gene cluster involved in antifungal activity expression in *Escherichia coli*. Appl. Environ. Microbiol. *74*, 723–730.

Cole, J., Chai, B., Farris, R., Wang, Q., Kulam, S., McGarrell, D., Garrity, G., and Tiedje, J. (2005). The Ribosomal Database Project (RDP-II): sequences and tools for high-throughput rRNA analysis. Nucleic Acids Res. *33*, D294–D296.

Compant, S., Reiter, B., Sessitsch, A., Nowak, J., Clement, C., and Ait Barka, E. (2005). Endophytic colonization of *Vitis vinfera* L. by plant growth-promoting bacterium *Burkholderia* sp. strain PsJN. Appl. Environ. Microbiol. *71*, 1685–1693.

Compant, S., Kaplan, H., Sessitsch, A., Nowak, J., Ait Barka, E., and Clement, C. (2008). Endophytic colonization of *Vitis vinifera* L. by *Burkholderia phytofirmans* strain PsJN: from the rhizosphere to inflorescence tissues. FEMS Microbiol. Ecol. *63*, 84–93.

Conn, V.M., and Franco, C.M.M. (2004a). Analysis of the endophytic actinobacterial population in the roots of wheat (*Triticum aestivum* L.) by terminal restriction fragment length polymorphism and sequencing of 16S rRNA genes. Appl. Environ. Microbiol. *70*, 1787–1794.

Conn, V.M., and Franco, C.M.M. (2004b). Effect of microbial inoculants on the indigenous actinobacterial endophyte population in the roots of wheat as determined by terminal restriction fragment length polymorphism. Appl. Environ. Microbiol. *70*, 6407–6413.

Daniel, R. (2005). The metagenomics of soil. Nat. Rev. Microbiol. *3*, 470–478.

Darsonval, A., Darrasse, A., Meyer, D., Demarty, M., Durand, K., Bureau, C., Manceau, C., and Jacques, M.A. (2008). The type III secretion system of *Xanthomonas fuscans* subsp *fuscans* is involved in the phyllosphere colonization process and in transmission to seeds of susceptible beans. Appl. Environ. Microbiol. *74*, 2669–2678.

de Weert, S., Dekkers, L.C., Kuiper, I., Bloemberg, G.V., and Lugtenberg, B.J.J. (2004). Generation of enhanced competitive root-tip-colonizing Pseudomonas bacteria through accelerated evolution. J Bacteriol. *186*, 3153–3159

DeSantis, T.Z., Hugenholtz, P., Larsen, N., Rojas, M., Brodie, E.L., Keller, K., Huber, T., Dalevi, D., Hu, P., and Andersen, G.L. (2006). Greengenes, a chimera-checked 16S rRNA gene database and workbench compatible with ARB. Appl. Environ. Microbiol. *72*, 5069–5072.

DeSantis, T.Z., Brodie, E.L., Moberg, J.P., Zubieta, I.X., Piceno, Y.M., and Andersen, G.L. (2007). High-density universal 16S rRNA microarray analysis reveals broader diversity than typical clone library when sampling the environment. Microb. Ecol. *53*, 371–383.

Dulla, G., and Lindow, S.E. (2008). Quorum size of *Pseudomonas syringae* is small and dictated by water availability on the leaf surface. Proc. Natl. Acad. Sci. U. S. A. *105*, 3082–3087.

el Zahar Haichar, F., Marol, C., Berge, O., Rangel-Castro, J.I., Prosser, J.I., Balesdent, J, Heulin, T., and Achouak, W. (2008). Plant host habitat and root exudates shape soil bacterial community structure. ISME J. *2*, 1221–1230.

Ercolani, G.L. (1991). Distribution of epiphytic bacteria on olive leaves and the influence of leaf age and sampling time. Microb. Ecol. *21*, 35–48.

Escoriza, M.F., Vanbriesen, J.M., Stewart, S., and Maier, J. (2006). Studying bacterial metabolic tates using Raman spectroscopy. Appl. Spectrosc. 60, 865, 971–976.

Espinosa-Urgel, M., Ramos, J-L. (2004). Cell density-dependent gene contributes to efficient seed colonization by *Pseudomonas putida* KT2440. Appl Environ Microbiol. 70, 5190–5198.

Feil, H., Feil, W.S., Chain, P., Larimer, F., DiBartolo, G., Copeland, A., Lykidis, A., Trong, S., Nolan, M., Goltsman, E., *et al.* (2005). Comparison of the complete genome sequences of *Pseudomonas syringae* pv. *syringae* B728a and pv. *tomato* DC3000. Proc. Natl. Acad. Sci. U. S. A. *102*, 11064–11069.

Feng, Y., Shen, D., and Song, W. (2006). Rice endophyte *Pantoea agglomerans* YS19 promotes host plant growth and affects allocations of host photosynthates. J. Appl. Microbiol. *100*, 938–945.

Filion, M., Hamelin, R.C., Bernier, L., and St-Arnaud, M. (2004). Molecular profiling of rhizosphere microbial communities associated with healthy and diseased black spruce (*Picea mariana*) seedlings grown in a nursery. Appl. Environ. Microbiol. *70*, 3541–3551.

Frias-Lopez, J., Shi, Y., Tyson, G.W., Coleman, M.L., Schuster, S.C., Chisholm, S.W., and DeLong, E.F. (2008). Microbial community gene expression in ocean surface waters. Proc. Natl. Acad. Sci. U. S. A. *105*, 3805–3810.

Fromin, N., Hamelin, J., Tarnawski, S., Roesti, D., Jourdain-Miserez, K., Forestier, N., Teyssier-Cuvelle, S., Gillet, F., Aragno, M., and Rossi, P. (2002). Statistical analysis of denaturing gel electrophoresis (DGE) fingerprinting patterns. Environ. Microbiol. *4*, 634–643.

Fürnkranz, M., Wanek, W., Richter, A., Abell, G., Rasche, F., and Sessitsch, A. (2008). Nitrogen fixation by phyllosphere bacteria associated with higher plants and their colonizing epiphytes of a tropical lowland rainforest of Costa Rica. ISME J. *2*, 561–570.

Gomes, N.C.M., Fagbola, O., Costa, R., Rumjanek, N.G., Buchner, A., Mendona-Hagler, L., and Smalla, K. (2003). Dynamics of fungal communities in bulk and maize rhizosphere soil in the tropics. Appl. Environ. Microbiol. *69*, 3758–3766.

Gourion, B., Rossignol, M., and Vorholt, J.A. (2006). A proteomic study of *Methylobacterium extorquens* reveals a response regulator essential for epiphytic growth. Proc. Natl. Acad. Sci. U. S. A. *103*, 13186–13191.

Gourion, B., Francez-Charlot, A., and Vorholt, J.A. (2008). PhyR is involved in the general stress response of *Methylobacterium extorquens* AM1. J. Bacteriol. *190*, 1027–1035.

Griffiths, R.I., Bailey, M.J., McNamara, N.P., and Whiteley, A.S. (2006). The functions and components of the Sourhope soil microbiota. Appl. Soil Ecol. *33*, 114–126.

Hallman, J., Quadt-Hallmann, A., Mahaffee, W.F., and Kloepper, J.W. (1997). Bacterial endophytes in agricultural crops. Can. J. Microbiol. *43*, 895–914.

Handelsman, J. (2004). Metagenomics: Application of genomics to uncultured microorganisms. Microbiol. Mol. Biol. Rev. *68*, 669–685.

Hawksworth, D.L., and Rossman, A.Y. (1997). Where are all the undescribed fungi? Phytopathology *87*, 888–891.

He, Z.L., Gentry, T.J., Schadt, C.W., Wu, L., Liebich, J., Chong, S.C., Wu, W., Jardine, P., Criddle, C., and Zhou, J.Z. (2007). GeoChip: A comprehensive microarray for investigating biogeochemical, ecological and environmental processes. ISME J. *1*, 67–77.

Heuser, T., and Zimmer, W. (2002). Quantitative analysis of phytopathogenic ascomycota on leaves of pedunculate oaks (*Quercus robur* L.) by real-time PCR. FEMS Microbiol. Lett. *209*, 295–299.

Heuser, T., and Zimmer, W. (2003). Genus- and isolate-specific real-time PCR quantification of *Erwinia* on leaf surfaces of English oaks (*Quercus robur* L.). Curr. Microbiol. *47*, 214–219.

Hiltner, L. (1904). Über neuere Erfahrungen und Probleme auf dem Gebiete der Bodenbakteriologie unter besonderer Berücksichtigung der Gründüngung und Brache. Arbeiten der Deutschen Landwirtschaftlichen Gesellschaft *98*, 59–78.

Holland, M.A., and Polacco, J.C. (1994). PPFMs and other covert contaminants: is there more to plant physiology than just plant? Annu. Rev. Plant Physiol. Plant Molec. Biol. *45*, 197–209.

Horner-Devine, M.C., Lage, M., Hughes, J.B., and Bohannan, B.J.M. (2004). A taxa-area relationship for bacteria. Nature *432*, 750–753.

Huang, W.E., Griffiths, R.I., Thompson, I.P., Bailey, M.J., and Whiteley, A.S. (2004). Raman microscopic analysis of single microbial cells. Anal. Chem. *76*, 4452–4458.

Huang, W.E., Bailey, M.J., Thompson, I.P., Whiteley, A.S., and Spiers, A.J. (2007a). Single-cell Raman spectral profiles of *Pseudomonas fluorescens* SBW25 reflects *in vitro* and *in planta* metabolic history. Microb. Ecol. *53*, 414–425.

Huang, W.E. Stoecker, K., Griffiths, R., Newbold, L., Daims, H., Whiteley, A.S., and Wagner, M. (2007b). Raman-FISH: combining stable-isotope Raman spectroscopy and fluorescence *in situ* hybridization for the single cell analysis of identity and function. Environ. Microbiol. *9*, 1878–1889.

Hugenholtz, P., Goebel, B.M., and Pace, N.R. (1998). Impact of culture-independent studies on the emerging phylogenetic view of bacterial diversity. J. Bacteriol. *180*, 4765–4774.

Idris, R., Trifonova, R., Puschenreiter, M., Wenzel, W.W., and Sessitsch, A. (2004). Bacterial communities associated with flowering plants of the Ni-hyperaccumulator *Thlaspi goesingense*. Appl. Environ. Microbiol. *70*, 2667–2677.

Iniguez, A.L., Dong, Y., Carter, H.D., Ahmer, B.M.M., Stone, J.M., and Triplett, E.W. (2005). Regulation of enteric endophytic bacterial colonization by plant defenses. Mol. Plant-Microbe Interact. *18*, 169–178.

Jacobs, J.L., Carroll, T.L., and Sundin, G.W. (2005). The role of pigmentation, ultraviolet radiation tolerance, and leaf colonization strategies in the epiphytic survival of phyllosphere bacteria. Microb. Ecol. *49*, 104–113.

Keller, M., and Zengler, K. (2004). Tapping into microbial diversity. Nat. Rev. Microbiol. *2*, 141–150.

Kielak, A., Pijl, A.S., van Veen, J.A., and Kowalchuk, G.A. (2008). Differences in vegetation composition and plant species identity lead to only minor changes in soil-borne microbial communities in a former arable field. FEMS Microbiol. Ecol. *63*, 372–382.

Kim, J.J., and Sundin, G.W. (2000). Regulation of the rulAB mutagenic DNA repair operon of Pseudomonas syringae by UV-B (290 to 320 nanometers) radiation and analysis of rulAB-mediated mutability in vitro and in planta. J. Bacteriol. *182*, 6137–6144.

Knauth, S., Hurek, T., Brar, D., and Reinhold-Hurek, B. (2005). Influence of different *Oryza* cultivars on expression of nifH gene pools in roots of rice. Environ. Microbiol. 7, 1725–1735.

Knief, C., Frances, L., Cantet, F., and Vorholt, J.A. (2008). Cultivation-independent characterization of *Methylobacterium* populations in the plant phyllosphere by automated ribosomal intergenic spacer analysis. Appl. Environ. Microbiol. *74*, 2218–2228.

Knoll, D., and Schreiber, L. (2000). Plant-microbe interactions: Wetting of ivy (*Hedera helix* L.) leaf surfaces in relation to colonization by epiphytic microorganisms. Microb. Ecol. *40*, 33–42.

Kowalchuk, G.A., Buma, D.S., de Boer, W., Klinkhamer, P.G.L., and van Veen, J.A. (2002). Effects of above-ground plant species composition and diversity on the diversity of soil-borne microorganisms. Antonie Van Leeuwenhoek *81*, 509–520.

Kowalchuk, G.A., Bruinsma, M., and Van Veen, J.A. (2003). Assessing soil ecosystem responses to GM plants. Trends Ecol. Evol. *18*, 403–410.

Kowalchuk, G.A., Speksnijder, A.G.C.L., Zhang, K., Goodman, R.M., and van Veen, J.A. (2007). Finding the needles in the meta-genome haystack. Microb. Ecol. *53*, 475–485.

Krause, A., Ramakumar, A., Bartels, D., Battistoni, F., Bekel, T., Boch, J., Bohm, M., Friedrich, F., Hurek, T., Krause, L., *et al.* (2006). Complete genome of the mutualistic, N^2-fixing grass endophyte *Azoarcus* sp. strain BH72. Nat. Biotech. *24*, 1385–1391.

Kürkcüoglu, S., Degenhardt, J., Lensing, J., Al-Masri, A.N., and Gau, A.E. (2007). Identification of differentially expressed genes in *Malus domestica* after application of the non-pathogenic bacterium *Pseudomonas fluorescens* Bk3 to the phyllosphere. J. Exp. Bot. *58*, 733–741.

Lambais, M.R., Crowley, D.E., Cury, J.C., Bull, R.C., and Rodrigues, R.R. (2006). Bacterial diversity in tree canopies of the Atlantic forest. Science *312*, 1917–1917.

Legendre, P., and Legendre, L. (1998). Numerical ecology, 2nd English edn (Amsterdam: Elsevier Science B.V.).

Leigh, M.B., Pellizari, V.H., Uhlik, O., Sutka, R., Rodrigues, J., Ostrom, N.E., Zhou, J.H., and Tiedje, J.M. (2007). Biphenyl-utilizing bacteria and their functional genes in a pine root zone contaminated with polychlorinated biphenyls (PCBs). ISME J. *1*, 134–148.

Leveau, J.H.J., and Gerards, S. (2008). Discovery of a bacterial gene cluster for catabolism of the plant hormone indole 3-acetic acid. FEMS Microbiol. Ecol. *65*, 238–250.

Leveau, J.H.J., and Lindow, S.E. (2002). Bioreporters in microbial ecology. Curr. Opin. Microbiol. *5*, 259–265.

Leveau, J.H.J., and Lindow, S.E. (2005). Utilization of the plant hormone indole-3-acetic acid for growth by *Pseudomonas putida* strain 1290. Appl. Environ. Microbiol. *71*, 2365–2371.

Libbert, E., and Risch, H. (1969). Interactions between plants and epiphytic bacteria regarding their auxin metabolism. V. Isolation and identification of the IAA-producing and -destroying bacteria from pea plants. Physiol. Plantarum *22*, 51–58.

Lindow, S.E. (1988). Lack of correlation of in vitro antibiosis with antagonism of ice nucleation active bacteria on leaf surfaces by non-ice nucleation active bacteria. Phytopathology *78*, 444–450.

Lindow, S.E., and Brandl, M.T. (2003). Microbiology of the phyllosphere. Appl. Environ. Microbiol. *69*, 1875–1883.

Lorenz, P., and Eck, J. (2005). Metagenomics and industrial applications. Nat. Rev. Microbiol. *3*, 510–516.

Lu, Y., Abraham, W.-R., and Conrad, R. (2007). Spatial variation of active microbiota in the rice rhizosphere revealed by in situ stable isotope probing of phospholipid fatty acids. Environ. Microbiol. *9*, 474–481.

Lugtenberg, B.J.J., Dekkers, L., and Bloemberg, G.V. (2001). Molecular determinants of rhizosphere colonization by *Pseudomonas*. Ann. Rev. Phytopath. 39, 461–490.

Manceau, C., and Kasempour, M.N. (2002). Endophytic versus epiphytic colonization of plants: what comes first? In Phyllosphere microbiology, S.E. Lindow, E.I. Hecht-Poinar, and V.J. Elliott, eds (St-Paul, MN: APS Press), pp. 115–123.

Manefield, M., Whiteley, A.S., Griffiths, R.I., and Bailey, M.J. (2002). RNA stable isotope probing: A novel means of linking microbial community function to phylogeny. Appl.Env. Microbiol. *68*, 5367–5373.

Marco, M.L., Legac, J., and Lindow, S.E. (2003). Conditional survival as a selection strategy to identify plant-inducible genes of *Pseudomonas syringae*. Appl. Environ. Microbiol. *69*, 5793–5801.

Marco, M.L., Legac, J., and Lindow, S.E. (2005). *Pseudomonas syringae* genes induced during colonization of leaf surfaces. Environ. Microbiol. 7, 1379–1391.

Mark, G.L., Dow J.M., Kiely, P.D., Higgins, H., Haynes, J., Baysse, C., Abbas, A., Foley, T., Franks, A., Morrissey, J., and O'gara, F. (2005). Transcriptome profiling of bacterial responses to root exudates identifies genes involved in microbe-plant interactions. Proc. Natl. Acad. Sci. U.S.A. 102,17454–17459.

Margulies, M., Egholm, M., Altman, W.E., Attiya, S., Bader, J.S., Bemben, L.A., Berka, J., Braverman, M.S., Chen, Y.J., Chen, Z.T., *et al.* (2005). Genome sequencing in microfabricated high-density picolitre reactors. Nature *437*, 376–380.

Marois, J.J., and Coleman, P.M. (1995). Ecological succession and biological control in the phyllosphere. Can. J. Bot. *73*, S76–S82.

Martiny, J.B.H., Bohannan, B.J.M., Brown, J.H., Colwell, R.K., Fuhrman, J.A., Green, J.L., Horner-Devine, M.C., Kane, M., Krumins, J.A., Kuske, C.R., *et al.* (2006). Microbial biogeography: putting microorganisms on the map. Nat. Rev. Microbiol. *4*, 102–112.

Mercier, J., and Lindow, S.E. (2000). Role of leaf surface sugars in colonization of plants by bacterial epiphytes. Appl. Environ. Microbiol. *66*, 369–374.

Miché, L., Battistoni, F., Gemmer, S., Belghazi, M., and Reinhold-Hurek, B. (2006). Upregulation of jasmonate-inducible defense proteins and differential colonization of roots of *Oryza sativa* cultivars with the endophyte *Azoarcus* sp. Mol. Plant-Microbe Interact. *19*, 502–511.

Mirete, S., de Figueras, C.G., and Gonzalez-Pastor, J.E. (2007). Novel nickel resistance genes from the rhizosphere metagenome of plants adapted to acid mine drainage. Appl. Environ. Microbiol. *73*, 6001–6011.

Monier, J.M., and Lindow, S.E. (2003). Differential survival of solitary and aggregated bacterial cells promotes aggregate formation on leaf surfaces. Proc. Natl. Acad. Sci. U. S. A. *100*, 15977–15982.

Monier, J.M., and Lindow, S.E. (2004). Frequency, size, and localization of bacterial aggregates on bean leaf surfaces. Appl. Environ. Microbiol. *70*, 346–355.

Monier, J.M., and Lindow, S.E. (2005a). Aggregates of resident bacteria facilitate survival of immigrant bacteria on leaf surfaces. Microb. Ecol. *49*, 343–352.

Monier, J.M., and Lindow, S.E. (2005b). Spatial organization of dual-species bacterial aggregates on leaf surfaces. Appl. Environ. Microbiol. *71*, 5484–5493.

Morris, C.E., Bardin, M., Berge, O., Frey-Klett, P., Fromin, N., Girardin, H., Guinebretiere, M.-H., Lebaron, P., Thiery, J.M., and Troussellier, M. (2002). Microbial biodiversity: Approaches to experimental design and hypothesis testing in primary scientific literature from 1975 to 1999. Microbiol. Mol. Biol. Rev. *66*, 592–616.

Morris, C.E., and Monier, J.M. (2003). The ecological significance of biofilm formation by plant-associated bacteria. Annu. Rev. Phytopathol. *41*, 429–453.

Nannipieri, P., Ascher, J., Ceccherini, M.T., Guerri, G., Renella, G., and Pietramellara, G. (2008). Recent Advances in Functional Genomics and Proteomics of Plant Associated Microbes. In Molecular Mechanisms of Plant and Microbe Coexistence Vol. 15, Nautiyal Shekhar, C., and Dion, P., eds. (Berlin Heidelberg: Springer), pp. 215–241.

Narasimhan, K., Basheer, C., Bajic, V.B., and Swarup, S. (2003) Enhancement of plant-microbe interactions using a rhizosphere metabolomics-driven approach and its application in the removal of polychlorinated biphenyls. Plant Physiol. *132*, 146–153.

Neufeld, J.D., Wagner, M., and Murrell, J.C. (2007). Who eats what, where and when? Isotope-labelling experiments of complex microbial communities are coming of age. ISME J. *1*, 103–110.

Okinaka, Y., Yang, C.H., Perna, N.T., and Keen, N.T. (2002). Microarray profiling of *Erwinia chrysanthemi* 3937 genes that are regulated during plant infection. Mol. Plant Microbe Interact. *15*, 619–629.

Ostle, N., Whiteley, A.S., Ostle, N., Whiteley, A.S., Bailey, M.J., Sleep, D., Ineson, P., and Manefield, M.J. (2003). Active microbial RNA turnover in a grassland soil estimated using a $^{13}CO^2$ spike. Soil Biol. BioChem. *35*, 877–885.

Pirttilä, A.M., Laukkanen, H., Pospiech, H., and Hohtola, A. (2000). Detection of intracellular bacteria in the buds of Scotch pine (*Pinus silvestris* L.) by in-situ hybridization. Appl. Environ. Microbiol. *66*, 3073–3077.

Poretsky, R.S., Bano, N., Buchan, A., LeCleir, G., Kleikemper, J., Pickering, M., Pate, W.M., Moran, M.A., and Hollibaugh, J.T. (2005). Analysis of microbial gene transcripts in environmental samples. Appl. Environ. Microb. *71*, 4121–4126.

Pühler, A., Arlat, M., Becker, A., Göttfert, M., Morrissey, J.P., and O'Gara, F. (2004) What can bacterial genome research teach us about bacteria–plant interactions? Curr. Opin. Plant Biol. 7, 137–147.

Radajewski, S., Ineson, P., Parekh, N.R., and Murrell, J.C. (2000). Stable-isotope probing as a tool in microbial ecology. Nature *403*, 646–649.

Rainey, P.B. (1999). Adaptation of *Pseudomonas fluorescens* to the plant rhizosphere. Environ. Microbiol. *1*, 243–257.

Ramette, A. (2007). Multivariate analyses in microbial ecology. FEMS Microbiol. Ecol. *62*, 142–160.

Ramey, B.E., Koutsoudis, M., von Bodman, S.B., and Fuqua, C. (2004). Biofilm formation in plant-microbe associations. Curr. Opin. Microbiol. 7, 602–609.

Rangel-Castro, J.I., Killham, K., Ostle, N., Nicol, G.W., Anderson, I.C., Scrimgeour, C.M., Ineson, P., Meharg, A., and Prosser, J.I. (2005). Stable isotope probing analysis of the influence of liming on root exudate utilization by soil microorganisms. Environ. Microbiol. 7, 828–838.

Rasche, F., Velvis, H., Zachow, C., Berg, G., Van Elsas, J.D., and Sessitsch, A. (2006). Impact of transgenic potatoes expressing anti-bacterial agents on bacterial endophytes is comparable with the effects of plant genotype, soil type and pathogen infection. J. Appl. Ecol. *43*, 555–566.

Rasche, F., Lueders, T., Schaefer, S., Buegger, F., Gattinger, A., Schloter, M., Hood-Nowotny, R.C., and Sessitsch, A. (2008). DNA stable isotope probing enables the identification of active bacterial endophytes in potato. New Phytologist *181*, 802–807.

Reiter, B., Bürgmann, H., Burg, K., and Sessitsch, A. (2003). Endophytic *nifH* gene diversity in African sweet potato. Can. J. Microbiol. *49*, 549–555.

Reiter, B., and Sessitsch, A. (2006). Bacterial endophytes of the wildflower *Crocus albiflorus* analyzed by characterization of isolates and by a cultivation-independent approach. Can. J. Microbiol. *52*, 140–149.

Rondon, M.R., August, P.R., Bettermann, A.D., Brady, S.F., Grossman, T.H., Liles, M.R., Loiacono, K.A., Lynch, B.A., MacNeil, I.A., Minor, C., Tiong, C.L., Gilman, M., Osburne, M.S., Clardy, J., Handelsman, J., and Goodman, R.M. (2000). Cloning the soil metagenome: a strategy for accessing the genetic and functional diversity of uncultured microorganisms. Appl. Environ. Microbiol. *66*, 2541–2547.

Roesch, L.F., Fulthorpe, R.R., Riva, A., Casella, G., Hadwin, A.K.M., Kent, A.D., Daroub, S.H., Camargo, F.A.O., Farmerie, W.G., and Triplett, E.W. (2007). Pyrosequencing enumerates and contrasts soil microbial diversity. ISME J. *1*, 283–290.

Rudrappa, T., Biedrzycki, M.L., and Bais, H.P. (2008). Causes and consequences of plant-associated biofilms. FEMS Microbiol. Ecol. *64*, 153–166.

Sanguin, H., Herrera, A., Oger-Desfeux, C., Dechesne, A., Simonet, P., Navarro, E., Vogel, T.M., Moenne-Loccoz, Y., Nesme, X., and Grundmann, G.L. (2006a). Development and validation of a prototype 16S rRNA-based taxonomic microarray for *Alphaproteobacteria*. Environ. Microbiol. *8*, 289–307.

Sanguin, H., Remenant, B., Dechesne, A., Thioulouse, J., Vogel, T.M., Nesme, X., Moenne-Loccoz, Y., and Grundmann, G.L. (2006b). Potential of a 16S rRNA-based taxonomic microarray for analyzing the rhizosphere effects of maize on *Agrobacterium* spp. and bacterial communities. Appl. Environ. Microbiol. *72*, 4302–4312.

Sankaran, K.V., Sutton, B.C., and Minter, D.W. (1995). A checklist of fungi recorded on *Eucalyptus*. Mycol. Pap. *170*, 1–376.

Schreiber, L., Krimm, U., Knoll, D., Sayed, M., Auling, G., and Kroppenstedt, R.M. (2005). Plant-microbe interactions: identification of epiphytic bacteria and their ability to alter leaf surface permeability. New Phytol. *166*, 589–594.

Sessitsch, A., Reiter, B., Pfeifer, U., and Wilhelm, E. (2002). Cultivation-independent population analysis of bacterial endophytes in three potato varieties based on eubacterial and *Actinomycetes*-specific PCR of 16S rRNA genes. FEMS Microbiol. Ecol. *39*, 23–32.

Sessitsch, A., Reiter, B., and Berg, G. (2004). Endophytic bacterial communities of field-grown potato plants and their plant-growth-promoting and antagonistic abilities. Can. J. Microbiol. *50*, 239–249.

Smalla, K., Wieland, G., Buchner, A., Zock, A., Parzy, J., Kaiser, S., Roskot, N., Heuer, H., and Berg, G. (2001). Bulk and rhizosphere soil bacterial communities studied by denaturing gradient gel electrophoresis: Plant-dependent enrichment and seasonal shifts revealed. Appl. Environ. Microbiol. *67*, 4742–4751.

Smith, S.E., and Read, D.J. (2008). Mycorrhizal symbiosis, 3rd edn (Academic Press).

Spaepen, S., Vanderleyden, J., and Remans, R. (2007). Indole-3-acetic acid in microbial and microorganism-plant signaling. FEMSS Microbiol. Rev. *31*, 425–448.

St-Arnaud, M., Hamel, C., Vimard, B., Caron, M., and Fortin, J.A. (1996). Enhanced hyphal growth and spore production of the arbuscular mycorrhizal fungus *Glomus intraradices* in an in vitro system in the absence of host roots. Mycol. Res. *100*, 328–332.

Streit, W.R., and Schmitz, R.A. (2004). Metagenomics - the key to the uncultured microbes. Curr. Opin. Microbiol. *7*, 492–498.

Sturz, A.V., Christie, B.R., and Nowak, J. (2000). Bacterial endophytes: potential role in developing sustainable systems of crop production. Crit. Rev. Plant Sci. *19*, 1–30.

Sun, L., Qiu, F.B., Zhang, X.X., Dai, X., Dong, X.Z., and Song, W. (2008). Endophytic bacterial diversity in rice (*Oryza sativa* L.) roots estimated by 16S rDNA sequence analysis. Microb. Ecol. *55*, 415–424.

Tan, Z., Hurek, T., and Reinhold-Hurek, B. (2003). Effect of N-fertilization, plant genotype and environmental conditions on *nifH* gene pools in roots of rice. Environ. Microbiol. *5*, 1009–1015.

Tarlera, S., Jangid, K., Ivester, A.H., Whitman, W.B., and Williams, M.A. (2008). Microbial community succession and bacterial diversity in soils during 77,000 years of ecosystem development. FEMS Microbiol. Ecol. *64*, 129–140.

Timmusk, S., Grantcharova, N., and Wagner, E.G.H. (2005). *Paenibacillus polymyxa* invades plant roots and forms biofilms. Appl. Environ. Microbiol. *71*, 7292–7300.

Thompson, I.P., Bailey, M.J., Fenlon, J.S., Fermor, T.R., Lilley, A.K., Lynch, J.M., McCormack, P.J., McQuilken, M.P. *et al.* (1993). Quantitative and qualitative seasonal changes in the microbial community from the phyllosphere of sugar beet (*Beta vulgaris*). Plant Soil *150*, 177–191.

Tringe, S.G., von Mering, C., Kobayashi, A., Salamov, A.A., Chen, K., Chang, H.W., Podar, M., Short, J.M., Mathur, E.J., Detter, J.C., *et al.* (2005). Comparative metagenomics of microbial communities. Science *308*, 554–557.

Tukey, H.B. (1970). Leaching of substances from plants. Ann. Rev. Plant Physiol. *21*, 305–324.

Turnbaugh, P.J., and Gordon, J.I. (2008). An invitation to the marriage of metagenomics and metabolomics. Cell *134*, 708–713.

Tyson, G.W., Chapman, J., Hugenholtz, P., Allen, E.E., Ram, R.J., Richardson, P.M., Solovyev, V.V., Rubin, E.M., Rokhsar, D.S., and Banfield, J.F. (2004). Community structure and metabolism through reconstruction of microbial genomes from the environment. Nature *428*, 37–43.

van der Wal, A., van Veen, J.A., Smant, W., Boschker, H.T.S., Bloem, J., Kardol, P., van der Putten, W.H., and de Boer, W. (2006). Fungal biomass development in a chronosequence of land abandonment. Soil Biol. Biochem. *38*, 51–60.

van Elsas, J.D., Costa, R. Jansson, J., Sjöling, S., Bailey, M.J., Nalin, R., Vogel, T.M., and van Overbeek, L. (2008). The metagenome of diseases supressive soils–experiences from the METACONTROL project. Trends Biotechnol. *26*, 591–601.

Van Sluys, M.A., Montero-Vitorello, C.B., Camargo, L.E.A., Menck, C.F.M., da Silva, A.C.R., Ferro, J.A., Oliveira, M.C., Setubal, J.C., Kitajiman, J.P., and Simpson, A.J. (2000). Comparative genomic analysis of plant-associated bacteria. Annu. Rev. Phytopathol. 40, 169–189.

Vandenkoornhuyse, P., Mahe, S., Ineson, P., Staddon, P., Ostle, N., Cliquet, J.B., Francez, A.J., Fitter, A.H., and Young, J.P.W. (2007). Active root-inhabiting microbes identified by rapid incorporation of plant-derived carbon into RNA. Proc. Natl. Acad. Sci. U. S. A. *104*, 16970–16975.

Verberkmoes, N.C., Russell, A.L., Shah, M., Godzik, A., Rosenquist, M., Halfvarson, J., Lefsrud, M.G., Apajalahti, J., Tysk, C., Hettich, R.L., and Jansson, J.K. (2009). Shotgun metaproteomics of the human distal gut microbiota. ISME J. *3*, 179–189.

Vujanovic, V., Hamelin, R.C., Bernier, L., Vujanovic, G., and St-Arnaud, M. (2007). Fungal diversity, dominance, and community structure in the rhizosphere of clonal *Picea mariana* plants throughout nursery production chronosequences. Microb. Ecol. *54*, 672–684.

Wacker, T.L., Safir, G.R., and Stephenson, S.N. (1990). Evidence for succession of mycorrhizal fungi in Michigan asparagus. Acta Horticulturae *271*, 273–278.

Wagner, M., Nielsen, P.H., Loy, A., Nielsen, J.L., and Daims, H. (2006). Linking microbial community structure with function: fluorescence *in situ* hybridization-microautoradiography and isotope arrays. Curr Opin Biotechnol 17, 1–9 .

Wagner, M., Smidt, H., Loy, A., and Zhou, J. (2007). Unravelling microbial communities with DNA-microarrays: challenges and future directions. Microb. Ecol. 53, 498–506.

Weidner, S., Arnold, W., Stackebrandt, E., and Puhler, A. (2000). Phylogenetic analysis of bacterial communities associated with leaves of the seagrass *Halophila stipulacea* by a culture-independent small-subunit rRNA gene approach. Microb. Ecol. *39*, 22–31.

Whipps, J.M., Hand, P., Pink, D., and Bending, G.D. (2008). Phyllosphere microbiology with special reference to diversity and plant genotype. J. Appl. Microbiol. *105*, 1744–1755.

Whiteley, A.S., Manefield, M., and Lueders, T. (2006). Unlocking the 'microbial black box' using RNA-based stable isotope probing technologies. Curr. Opin. Biotechnol. 17, 67–71.

Wilmes, P., and Bond, P.L. (2004). The application of two-dimensional polyacrylamide gel electrophoresis and downstream analyses to a mixed community of prokaryotic microorganisms. Environ. Microbiol. *6*, 911–920.

Wilmes, P., and Bond, P.L. (2006). Metaproteomics: studying functional gene expression in microbial ecosystems. Trends Microbiol. *14*, 92–97.

Yang, C.H., Crowley, D.E., Borneman, J., and Keen, N.T. (2001). Microbial phyllosphere populations are more complex than previously realized. Proc. Natl. Acad. Sci. U. S. A. *98*, 3889–3894.

Yergeau, E., Sommerville, D.W., Maheux, E., Vujanovic, V., Hamel, C., Whalen, J.K., and St-Arnaud, M. (2006a). Relationships between *Fusarium* population structure, soil nutrient status and disease incidence in field-grown asparagus. FEMS Microbiol. Ecol. *58*, 394–403.

Yergeau, E., Vujanovic, V., and St-Arnaud, M. (2006b). Changes in communities of *Fusarium* and arbuscular-mycorrhizal fungi as related to different asparagus cultural factors. Microb. Ecol. 52, 104–113.

Yergeau, E., Newsham, K.K., Pearce, D.A., and Kowalchuk, G.A. (2007). Patterns of bacterial diversity across a range of Antarctic terrestrial habitats. Environ. Microbiol. *9*, 2670–2682.

Yergeau, E., Schoondermark-Stolk, S.A., Brodie, E.L., Déjean, S., DeSantis, T.Z., Gonçalves, O., Piceno, Y.M., Andersen, G.L., and Kowalchuk, G.A. (2009). Environmental microarray analyses of Antarctic soil microbial communities. ISME J. *3*, 340–351.

Zhang, L., Hurek, T., and Reinhold-Hurek, B. (2007). A *nifH*-based oligonucleotide microarray for functional diagnostics of nitrogen-fixing microorganisms. Microb. Ecol. 53, 456–470.

Zhang, X-X., Lilley, A.K., Bailey, M.J., and Rainey, P.B. (2004). Functional and phylogenetic analysis of a plant-inducible oligoribonuclease(*orn*) gene from an indigenous *Pseudomonas* plasmid. Microbiol. (UK) 150, 2889–2898.

Marine Environments

8

Alexander H. Treusch, Ulrich Stingl and Stephen J. Giovannoni

Abstract

Ocean microbial communities play important roles in global geochemical cycles. From the earliest cultivation experiments to today's metagenomic analyses, most of the major discoveries in this field were driven by applications of novel methods. Molecular ecology had a major impact by revealing the true scope of microbial diversity and providing genetic markers that could be used to track important species when cultures were unavailable. In some cases, metagenomics provided insights into the biochemical adaptations of these organisms. A renaissance in culturing techniques led to isolates of many abundant ocean microbes that subsequently could be studied in a laboratory setting. Today a consortium of approaches that span scales from molecules to ocean basins are being applied to ocean ecosystems, with the result that marine microbiology is becoming a highly integrated science.

Introduction

What sets the Earth apart from every other planet we know of is the enormous amount of water available in liquid form, most of which is stored in global oceans that cover roughly 70% of the Earth's surface. Thus, it is no surprise that life is suspected to have originated in the oceans and today they are recognized as a major source of biological diversity. Elemental cycles, and particularly the carbon cycle, are heavily influenced by life in the oceans. Roughly half of global photosynthesis takes place in the oceans, and they contain almost as much carbon in the form of dissolved organic matter as there is carbon in atmospheric CO_2.

Because of their size and importance, the oceans have long attracted the interest of microbiologists, and have played a particularly important role in the development of environmental molecular biology. One of the reasons for the rapid rise of molecular approaches in marine microbiology is practical: the cells in aquatic systems are suspended and can be collected easily. Sediment and soil microbiologists face much greater challenges extracting cells and nucleic acids from their samples (see Chapter 6). So, for good reasons, the oceans have often been chosen for testing new methods that have later been applied elsewhere.

However, studying microbial life in the oceans also posseses unique challenges. The sheer size of the oceans and the fact that large parts of this ecosystem are not readily accessible mean that sampling can often only be achieved by organizing expensive expeditions, and even then the diversity of ocean ecosystems, spanning so much latitude and depth, can make it risky to generalize without many samples.

The early days of marine microbiology

Some early microbiologists questioned the existence of marine bacteria because of the salinity, ultraviolet radiation at the surface, high pressures in the deep sea and the threat of predation by abundant enemies. Even after the first reports of bacteria in coastal waters, some supposed them to be land-borne contaminants.

One of the first bacterial species that was accurately described, *Spirochaeta plicatilis*, was isolated from seawater (Ehrenberg, 1838). Other marine bacterial species were isolated and described in the following years, like *Beggiatoa mirabilis* (Cohn, 1865), *Beggiatoa minima* (Warming, 1875), *Thiospirillum violaceum*, *Thiospirillum rosenbergii* and *Achromatium mulleri* (all: Warming, 1876). Despite this early work with bacteria collected from the shoreline, the first systematic approach to isolate and describe marine species was not until 1884, when Certes (1884) collected 100 marine water and sediment samples, some from great depths. Employing aerobic culturing conditions, he obtained cultures from nearly all the samples, thus proving that there are viable bacteria throughout the water column and some can survive the pressure of great depths and decompression, to grow at the surface. Bernhard Fischer published a study in 1886 in which he showed that, in contrast to coastal areas, the air over the open ocean contains nearly no land-borne bacteria. Thus, he concluded that the bacteria found in open ocean waters could not originate from land. He further isolated 14 different marine species that grew at 0°C (Fischer, 1888) and described a water sampler for bacteria that can be used for samples from every depth (Fischer, 1893). In another study, for which he collected the data during a cruise across the Atlantic Ocean, Fischer reported viable cells in all the 175 samples he analysed as well as new isolates of photogenic bacteria and some interesting ecologic relationships (Fischer, 1894). Russell (1891) found viable bacteria in waters and sediment samples in the Gulf of Naples. He showed that some of the bacteria found in the mud could not be detected in the overlying water, indicating to him that this is their native habitat. The pioneering work of Certes, Fischer and Russell set the stage for several cruises that followed in which researchers studied multiple water and sediment samples by plate counting and isolated many additional species. It was concluded that many of the bacteria found in marine samples originated in the marine environment and were not contaminants from land (see an overview of these studies in ZoBell, 1946). However, the marine environment was not the main focus of microbiology at that time, which is why the 1939 edition of *Bergey's Manual of Determinative Bacteriology* lists only 86 marine species among a total of 1335 described species, and most of them weren't even fully described. Nevertheless, the pioneering studies established the presence of bacteria in the oceans, both in the water column and in sediments. They showed that the bacteria occurred naturally in these environments, divided there, and were not land-borne contaminants.

After the early phase of exploration and cataloguing the focus turned to understanding bacteria in the oceans as biochemical agents that participate in the flux of elements. One of the first overviews of the field of marine microbiology that stresses this fact was written by Issatchenko (1914) in Russian. He summarized his and other researcher's findings while publishing a study of Arctic seawater and mud, where he found bacterial communities comprising nitrogen-fixers, nitrifiers, denitrifiers, ammonifiers, sulphate-reducers and other organisms. In 1946 Claude ZoBell published his *Marine Microbiology, a Monograph on Hydrobacteriology*, which is considered a landmark in the field today because it summarizes the knowledge about marine microorganisms and their ecology as well as the techniques and media needed to study them.

Because the early studies were done by cultivation and plate counts, it was not surprising that the researchers selected for a subset of organisms that could replicate in agar media. Nonetheless, they described astonishingly many of the groups of bacteria that are involved in global element cycles, although they miscalculated their abundance.

Marine microbiology changed significantly with the introduction of molecular techniques that enabled people to overcome the limits of what later was known as the 'plate count anomaly' (Staley and Konopka, 1985). Now organisms that could not be grown under classic culturing conditions could be detected, studied and targeted by advanced cultivation approaches that made use of environmental molecular data, and cultivation independent methods that rely on reconstructing microbial ecology from nucleic acid sequences. This gave new impetus to the field and led to the cultivation of dominant bacterial species (Rappe *et al.*, 2002) and large metagenomic datasets (Venter *et al.*, 2004).

The remaining parts of this chapter will cover this recent history of molecular studies of marine microorganisms and describe new discoveries in more detail. Although marine environments have many facets, in this chapter we focus on planktonic microorganisms that live in the water column. We do not cover the microbiology of marine sediments, or marine hydrothermal vents.

The molecular biology renaissance

Who is out there? Molecular phylogeny reveals microbial diversity in the oceans

Marine bacteria are major players in global nutrient cycles. While 50% of the global primary production is created by photoautotrophic picoplankton (cyanobacteria), the majority of bacteria in most oceanic environments are assumed to be heterotrophic and degrade organic carbon. Most of the organic carbon in the oceans that is accessible to heterotrophic bacteria is in the form of dissolved organic matter (DOM), which is mainly created by excretion of dissolved organic carbon (DOC) by living cells, and the decay of dead organisms from all trophic levels. This microbial recycling of carbon within the classic food chain, which leads from photoautotrophic algae to zooplankton and higher trophic levels, was coined the microbial loop hypothesis (Azam *et al.*, 1983) and is based on the ideas of Pomeroy (1974). At the time this hypothesis was built, nearly nothing about the identity of the microbial community was known. This missing information started to appear with the first sequencing of bacterial ribosomal genes from marine environments (Giovannoni *et al.*, 1990). Culture-independent studies based on the analysis of 16S rRNA genes in the last 15 years have provided a reasonably thorough picture of the types of bacteria present in bacterioplankton communities. Though coastal sites can differ markedly from oligotrophic open-ocean sites in that they have quite different trophic structures and higher productivity and are influenced by freshwater inputs, bacterioplankton communities in coastal regions and open oceans appear to harbour many closely related microorganisms (Rappe *et al.*, 2000; Acinas *et al.*, 2004; Morris *et al.*, 2006). Nearly all of the major marine bacterioplankton groups (Fig. 8.1) are consistently recovered in environmental gene and metagenome libraries from coastal and open-ocean surface samples, including the SAR86 cluster, SAR11 cluster, SAR116 cluster, marine *Actinobacteria* cluster, Marine Group A, *Roseobacter* clade, and marine *Archaea*. While it is important to note that the recovery of closely related phylotypes from the temperate waters of the Oregon coast and subtropical waters off of Hawaii in the Pacific Ocean supports the notion that the major clades of marine bacteria are cosmopolitan in distribution, they are often reproducibly not identical. This hypothesis is corroborated by a recent study of Pommier *et al.* (2007), who examined ribosomal sequences from nine coastal locations around the world and saw that only two operational taxonomic units (OTU, based on a 97% sequence similarity cut-off) were present in all clone libraries. This can be interpreted as indicating a high degree of specific adaptation of the major bacterioplankton groups to divergent marine environments.

With a phylogenetic outline of marine microbial diversity in hand, and data available for the design of probes that could be used for detection, research focused on learning more about the important groups of marine microorganisms. The genomes of cultivated representatives of important marine microbial clades are being sequenced and show a wide variety of genome sizes and potential metabolic capabilities (see Fig. 8.1). Some important microbial groups have so far resisted cultivation. Information about these organisms comes from methods like labelling with BrdU (Mou *et al.*, 2007), stable isotope probing (Neufeld *et al.*, 2007) and metagenomics (Rusch *et al.*, 2007). In Fig. 8.1 these groups are highlighted by asterisks (see Chapters 2 and 4 for a description of these methods).

The importance of Archaea in marine ecosystems was overlooked for a long time because of the incorrect perception that they are all extremophiles or have very special metabolism (methanogens) and therefore cannot exist in mesophilic habitats like the ocean. This changed with the discovery of small subunit rRNA genes of both major archaeal lineages, Cren- and Euryarchaeota, in seawater by molecular ecological methods (DeLong, 1992; Fuhrman

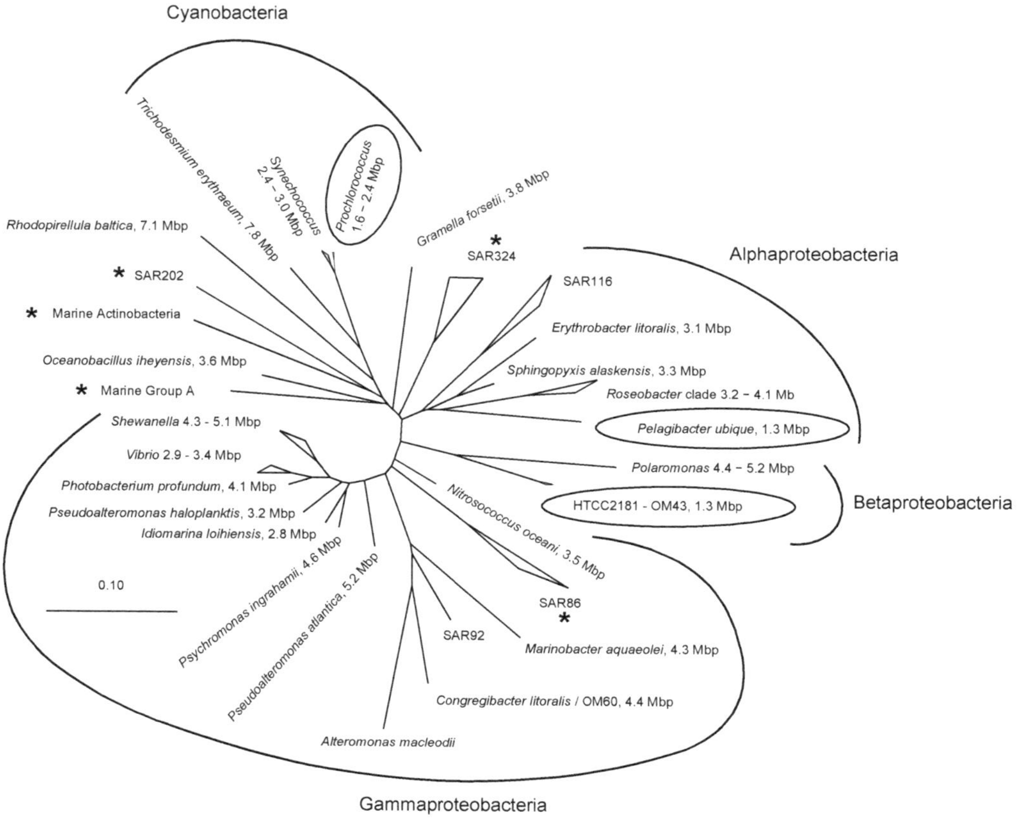

Figure 8.1 Phylogeny of selected marine bacteria based on their small subunit rRNA genes. Asterisks indicate major groups of which no cultivated species has been reported. Where available, the genome size of the species is shown. Species with unusual small genomes are shown framed.

et al., 1992). Consecutive studies showed the widespread distribution of marine Crenarchaea in different parts and depths of the ocean (Stein and Simon, 1996; Karner *et al.*, 2001; Herndl *et al.*, 2005; Ingalls *et al.*, 2006) as well as symbionts in marine sponges (Preston *et al.*, 1996). Close relatives of the marine Crenarchaeota have been detected in many soil environments as well (e.g. Jurgens *et al.*, 1997; Ochsenreiter *et al.*, 2003).

A second group of Archaea found in mesophilic environments are the marine pelagic Euryarchaeota that form at least three distinct phylogenetic groups, with the marine group II euryarchaeotes being the most abundant clade (DeLong, 1992). In surface waters of the oceans, group II Euryarchaeota normally are more abundant than the marine Crenarchaeota (DeLong, 1992; Massana *et al.*, 1997; Massana *et al.*, 2000). Blooms of these microorganisms have been described in the North Sea (Pernthaler *et al.*, 2002), but still little is known about the lifestyle and metabolism of these Archaea.

Patterns in time and space

Stratification

One of the early observations to emerge from rRNA probe hybridization studies was that microbial populations in the ocean surface layer are highly stratified. Often the first evidence of stratification to emerge was indications from environmental 16S rRNA (Fuhrman *et al.*, 1992, 1993; Giovannoni *et al.*, 1996; Gordon and Giovannoni, 1996; Fuhrman *et al.*, 1997; Wright *et al.*, 1997), or later BAC, and fosmid clone libraries (e.g. Beja *et al.*, 2000b), which recovered unusual taxa and genes from different depths, but suffered from the usual problem with environmental genomics, for example undersampling, and lacked sufficient coverage to draw statistically robust conclusions (DeLong *et al.*, 2006). RNA probe hybridization methods, using either bulk environmental rRNA targets (Giovannoni *et al.*, 1996; Gordon and Giovannoni, 1996; Massana *et al.*, 1997; Wright *et al.*, 1997) or FISH protocols with whole cells (Massana *et al.*, 1997; Morris *et al.*, 2002; Morris *et al.*, 2004), produced convincing and more detailed evidence of community stratification.

Two kinds of community stratification have been described. Some important plankton clades are relegated to one part of the water column (Giovannoni *et al.*, 1996; Gordon and Giovannoni, 1996; Massana *et al.*, 1997; Wright *et al.*, 1997; Cho *et al.*, 2004), and in other cases the clade is found throughout the water column, but consists of multiple ecotypes that have depth-specific distributions (Field *et al.*, 1997; Rocap *et al.*, 2002).

The biggest distinction in microbial communities occurs at the boundary of the photic and aphotic zone. Microbial groups that appear to be much more abundant in the aphotic zone include the SAR202 (GNS) clade (Giovannoni *et al.*, 1996; Morris *et al.*, 2004), the SAR324 (Marine Group B) clade of Deltaproteobacteria (Wright *et al.*, 1997), Marine Group A (Gordon and Giovannoni, 1996), and the Group I Archaea (Massana *et al.*, 1997).

DeLong and coworkers undertook metagenomic analyses of seven samples spanning the surface to 4000 meters in the North Pacific Subtropical Gyre. The major drawback of the metagenomic approach they undertook was that it under-sampled, and produced anomalies that made it difficult to draw conclusions about the significance of their findings. For example, they recovered no SAR11 16S rRNA genes in samples from 10–500 m, although SAR11 has been shown to be the most abundant clade in this region of the water column. Secondly, they recovered genes from oxygenic phototrophs – *Procholorcoccus* and uncultured cyanobacteria – from 770 m, well below the photic zone.

Seasonal cycles in bacterioplankton communities

Seasonal cycles in phyoplankton production were well known before the introduction of molecular methods to oceanography (Raymont, 1980) and were eventually observed with great clarity from space by satellites sensing chlorophyll (Hurrell, 1995). Time-series measurements that employed molecular methods to measure seasonal variation in the abundance of uncultured microbial clades have since shown that many important microbial groups undergo annual oscillations in abundance (Morris *et al.*, 2002; Morris *et al.*, 2005; Fuhrman *et al.*, 2006).

Perhaps the best understood marine bacterium is the cyanobacterial species *Prochlorococcus* and its close relative *Synechococcus*. Fig 8.2 shows a cryo-electron tomography picture of a *Prochlorococcus* cell in which the large, pigment-containing, intracytoplasmic lamellae are visible. These organisms were among the first important marine bacteria whose significance was detected and who were cultivated. Progress in the study of those cyanobacteria was rapid early on because fluorescence from their pigments made it possible to count and image living cells. For example, DuRand *et al.* (2001) counted the cyanobacteria alongside the phytoplankton in monthly samples from the Bermuda Atlantic Time-series Study (BATS) using flow cytometry. They could show that *Synechococcus* was most abundant during spring blooms in which the water column undergoes deep mixing events. *Prochlorococcus* was more abundant than *Synechococcus* and showed a deeper subsurface maximum concentration during the summer and fall when the water column is stratified. Other important bacterioplankton, lacking such convenient optical signatures for detection, were not recognized, counted and cultured until after nucleic acid probes were developed for their identification.

Metagenomics, a search for function

Marine ecology is fundamentally simple. The oceans support life that is driven by power from light, and limited by nutrients, particularly the macronutrients phosphate and nitrogen, but also iron. The cycle is fast, with little living biomass accumulating, in contrast to land-based systems. Photosynthesis and heterotrophy are the dominant modes of metabolism.

However simple the basics facts, the specific adaptive strategies that marine microorganisms employ to eke a living in ocean environments are complex and elusive. In search of information

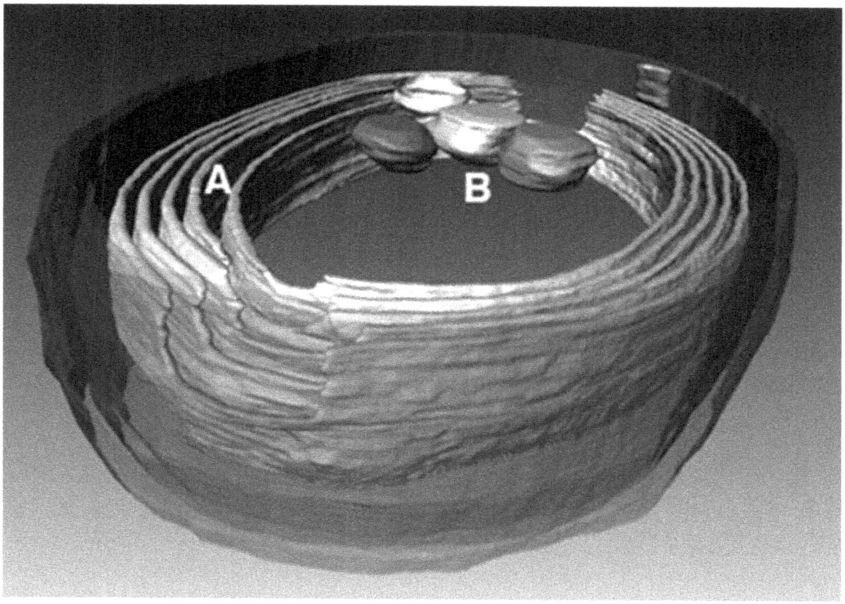

Figure 8.2 Surface-rendered model of Prochlorococcus strain MIT9313 generated from images taken with cryo-electron tomography. Highly visible are the large intracytoplasmic membranes harbouring **A** the photosynthetic pigments and **B** some carboxysomes. Reproduced with permission from Ting *et al*. (2007).

about the ecology of species marine microbiologists turned to metagenomics – sequencing DNA from the environment (see Chapter 2) – to reveal the unseen mechanisms of microbial life in the oceans. In surface waters harvesting sunlight proved to be a more important, and varied, activity than had been recognized.

Proteorhodopsin: a new kind of phototrophy

Metagenomics led directly to the discovery that rhodopsin-based proton pumps are present in marine systems, providing a third type of phototrophy that captures energy from light in the ocean's euphotic zone. These retinal-binding membrane proteins were first discovered in halophilic archaea, and are called bacteriorhodopsins (Oesterhelt and Stoeckenius, 1971). Additional members of the rhodopsin family of proteins are the chloride ion pumps (halorhodopsins) and photosensory rhodopsins (sensory rhodopsins) from halophilic Archaea. This family has been further enlarged by the identification of the so-called proteorhodopsin in uncultivated marine Gammaproteobacteria during an environmental genomic study (Beja *et al.*, 2000a). Biochemical and spectroscopic characterization showed that this protein is indeed a light-driven proton pump and not a sensory rhodopsin (Beja *et al.*, 2000a). In later studies it was shown that proteorhodopsins functioning as proton pumps could potentially support a large part of the energy requirements of these cells (Beja *et al.*, 2001), thus raising the hypothesis that light energy can be used by proteorhodopsin containing microorganisms to create an additional proton motive force without the cost of oxidizing carbon compounds (Beja *et al.*, 2000a). This would theoretically lead to a more effective use of organic carbon, which may be at limiting concentrations in most marine environments (Arrigo, 2005).

Spectral fine-tuning of proteorhodopsins is an ecologically interesting adaption to the different light intensities in marine habitats (2001). Variants of proteorhodopsin are divided into two groups that have absorption maxima spanning a range from 490 nm (blue) to 540 nm (green), with additional small scale changes in absorption maxima within the green proteorhodopsin group (505–540 nm) (Man *et al.*, 2003). The amino acid residue 105, located inside the retinal binding pocket, was identified as the major factor responsible for spectral tuning of proteorhodopsins

(Man *et al.*, 2003). Additional residues, predicted to be outside of the retinal binding pocket at positions 65 and 70, probably are responsible for the fine-tuning of the absorption maxima (Man-Aharonovich *et al.*, 2004), proving the bioinformatic prediction of Bielawski *et al.* (2004) that spectral tuning is a combined effect of the amino acids that directly interact with retinal and other amino acids that change the conformation of the retinal binding pocket.

The ecological significance of proteorhodopsin phototrophy has been shown with culture-independent surveys that reveal a widespread distribution of proteorhodopsin-bearing bacteria in the oceans. They are found in the Antarctic and Pacific Ocean (Beja *et al.*, 2001; de la Torre *et al.*, 2003) as well as in the Mediterranean Sea and Red Sea (Man *et al.*, 2003; Sabehi *et al.*, 2003). Proteorhodopsins are not specific for Gammaproteobacteria and are widely distributed within the Proteobacteria (de la Torre *et al.*, 2003). Using an environmental shotgun sequencing approach Venter *et al.* (2004) identified more than 780 putative genes coding for proteins with homologies to proteorhodopsin in the Sargasso Sea, increasing the sequence diversity of rhodopsins. An analysis of the phylogenetic marker genes encoded on the same scaffolds as the rhodopsin genes demonstrated that rhodopsins are widespread in marine planktonic bacteria, including phyla other than the Proteobacteria.

However, the benefit of proteorhodopsins has yet to be convincingly demonstrated for the organisms that harbour these genes. While the original work on proteorhodopsin was performed with heterologously expressed proteorhodopsin genes in *Escherichia coli* (Beja *et al.*, 2000a; Beja *et al.*, 2001), more and more cultures of marine bacteria containing proteorhodopsin are being described. No visible effect of light on growth of the proteobacterium *Pelagibacter ubique* and HTCC2207 was observed (Giovannoni *et al.*, 2005; Stingl *et al.*, 2007b), but Gómez-Consarnau *et al.* (2007) reported a significant enhancement of both growth rate and yield in a marine *Bacteroidetes* species.

Another twist in the proterhodopsin story came when it was shown that the Group II mesophilic marine Euryarchaeota harbour this gene (Frigaard *et al.*, 2006). Other than the presence of these mesophilic Archaea in surface waters of the Ocean, little was known about their physiology. Most of the insight came from metagenomic analyses. One study identified a 60-kb bacterial artificial chromosome clone coding for a 5S and a 23S rRNA gene of the group II Euryarchaeota and 38 protein-encoding genes (Beja *et al.*, 2000b). The G+C content of the clone was 45% and the 23S rRNA gene was not linked to a 16S rRNA gene, similar to the closest cultivated phylogenetic neighbours in the order *Thermoplasmales*. Moreira *et al.* (2004) analysed a 39.5-kb clone from a deep-sea group II pelagic euryarchaeote that was isloated from a cosmid library constructed from waters sampled at 500 m depth. Interestingly, the clone exhibited a G+C content of 57%. Both studies identified mainly house-keeping genes but couldn't reveal any genes giving hints to specific metabolic properties. This changed with the recent identification of proterhodopsin genes encoding euryarchaeal clones in fosmid libraries (Frigaard *et al.*, 2006). The phylogenetic relationship of the newly discovered proteorhodopsin genes suggested a horizontal gene transfer between planktonic Bacteria and Archaea. 10% of the marine group II euryarchaeotes in the photic zone had the archaeal proteorhodopsin gene linked to a group II euryarchaeal 16S rRNA gene. Additionally, the archaeal-type proteorhodopsin gene was identified in other genomic contexts as well. The reported ratio between the observed marine group II euryarchaeal 16S rRNA genes and archaeal-type proteorhodopsin genes was roughly 1:1 in the photic zone hinting that most of the euryarchaeota present may encode this gene.

Ammonia oxidation by Archaea

The initial discovery of the Group I Marine Crenarchaea was rapidly followed by proof of their global distribution in the oceans, but evidence of their physiology and contribution to geochemistry have appeared slowly and remain incomplete. Some studies reported the Crenarchaeota taking up amino acids (Ouverney and Fuhrman, 2000; Herndl *et al.*, 2005), which would indicate heterotrophy. Other studies suggested that some pelagic Crenarchaeota may be autotrophs (Hoefs *et al.*, 1997; Pearson *et al.*, 2001) capable of light-independent bicarbonate fixation (Wuchter *et al.*, 2003; Herndl *et al.*, 2005). Evidence from a soil fosmid clone that linked a crenarchaeal rRNA operon to novel ammonia monooxygenase genes pointed out that these organisms might be involved in nitrification processes (Treusch *et al.*, 2005). These novel genes are widespread in marine and soil environments (Venter *et al.*, 2004; Francis *et al.*, 2005) and in soils it was shown that they are expressed (Leininger *et al.*, 2006). The cultivation of *Nitrosopumilus maritimus* finally proved that at least some of the mesophilic Crenarchaeota are nitrifiers (Konneke *et al.*, 2005). Additionally, *N. maritimus* grows in the absence of organic carbon, showing that it has a pathway to fix inorganic carbon (Konneke *et al.*, 2005). It was already suggested by metagenomic studies that marine Crenarchaeota fix carbon using a modified 3-hydroxypropionate cycle (Hallam *et al.*, 2006). A recent study showed a new carbon fixation pathway in the thermophilic archaeon *Metallosphaera sedula* by genome comparison and biochemical testing (Berg *et al.*, 2007). The enzymes for this 3-hydroxypropionate/4-hydroxybutyrate pathway are present in the genome of *Cenarchaeum symbiosum* and in the datasets of the Global Ocean Survey (GOS) studies (Rusch *et al.*, 2007; Yooseph *et al.*, 2007). If further evidence supports extrapolation of these findings to the abundant marine Crenarchaeota, then they are likely to have a significant impact on global cycles of nitrogen and carbon.

Aerobic anoxic photosynthesis

Aerobic anoxygenic phototrophic bacteria (AAP) (Yurkov and Beatty, 1998), that use bacteriochlorophyll to support phototrophic electron transport, were first reported to be important in the oceans in 2000 and are now believed to be widely distributed in seawater (Kolber *et al.*, 2000, 2001). They are reported to account for up to 11% of the microbial community in ocean surface waters (Kolber *et al.*, 2001).

Several AAP strains belonging to the alpha-3 and alpha-4 Proteobacteria were isolated from marine habitats long before the discovery that these organisms could be abundant and impact geochemical cycles in the oceans. Most of the strains isolated in early studies grow optimally in organic rich-media containing peptone and yeast extract (Allgaier *et al.*, 2003; Koblizek *et al.*, 2003). All these easily culturable marine AAP Alphaproteobacteria mainly belong to the genera *Roseobacter* and *Erythrobacter* (Yurkov and Beatty, 1998), and have been the subjects of physiological studies. Recently, it has been shown that their photosynthetic apparatus is functional (Kolber *et al.*, 2001), although they are not capable of strict photoautotrophic growth (Koblizek *et al.*, 2003). The genes required for the bacteriochlorophyll-based photosystem are clustered contiguously in a 45-kb chromosomal region (Alberti *et al.*, 1995), that also can be located on a linear plasmid (Pradella *et al.*, 2004). This superoperon consists of genes coding for the bacteriochlorophyll biosynthetic pathway (*bch*), genes for subunits of the light-harvesting complex and the reaction centre (*puf*), as well as genes coding for the carotenoid biosynthetic pathway (*crt*). Environmental genomic studies demonstrated this superoperon of AAP genes on genomic material cloned into BAC vectors that phylogenetically belonged to organisms of the Alpha, Beta and Gamma Proteobacteria (Beja *et al.*, 2002).

Using *puf*M, coding for one subunit of the reaction centre, and genes of the bacteriochlorophyll biosynthetic pathway as phylogenetic markers, it has been shown that *Roseobacter* and *Roseobacter*-like bacteria constitute a significant proportion of the aerobic anoxygenic phototroph community in the Pacific Ocean (Beja *et al.*, 2002) and the Red and Mediterranean Sea (Oz *et al.*, 2005). Recent results suggest a dominance of these groups in the Pacific Ocean as they account for up to 40% of the bacterial population in the Monterey Bay area (Suzuki *et al.*, 2001). They also constitute 26% of all rRNA genes identified in BAC libraries constructed from surface water samples of the same area (Suzuki *et al.*, 2004). This shows their significance and hints to their importance for phototrophy in the oceans.

After the metagenomic evidence already indicated that uncultured gammaproteobacterial AAPs are important components of bacterioplankton (Beja *et al.*, 2002), they were later identified within members of the OM60 clade (Cho *et al.*, 2007). The OM60 clade was first identified by phylogenetic analysis of environmental 16S rRNA clones from coastal seawater samples (Rappe *et al.*, 1997). The first isolate from this clade, KT71, was described in 2001 (Eilers *et al.*,

2001). More reports of isolated strains from this clade followed (Connon and Giovannoni, 2002; Brinkmeyer *et al.*, 2003; Cho and Giovannoni, 2004; Agogue *et al.*, 2005). Studies employing rRNA probes to detect the OM60 clade by fluorescence *in situ* hybridization in samples from the North Sea revealed cell abundances ranging from 6% to 11% of total bacterial counts. However, an interplay of new technologies revealed the OM60 clade strain HTCC2080 to be a cultured representative of the marine gammproteobacterial AAPs when proteomic data from this culture showed an exact match to a metagenomic fragment recovered from BAC clone 65D09 encoding the *puf* operon (Beja *et al.*, 2002). The proteomic study revealed the presence of the *puf* genes and also showed that they were actually expressed (Cho *et al.*, 2007) (see Fig. 8.3).

This example nicely demonstrated that a combination of culture-based and culture-independent technologies can answer questions of ecologic relevance that the individual methods failed to reveal.

Metagenomic undersampling

Rapidly dropping sequencing costs are making metagenomic data widely available, but most metagenomic datasets face the problem of undersampling because of the extraordinarily high diversity of DNA in most natural ecosystems. Handelsman *et al.* (2002) calculated, that for a soil sample that contained 4000 different genomes of the size of *E. coli* it would be necessary to obtain 250,000 clones with 80-kb inserts for single fold coverage. For adequate representation of the community and its diversity 1–10 million clones would be necessary. Along similar lines, Venter *et al.* (2004) calculated that their study of the Sargasso Sea surface water, with roughly 1800 species by their estimates, would have needed 12-fold deeper sampling to obtain 95% of the unique sequences. However, this first marine

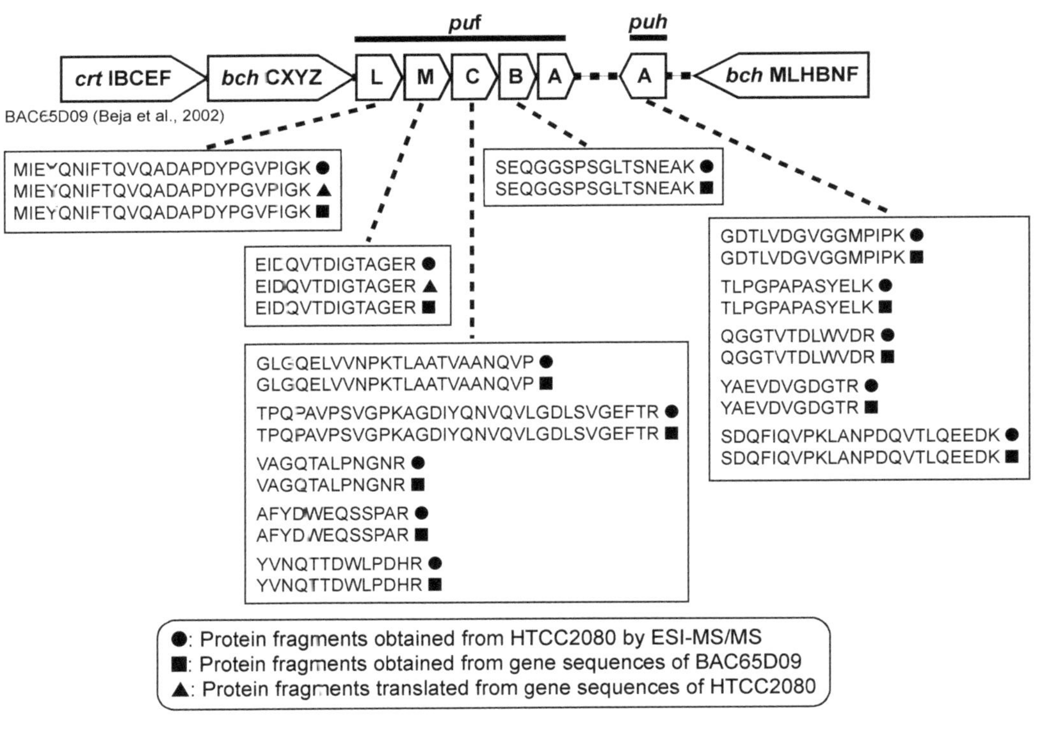

Figure 8.3 Protein profile of the photosynthetic reaction center of HTCC2080. The figure shows the genes in the order they were discovered on a metagenomic BAC clone (Beja *et al.* 2002). Later, these genes were identified in the new-cultivated strain HTCC2080 and a proteomic analysis showed that the genes are actually expressed (Cho *et al.*, 2007).

dataset had a higher coverage than all the following studies of the Global Ocean Survey (GOS) (Rusch *et al.*, 2007; Yooseph *et al.*, 2007), making comparisons between these samples problematic. New approaches to bioinformatics and statistics are being undertaken to overcome these limitations (e.g. Huson *et al.*, 2007; McHardy *et al.*, 2007), but higher coverage would be the best solution.

Another problem that confuses direct comparisons between different metagenomic samples is natural variation in oceanographic conditions. For example when samples are collected from different sites in different seasons, seasonal and geographical trends in the dataset may be impossible to distinguish. Rusch *et al.* (2007) observed a correlation between the similarity of sequences and the temperature and salinity of the marine environment the samples had been taken from, suggesting that much geographical variation might originate either from seasonal cycles in community composition or from latitudinal variation.

Using metagenomic approaches to profile microbial communities over time and space has recently been discussed, but seems to have, from a current point of view, major drawbacks. First, there is the massive undersampling with this approach that can arguably be solved by using the 454 sequencing technology (Margulies *et al.*, 2005). Second, the costs for this technology are, although decreasing, still high and make other approaches more economic. We did a comparison of terminal restriction length polymorphism (T-RFLP), a widely used community profiling technique (e.g. Morris *et al.*, 2005), a 454 metagenomic approach and a 454 sequencing of the variable region 3 of the 16S rRNA gene approach (Table 8.1) (see Chapter 4 for a description of these methods). When comparing the costs to the potential results it becomes clear that a straightforward metagenomic approach might not be the best solution if you are interested in profiling the community for many samples. Classic T-RFLP analyses or modern approaches like the 454-based analysis of the V3 region give a higher resolution with significantly lower costs. Depending on the questions asked and the experimental design each method has its strengths and weaknesses that need to be considered.

Culturing rediscovered

After it became obvious that many marine microorganisms were overlooked by classic culturing this problem was addressed with new approaches and by the refinement of methods that had been traditional mainstays of microbial cultivation. The biggest problem with conventional cultivation techniques is their selectivity and bias towards fast growing organisms (Ferguson *et al.*, 1984; Eilers *et al.*, 2000). All of the new approaches use low nutrient media and in parallel try to separate the cells of different microorganisms before inoculation to avoid competition between fast and slow growing organisms. Additionally, some approaches try to simulate the physical and chemical conditions in the natural environment.

Button and colleagues (1993) published an approach in which they used unamended sterilized seawater to grow the cultures. To isolate oligotrophic marine microorganisms they diluted natural samples so that tubes of sterile medium were inoculated with an average number of cells that ranged from 0.1 to 1000. They reported a range of 2–60% of culture tubes with growth for inocula from marine waters from Alaska and the Netherlands, as determined by flow cytometry. With this approach they isolated and described *Sphingomonas alaskensis* and *Cycloclasticus oligotrophus* (Schut *et al.*, 1993; Wang *et al.*, 1996; Button *et al.*, 1998; Vancanneyt *et al.*, 2001). The stochastic nature of dilution-to-extinction meant that it could not be exploited without dramatically increasing the number of cultures that could be obtained and identified. With this motivation, high-throughput versions of this technology were developed by Connon and Giovannoni (2002). Their approaches also made use of new media that were prepared using procedures designed to minimize contamination from metals and detergents. The high-throughput cultivation (HTC) method gave 14- to 1400-fold higher cultivation efficiencies from coastal seawater in comparison with classic culturing techniques. Among the resulting cultures were species related to previous uncultured lineages like SAR11, SAR92, OM43 and OM60/OM241. The technology provided a breakthrough with the cultivation of '*Candidatus* Pelagibacter ubique', a member of the ubiquitous marine bacterioplankton clade SAR11 (Rappe *et al.*, 2002). In the following years HTC technology

Table 8.1 Comparison of characteristics of three molecular methods used to assess microbial community composition

	T-RFLP	454 V3 region[1]	454 Metagenomics
DNA collection/isolation/preparation	Identical	Identical	Identical
PCR	\$130[2]	\$40[3]	N/A
Restriction digest	\$1	N/A	N/A
Library preparation	N/A	N/A	\$1,600
Assay (per sample)	\$1	\$750[4]	\$12,000[5]
Computing resources	Desktop (50Mb)	Server (5Gb)	Server (80Gb)
Time for analysis	Hours	Days	Weeks
Cost for 16 samples	\$162	\$12,040	\$217,600
Population abundance required for statistically significant coverage	*ca* 1% of population[6]	*ca* 0.1% of population[7]	*ca* 10% of population[8]

1 Variable region 3 of the 16S rRNA gene.

2 \$10 for a 20mer primer + \$120 for a 20mer primer with a fluorescent tag.

3 \$20 each for two 40mer primers, including 454 specific priming sites.

4 Assuming 16 sectors per chip.

5 Assuming one chip for each sample.

6 Peaks tend to correlate with phylogenetic groups but actual species identities and polymorphisms are unknown; major clades detected.

7 Most species identified including low abundance species.

8 Based on recent results; one environmental sample on half of a chip yielded 25 16S rRNA sequences in the V3 region.

proved to be very successful and was an essential tool to culture numerous new strains (e.g. Cho and Giovannoni, 2003a–c, 2004; Cho *et al.*, 2004).

Another way to get improved culturability is to simulate the natural environment in more detail than it can be done in Petri dishes or liquid cultures. The presumption behind this approach is that nearly every microbial species should grow in culture if all chemical components of the natural environment are present. This includes metabolites exchanged by different species and signalling molecules. Kaeberlein *et al.* (2002) got pure cultures of new marine species using diffusion growth chambers. They embedded dilutions of extracted cells from a marine sandy sediment with agar in the chambers and sealed them with 0.03 μm pore-size membranes. Incubation was done in an aquarium filled with seawater on a sample from the same marine sediment. Also trying to allow the exchange of metabolites and signalling molecules between different cells without the risk of overgrowth, Zengler *et al.* (2002) developed a method based on the encapsulation of single cells into gel microdroplets (GMDs). By culturing those encapsulated cells together in growth columns with continuous flow of fresh media, the natural environment was simulated. Flow sorting of GMDs in which microcolonies had been grown resulted in the isolation of novel strains from the phyla Planctomycetes, Cytophaga–Flavobacterium–Bacteroides and Alphaproteobacteria. Even more remarkable was the fact that only 47% of the isolated cultures fell into the nine common bacterioplankton clades, in contrast to 99% in a 16S rRNA gene clone library prepared from the same sample before encapsulation. This showed that less abundant members of the bacterioplankton can be isolated without being outcompeted.

These novel approaches showed that it is possible to get representative isolates for previously uncultured lineages of microorganisms that were only identified by molecular methods. However, it is noteworthy that no single technique and medium will be suitable for all microorganisms in one sample as it was demonstrated by Kopke *et al.* (2005). Targeted cultivation to suit the organisms of interest or massive parallel approaches with different culturing techniques and media to cover the full diversity of one sample will most likely be the future in cultivation approaches.

Future trends and web resources

When one looks to the future of marine microbiology it is apparent that the answers to many questions will be found in the details, such as complex co-evolved interactions between organisms, unforeseen forms of metabolism revealed by genomics and metagenomics, and spatial and temporal patterns that emerge from high-resolution ecological studies. New advances will come from old and new methods, but also from new cultured isolates that have a well-established ecological significance and stimulate post genomic investigations in a controlled laboratory setting.

The classic molecular approach of sequencing amplified and cloned 16S rRNA genes will continue to be a major contributor to determine diversity in any given sample. Extensive studies are now focusing more on analysing hundreds of clones and comparing samples from different sampling locations (e.g. Pommier *et al.*, 2007). New technologies like phylochips (e.g. Wagner *et al.*, 2007) (see Chapter 5) and 454 pyrosequencing may enable high resolution analyses of even more samples. As problems with these technologies are solved and prices drop they will most likely replace classic community profiling methods like T-RFLP and amplified ribosomal intergenic spacer analysis (ARISA) (e.g. Brown *et al.*, 2005). Time-series studies, depth profiles and transects with high-density sampling may become commonplace in molecular marine microbiology, matching in their detail chemical and physical oceanographic studies.

The emergence of new sequencing technologies and the large metagenomic sequence datasets they provide will have a dramatic effect. Already, as a consequence of the new sequencing technologies, bioinformatic data processing has become the bottleneck in analysing the sequence data. The new technologies also involve some challenges for the processing of data, because of the short reads, although the 454 pyrosequencing read lengths have steadily increased with improvements in the technology. As we demonstrated above, even with the massively parallel sequencing technologies that have recently become available, it remains impractical to implement metagenomic approaches with experimental designs that include many samples. In the short term, metagenomics will provide statistically robust data when it is applied to specific questions about numerically dominant taxa, particularly when the experimental design demands only a small number of samples. Some of the sampling problems implicit in metagenomics may be overcome by the use of enrichment cultures, physical cell separation by flow sorting and single cell genomics to provide samples with reduced diversity.

Culturing will continue to be important, particularly as high-throughput approaches that are suitable for working with many variables are used to capture more diversity. Likewise, targeted cultivation approaches that begin by inferring an uncultured organism's physiological properties from environmental, phylogenetic and metagenomic data are also likely to produce new cultures. The cultivation of strains closely related to previous isolates, as has been done with *Prochlorococcus/Synechococcus* (Ahlgren and Rocap, 2006) and SAR11 (Stingl *et al.*, 2007a), will enable microbial oceanographers to identify ecotypes and the physiological adaptations that enable them to specialize.

An important new area of research in molecular marine microbiology is focused on understanding microbial responses to their environment. Metatranscriptomics and metaproteomics are two approaches that potentially can reveal the status of the regulatory networks that cells use to adapt to changing environmental conditions (e.g. Giovannoni *et al.*, 2005) (see Chapter 3 for a discussion of metaproteomics). The nice thing about these new methods is that, in contrast to genomics and metagenomics, they show what features of an organisms metabolic potential are expressed in the environment. For example, Short and Zehr showed that only few of the types of the nitrogenase gene *nif*H that were detected in environmental DNA were actually expressed in the Chesapeake Bay estuary (Short and Zehr, 2007).

In summary, marine microbial ecologists have been rewarded with very rapid progress in recent years because of the availability of new technologies, but many exciting challenges remain. Some of these challenges result from the uncertain affects of climate change and the need to produce accurate predictions from complex systems in which microorganisms play a catalytic role. It seems certain that these challenges and continued technological progress will make the years ahead exciting for those fascinated with the microbiology of ocean ecosystems.

Websites of interest

Cultures and culturing

http://www.dsmz.de/ (German type culture collection. Online search for strains and growth media of deposited strains).

http://www.atcc.org/ (American type culture collection. Online search for strains and growth media of deposited strains).

Microbial observatories and oceanographic data

http://bats.bios.edu/index.html (homepage of the Bermuda Atlantic Time-series study site).

http://hahana.soest.hawaii.edu/hot/hot_jgofs.html (homepage of the Hawaii Ocean Time-series study site).

http://www.noaa.gov/ (homepage of the National Oceanic and Atmospheric Administration. Satellite and oceanographic data available for download: http://www.ngdc.noaa.gov/dmsp/download.html).

Microbial diversity

http://www.arb-home.de/ (software to reconstruct phylogenetic relationships between RNA, DNA and protein sequences. Mainly used for 16S and 23S rRNA phylogenies).

http://www.arb-silva.de/ (comprehensive 16S and 23S rRNA databases that are updated on a regular basis).

http://greengenes.lbl.gov/cgi-bin/nph-index.cgi (homepage dedicated to the analysis of 16S rRNA sequences).

http://egg.umh.es/micromar/index.php (a database of DNA sequences from marine microorganisms accompanied by the locations they were sampled from and oceanographic data).

Microbial genomics and metagenomics

http://www.moore.org/microgenome/ (an initiative to sequence the genomes of 155 marine microorganisms funded by the Gordon and Betty Moore foundation).

http://camera.calit2.net/ (homepage hosting the metagenomic datasets from the Global Ocean Survey and bioinformatic tools to work with it).

References

Acinas, S.G., Klepac-Ceraj, V., Hunt, D.E., Pharino, C., Ceraj, I., Distel, D.L., and Polz, M.F. (2004). Fine-scale phylogenetic architecture of a complex bacterial community. Nature *430*, 551–554.

Agogue, H., Casamayor, E.O., Bourrain, M., Obernosterer, I., Joux, F., Herndl, G.J., and Lebaron, P. (2005). A survey on bacteria inhabiting the sea surface microlayer of coastal ecosystems. FEMS Microbiol. Ecol. *54*, 269–280.

Ahlgren, N.A., and Rocap, G. (2006). Culture isolation and culture-independent clone libraries reveal new marine Synechococcus ecotypes with distinctive light and N physiologies. Appl. Environ. Microbiol. *72*, 7193–7204.

Alberti, M., Burke, D.H., and Hearst, J.E. (1995). Structure and sequence of the photosynthesis gene cluster. In Anoxygenic Photosynthetic Bacteria, C.E. Bauer, ed. (Dordrecht: Kluwer Academic Publishers), pp. 1083–1106.

Allgaier, M., Uphoff, H., Felske, A., and Wagner-Dobler, I. (2003). Aerobic anoxygenic photosynthesis in Roseobacter clade bacteria from diverse marine habitats. Appl. Environ. Microbiol. *69*, 5051–5059.

Arrigo, K.R. (2005). Marine microorganisms and global nutrient cycles. Nature *437*, 349–355.

Azam, F., Fenchel, T., Field, J.G., Gray, J.S., Meyerreil, L.A., and Thingstad, F. (1983). The ecological role of water-column microbes in the sea. Marine Ecology-Progress Series *10*, 257–263.

Béjà, O., Aravind, L., Koonin, E.V., Suzuki, M.T., Hadd, A., Nguyen, L.P., Jovanovich, S.B., Gates, C.M., Feldman, R.A., Spudich, J.L., Spudich, E.N., and DeLong, E.F. (2000a). Bacterial rhodopsin: evidence for a new type of phototrophy in the sea. Science *289*, 1902–1906.

Béjà, O., Suzuki, M.T., Koonin, E.V., Aravind, L., Hadd, A., Nguyen, L.P., Villacorta, R., Amjadi, M., Garrigues, C., Jovanovich, S.B., Feldman, R.A., and DeLong, E.F. (2000b). Construction and analysis of bacterial artificial chromosome libraries from a marine microbial assemblage. Environ. Microbiol. *2*, 516–529.

Béjà, O., Spudich, E.N., Spudich, J.L., Leclerc, M., and DeLong, E.F. (2001). Proteorhodopsin phototrophy in the ocean. Nature *411*, 786–789.

Béjà, O., Suzuki, M.T., Heidelberg, J.F., Nelson, W.C., Preston, C.M., Hamada, T., Eisen, J.A., Fraser, C.M., and DeLong, E.F. (2002). Unsuspected diversity among marine aerobic anoxygenic phototrophs. Nature *415*, 630–633.

Berg, I.A., Kockelkorn, D., Buckel, W., and Fuchs, G. (2007). A 3-hydroxypropionate/4-hydroxybutyrate autotrophic carbon dioxide assimilation pathway in Archaea. Science *318*, 1782–1786.

Bielawski, J.P., Dunn, K.A., Sabehi, G., and Beja, O. (2004). Darwinian adaptation of proteorhodopsin to different light intensities in the marine environment. Proc. Natl. Acad. Sci. U.S.A. *101*, 14824–14829.

Brinkmeyer, R., Knittel, K., Jurgens, J., Weyland, H., Amann, R., and Helmke, E. (2003). Diversity and structure of bacterial communities in Arctic versus Antarctic pack ice. Appl. Environ. Microbiol. *69*, 6610–6619.

Brown, M.V., Schwalbach, M.S., Hewson, I., and Fuhrman, J.A. (2005). Coupling 16S-ITS rDNA clone libraries and automated ribosomal intergenic spacer analysis to show marine microbial diversity: development and application to a time series. Environ. Microbiol. *7*, 1466–1479.

Button, D.K. (1993). Nutrient-limited microbial growth kinetics: overview and recent advances. Antonie Van Leeuwenhoek *63*, 225–235.

Button, D.K., Robertson, B.R., Lepp, P.W., and Schmidt, T.M. (1998). A small, dilute-cytoplasm, high-affinity, novel bacterium isolated by extinction culture and having kinetic constants compatible with growth at ambient concentrations of dissolved nutrients in seawater. Appl. Environ. Microbiol. 64,: 4467–4476.

Certes, A. (1884). Sur la culture, à l'abri des germes atmosphériques, des eaux et des sédiments rapportés par les expéditions du Travailleur et du Talisman; 1882–1883. Compt. Rend. Acad. Sci. *98*, 690–693.

Cho, J.C., and Giovannoni, S.J. (2003a). Fulvimarina pelagi gen. nov., sp. nov., a marine bacterium that forms a deep evolutionary lineage of descent in the order 'Rhizobiales'. Int. J. Syst. Evol. Microbiol. *53*, 1853–1859.

Cho, J.C., and Giovannoni, S.J. (2003b). Parvularcula bermudensis gen. nov., sp. nov., a marine bacterium that forms a deep branch in the alpha-Proteobacteria. Int. J. Syst. Evol. Microbiol. 53, 1031–1036.

Cho, J.C., and Giovannoni, S.J. (2003c). *Croceibacter atlanticus* gen. nov., sp. nov., a novel marine bacterium in the family Flavobacteriaceae. Syst. Appl. Microbiol. *26*, 76–83.

Cho, J.C., and Giovannoni, S.J. (2004). Cultivation and growth characteristics of a diverse group of oligotrophic marine Gammaproteobacteria. Appl. Environ. Microbiol. *70*, 432–440.

Cho, J.C., Vergin, K.L., Morris, R.M., and Giovannoni, S.J. (2004). Lentisphaera araneosa gen. nov., sp. nov, a transparent exopolymer producing marine bacterium, and the description of a novel bacterial phylum, Lentisphaerae. Environ. Microbiol. *6*, 611–621.

Cho, J.C., Stapels, M.D., Morris, R.M., Vergin, K.L., Schwalbach, M.S., Givan, S.A., Barofsky, D.F., and Giovannoni, S.J. (2007). Polyphyletic photosynthetic reaction centre genes in oligotrophic marine Gammaproteobacteria. Environ. Microbiol. *9*, 1456–1463.

Cohn, F. (1865). Zwei neue Beggiatoen. Hedwigia *4*, 81–84.

Connon, S.A., and Giovannoni, S.J. (2002). High-throughput methods for culturing microorganisms in very-low-nutrient media yield diverse new marine isolates. Appl. Environ. Microbiol. *68*, 3878–3885.

de la Torre, J.R., Christianson, L.M., Beja, O., Suzuki, M.T., Karl, D.M., Heidelberg, J., and DeLong, E.F. (2003). Proteorhodopsin genes are distributed among divergent marine bacterial taxa. Proc. Natl. Acad. Sci. U.S.A. *100*, 12830–12835.

DeLong, E.F. (1992). Archaea in coastal marine environments. Proc. Natl. Acad. Sci. U.S.A. *89*, 5685–5689.

DeLong, E.F., Preston, C.M., Mincer T, Rich V, Hallam, S.J., Frigaard, N.U., Martinez A, Sullivan, M.B., Edwards R, Brito, B.R., Chisholm, S.W., and Karl, D.M. (2006). Community genomics among stratified microbial assemblages in the ocean's interior. Science *311*, 496–503.

DuRand, M.D., Olson, R.J., and Chisholm, S.W. (2001). Phytoplankton population dynamics at the Bermuda Atlantic Time-series station in the Sargasso Sea. Deep-Sea Research Part II – Topical Studies in Oceanography *48*, 1983–2003.

Ehrenberg, C.G. (1838). Die Infusionsthierchen als vollkommende Organismen. Ein Blick in das tiefere organische Leben der Natur. Folio, Leipzig, 547 pp.

Eilers, H., Pernthaler, J., and Amann, R. (2000). Succession of pelagic marine bacteria during enrichment: a close look at cultivation-induced shifts. Appl. Environ. Microbiol. *66*, 4634–4640.

Eilers, H., Pernthaler, J., Peplies, J., Glockner, F.O., Gerdts, G., and Amann, R. (2001). Isolation of novel pelagic bacteria from the German bight and their seasonal contributions to surface picoplankton. Appl. Environ. Microbiol. *67*, 5134–5142.

Ferguson, R.L., Buckley, E.N., and Palumbo, A.V. (1984). Response of marine bacterioplankton to differential filtration and confinement. Appl. Environ. Microbiol. *47*, 49–55.

Field, K.G., Gordon, D., Wright, T., Rappe, M., Urback, E., Vergin, K., and Giovannoni, S.J. (1997). Diversity and depth-specific distribution of SAR11 cluster rRNA genes from marine planktonic bacteria. Appl. Environ. Microbiol. *63*, 63–70.

Fischer, B. (1888). Berichte über Bakterienwachsthum bei 0° C sowie über das Photographiren von Kulturen leuchtender Bakterien in ihrem eigenen Lichte. Centralbl. f. Bakt. *4*, 89–92.

Fischer, B. (1893). Ueber das Grundwasser von Kiel mit besonderer Berücksichtigung seines Eisengehaltes und über Versuche zur Entfernung des Eisens aus demselben. Zeitschr. f. Hyg. *13*, 251–335.

Fischer, B. (1894). Die Bakterien des Meeres nach den Untersuchungen der Plankton-Expedition unter gleichzeitiger Berücksichtigung einiger älterer und neuerer Untersuchungen. Ergebnisse der Plankton-Expedition der Humboldt-Stiftung *4*, 1–83.

Francis, C.A., Roberts, K.J., Beman, J.M., Santoro, A.E., and Oakley, B.B. (2005). Ubiquity and diversity of ammonia-oxidizing archaea in water columns and sediments of the ocean. Proc. Natl. Acad. Sci. U.S.A. *102*, 14683–14688.

Frigaard, N.U., Martinez, A., Mincer, T.J., and DeLong, E.F. (2006). Proteorhodopsin lateral gene transfer between marine planktonic Bacteria and Archaea. Nature *439*, 847–850.

Fuhrman, J.A., McCallum, K., and Davis, A.A. (1992). Novel major archaebacterial group from marine plankton. Nature *356*, 148–149.

Fuhrman, J.A., McCallum, K., and Davis, A.A. (1993). Phylogenetic diversity of subsurface marine microbial communities from the Atlantic and Pacific Oceans. Appl. Environ. Microbiol. *59*, 1294–1302.

Fuhrman, J.A., McCallum, K., and Davis, A.A. (1997). Widespread Archaea and novel bacteria from the deep sea as shown by 16S rRNA gene sequences. Mar. Ecol. Prog. Ser. *150*, 275–285.

Fuhrman, J.A., Hewson, I., Schwalbach, M.S., Steele, J.A., Brown, M.V., and Naeem, S. (2006). Annually reoccurring bacterial communities are predictable from ocean conditions. Proc. Natl. Acad. Sci. U.S.A. *103*, 13104–13109.

Giovannoni, S., and Stingl, U. (2007). The importance of culturing bacterioplankton in the 'omics' age. Nat Rev Microbiol *5*, 820–826.

Giovannoni, S.J., Britschgi, T.B., Moyer, C.L., and Field, K.G. (1990). Genetic diversity in Sargasso Sea bacterioplankton. Nature *345*, 60–63.

Giovannoni, S.J., Rappe, M.S., Vergin, K.L., and Adair, N.L. (1996). 16S rRNA genes reveal stratified open ocean bacterioplankton populations related to the Green Non-Sulfur bacteria. Proc. Natl. Acad. Sci. U.S.A. *93*, 7979–7984.

Giovannoni, S.J., Bibbs, L., Cho, J.C., Stapels, M.D., Desiderio, R., Vergin, K.L., Rappé, M.S., Laney, S., Wilhelm, L.J., Tripp, H.J., Mathur, E.J., and Barofsky, D.F. (2005). Proteorhodopsin in the ubiquitous marine bacterium SAR11. Nature *438*, 82–85.

Gomez-Consarnau, L., Gonzalez, J.M., Coll-Llado, M., Gourdon, P., Pascher, T., Neutze, R., *et al.* (2007). Light stimulates growth of proteorhodopsin-containing marine Flavobacteria. Nature *445*, 210–213.

Gordon, D.A., and Giovannoni, S.J. (1996). Detection of stratified microbial populations related to Chlorobium and Fibrobacter species in the Atlantic and Pacific oceans. Appl. Environ. Microbiol. *62*, 1171–1177.

Hallam, S.J., Mincer, T.J., Schleper, C., Preston, C.M., Roberts, K., Richardson, P.M., and DeLong, E.F. (2006). Pathways of carbon assimilation and ammonia oxidation suggested by environmental genomic analyses of marine Crenarchaeota. PLoS Biol *4*, e95.

Handelsman, J., Liles, M., Mann, D., Riesenfeld, C., and Goodman, R.M. (2002). Cloning the metagenome: culture-independent access to the diversity and functions of the uncultivated microbial world. In Methods in Microbiology-Functional Microbial Genomics, Wren, B., and Dorell, N., eds (Academic Press), pp. 241–255.

Herndl, G.J., Reinthaler, T., Teira, E., van Aken, H., Veth, C., Pernthaler, A., and Pernthaler, J. (2005). Contribution of Archaea to total prokaryotic production in the deep Atlantic Ocean. Appl. Environ. Microbiol. *71*, 2303–2309.

Hoefs, M., Schouten, S., De Leeuw, J.W., King, L.L., Wakeham, S.G., and Damste, J. (1997). Ether Lipids of Planktonic Archaea in the Marine Water Column. Appl. Environ. Microbiol. *63*, 3090–3095.

Hurrell, J.W. (1995). Decadal trends in the North Atlantic Oscillation: regional temperatures and precipitation. Science *269*, 676–679.

Huson, D.H., Auch, A.F., Qi, J., and Schuster, S.C. (2007). MEGAN analysis of metagenomic data. Genome Res. *17*, 377–386.

Ingalls, A.E., Shah, S.R., Hansman, R.L., Aluwihare, L.I., Santos, G.M., Druffel, E.R., and Pearson, A. (2006). Quantifying archaeal community autotrophy in the mesopelagic ocean using natural radiocarbon. Proc. Natl. Acad. Sci. U.S.A. *103*, 6442–6447.

Issatschenko, B.L. (1914). Investigations on the bacteria of the glacial Arctic Ocean. Petrograd: Monograph.

Jurgens, G., Lindstrom, K., and Saano, A. (1997). Novel group within the kingdom Crenarchaeota from boreal forest soil. Appl. Environ. Microbiol. *63*, 803–805.

Kaeberlein, T., Lewis, K., and Epstein, S.S. (2002). Isolating 'uncultivable' microorganisms in pure culture in a simulated natural environment. Science *296*, 1127–1129.

Karner, M.B., DeLong, E.F., and Karl, D.M. (2001). Archaeal dominance in the mesopelagic zone of the Pacific Ocean. Nature *409*, 507–510.

Koblizek, M., Beja, O., Bidigare, R.R., Christensen, S., Benitez-Nelson, B., Vetriani, C., *et al.* (2003). Isolation and characterization of Erythrobacter sp. strains from the upper ocean. Arch. Microbiol. *180*, 327–338.

Kolber, Z.S., Van Dover, C.L., Niederman, R.A., and Falkowski, P.G. (2000). Bacterial photosynthesis in surface waters of the open ocean. Nature *407*, 177–179.

Kolber, Z.S., Plumley, F.G., Lang, A.S., Beatty, J.T., Blankenship, R.E., VanDover, C.L., Vetriani, C., Koblizek, M., Rathgeber, C., and Falkowski, P.G. (2001). Contribution of aerobic photoheterotrophic bacteria to the carbon cycle in the ocean. Science *292*, 2492–2495.

Konneke, M., Bernhard, A.E., de la Torre, J.R., Walker, C.B., Waterbury, J.B., and Stahl, D.A. (2005). Isolation of an autotrophic ammonia-oxidizing marine archaeon. Nature *437*, 543–546.

Kopke, B., Wilms, R., Engelen, B., Cypionka, H., and Sass, H. (2005). Microbial diversity in coastal subsurface sediments: a cultivation approach using various electron acceptors and substrate gradients. Appl. Environ. Microbiol. *71*, 7819–7830.

Leininger, S., Urich, T., Schloter, M., Schwark, L., Qi, J., Nicol, G.W., Prosser, J.I., Schuster, S.C., and Schleper, C. (2006). Archaea predominate among ammonia-oxidizing prokaryotes in soils. Nature *442*, 806–809.

Man, D., Wang, W., Sabehi, G., Aravind, L., Post, A.F., Massana, R., Spudich, EN., Spudich, J.L., and Béjà, O. (2003). Diversification and spectral tuning in marine proteorhodopsins. EMBO J. *22*, 1725–1731.

Man-Aharonovich, D., Sabehi, G., Sineshchekov, O.A., Spudich, E.N., Spudich, J.L., and Beja, O. (2004). Characterization of RS29, a blue-green proteorhodopsin variant from the Red Sea. Photochem. Photobiol. Sci. *3*, 459–462.

Margulies, M., Egholm, M., Altman, W.E., Attiya, S., Bader, J.S., Bemben, L.A., Berka, J., Braverman, M.S., Chen, Y.J., Chen, Z., *et al.* (2005). Genome sequencing in microfabricated high-density picolitre reactors. Nature *437*, 376–380.

Massana, R., Murray, A.E., Preston, C.M., and DeLong, E.F. (1997). Vertical distribution and phylogenetic characterization of marine planktonic Archaea in the Santa Barbara Channel. Appl. Environ. Microbiol. *63*, 50–56.

Massana, R., DeLong, E.F., and Pedros-Alio, C. (2000). A few cosmopolitan phylotypes dominate planktonic archaeal assemblages in widely different oceanic provinces. Appl. Environ. Microbiol. *66*, 1777–1787.

McHardy, A.C., Martin, H.G., Tsirigos, A., Hugenholtz, P., and Rigoutsos, I. (2007). Accurate phylogenetic classification of variable-length DNA fragments. Nat. Methods *4*, 63–72.

Moreira, D., Rodriguez-Valera, F., and Lopez-Garcia, P. (2004). Analysis of a genome fragment of a deep-sea uncultivated Group II euryarchaeote containing 16S rDNA, a spectinomycin-like operon and several energy metabolism genes. Environ. Microbiol. *6*, 959–969.

Morris, R.M., Rappe, M.S., Connon, S.A., Vergin, K.L., Siebold, W.A., Carlson, C.A., and Giovannoni, S.J. (2002). SAR11 clade dominates ocean surface bacterioplankton communities. Nature *420*, 806–810.

Morris, R.M., Rappe, M.S., Urbach, E., Connon, S.A., and Giovannoni, S.J. (2004). Prevalence of the Chloroflexi-related SAR202 bacterioplankton cluster throughout the mesopelagic zone and deep ocean. Appl. Environ. Microbiol. *70*, 2836–2842.

Morris, R.M., Vergin, K.L., Cho, J.C., Rappe, M.S., Carlson, C.A., and Giovannoni, S.J. (2005). Temporal and spatial response of bacterioplankton lineages to annual convective overturn at the Bermuda Atlantic Time-series Study site. Limnology and Oceanography *50*, 1687–1696.

Morris, R.M., Longnecker, K., and Giovannoni, S.J. (2006). Pirellula and OM43 are among the dominant lineages identified in an Oregon coast diatom bloom. Environ. Microbiol. *8*, 1361–1370.

Mou, X., Hodson, R.E., and Moran, M.A. (2007). Bacterioplankton assemblages transforming dissolved organic compounds in coastal seawater. Environ. Microbiol. *9*, 2025–2037.

Neufeld, J.D., Schafer, H., Cox, M.J., Boden, R., McDonald, I.R., and Murrell, J.C. (2007). Stable-isotope probing implicates Methylophaga spp and novel Gammaproteobacteria in marine methanol and methylamine metabolism. ISME J. *1*, 480–491.

Ochsenreiter, T., Selezi, D., Quaiser, A., Bonch-Osmolovskaya, L., and Schleper, C. (2003). Diversity and abundance of Crenarchaeota in terrestrial habitats studied by 16S RNA surveys and real time PCR. Environ. Microbiol. *5*, 787–797.

Oesterhelt, D., and Stoeckenius, W. (1971). Rhodopsin-like protein from the purple membrane of *Halobacterium halobium*. Nat. New Biol. *233*, 149–152.

Ouverney, C.C., and Fuhrman, J.A. (2000). Marine planktonic archaea take up amino acids. Appl. Environ. Microbiol. *66*, 4829–4833.

Oz, A., Sabehi, G., Koblizek, M., Massana, R., and Beja, O. (2005). Roseobacter-like bacteria in red and mediterranean sea aerobic anoxygenic photosynthetic populations. Appl. Environ. Microbiol. 71,: 344–353.

Pearson, A., McNichol, A.P., Benitez-Nelson, B.C., Hayes, J.M., and Eglinton, T.I. (2001). Origins of lipid biomarkers in Santa Monica Basin surface sediment: A case study using compound-specific Delta C-14 analysis. Geochim. Cosmochim. Acta *65*, 3123–3137.

Pernthaler, A., Preston, C.M., Pernthaler, J., DeLong, E.F., and Amann, R. (2002). Comparison of fluorescently labeled oligonucleotide and polynucleotide probes for the detection of pelagic marine bacteria and archaea. Appl. Environ. Microbiol. *68*, 661–667.

Pomeroy, L.R. (1974). Oceans Food Web, a Changing Paradigm. Bioscience *24*, 499–504.

Pommier, T., Canbäck, B., Riemann, L., Boström, K.H., Simu, K., Lundberg, P., and Tunlid, A. (2007). Global patterns of diversity and community structure in marine bacterioplankton. Mol. Ecol. *16*, 867–880.

Pradella, S., Allgaier, M., Hoch, C., Pauker, O., Stackebrandt, E., and Wagner-Dobler, I. (2004). Genome organization and localization of the pufLM genes of the photosynthesis reaction center in phylogenetically diverse marine Alphaproteobacteria. Appl. Environ. Microbiol. *70*, 3360–3369.

Preston, C.M., Wu, K.Y., Molinski, T.F., and DeLong, E.F. (1996). A psychrophilic crenarchaeon inhabits a marine sponge: Cenarchaeum symbiosum gen. nov., sp. nov. Proc. Natl. Acad. Sci. U.S.A. *93*, 6241–6246.

Rappe, M.S., Kemp, P.F., and Giovannoni, S.J. (1997). Phylogenetic diversity of marine coastal picoplankton 16S rRNA genes cloned from the continental shelf off Cape Hatteras, North Carolina. Limnology and Oceanography *42*, 811–826.

Rappe, M.S., Vergin, K., and Giovannoni, S.J. (2000). Phylogenetic comparisons of a coastal bacterioplankton community with its counterparts in open ocean and freshwater systems. FEMS Microbiol. Eco.l 33, 219–232.

Rappe, M.S., Connon, S.A., Vergin, K.L., and Giovannoni, S.J. (2002). Cultivation of the ubiquitous SAR11 marine bacterioplankton clade. Nature 418,630–633.

Raymont, J.E.G. (1980). Plankton and Productivity in the Oceans, Vol. 1, Phytoplankton, 2nd edn (Oxford: Pergamon Press).

Rocap, G., Distel, D.L., Waterbury, J.B., and Chisholm, S.W. (2002). Resolution of Prochlorococcus and Synechococcus ecotypes by using 16S-23S ribosomal DNA internal transcribed spacer sequences. Appl. Environ. Microbiol. *68*, 1180–1191.

Rusch, D.B., Halpern, A.L., Sutton, G., Heidelberg, K.B., Williamson, S., Yooseph, S., Wu, D., Eisen, J.A., Hoffman, J.M., Remington, K., *et al.* (2007). The Sorcerer II Global Ocean Sampling expedition: northwest Atlantic through eastern tropical Pacific. PLoS Biol *5*, e77.

Russell, H.L. (1891). Untersuchungen über im Golf von Neapel lebende Bakterien. Zeitschr. f. Hyg. *11*, 165–206.

Sabehi, G., Massana, R., Bielawski, J.P., Rosenberg, M., Delong, E.F., and Beja, O. (2003). Novel Proteorhodopsin variants from the Mediterranean and Red Seas. Environ. Microbiol. *5*, 842–849.

Schut, F., de Vries, E.J., Gottschal, J.C., Robertson, B.R., Harder, W., Prins, R.A., and Button, D.K. (1993). Isolation of typical marine bacteria by dilution culture: growth, maintenance, and characteristics of isolates under laboratory conditions. Appl. Environ. Microbiol. *59*, 2150–2160.

Short, S.M., and Zehr, J.P. (2007). Nitrogenase gene expression in the Chesapeake Bay Estuary. Environ. Microbiol. *9*, 1591–1596.

Staley, J.T., and Konopka, A. (1985). Measurement of *in situ* activities of nonphotosynthetic microorganisms in aquatic and terrestrial habitats. Annu. Rev. Microbiol. *39*, 321–346.

Stein, J.L., and Simon, M.I. (1996). Archaeal ubiquity. Proc. Natl. Acad. Sci. U.S.A. *93*, 6228–6230.

Stingl, U., Tripp, H.J., and Giovannoni, S.J. (2007a). Improvements of high-throughput culturing yielded novel SAR11 strains and other abundant marine bacteria from the Oregon coast and the Bermuda Atlantic Time Series study site. ISME J. *1*, 361–371.

Stingl, U., Desiderio, R.A., Cho, J.C., Vergin, K.L., and Giovannoni, S.J. (2007b). The SAR92 clade: An abundant coastal clade of culturable marine bacteria possessing proteorhodopsin. Appl. Environ. Microbiol. *73*, 2290–2296.

Suzuki, M.T., Preston, C.M., Chavez, F.P., and DeLong, E.F. (2001). Quantitative mapping of bacterioplankton populations in seawater: field tests across an upwelling plume in Monterey Bay. Aquat. Microbiol. Ecol. *24*, 117–127.

Suzuki, M.T., Preston, C.M., Beja, O., de la Torre, J.R., Steward, G.F., and DeLong, E.F. (2004). Phylogenetic screening of ribosomal RNA gene-containing clones in Bacterial Artificial Chromosome (BAC) libraries from different depths in Monterey Bay. Microb Ecol *48*, 473–488.

Ting, C.S., Hsieh, C., Sundararaman, S., Mannella, C., and Marko, M. (2007). Cryo-electron tomography reveals the comparative three-dimensional architecture of Prochlorococcus, a globally important marine cyanobacterium. J. Bacteriol. *189*, 4485–4493.

Treusch, A.H., Leininger, S., Kletzin, A., Schuster, S.C., Klenk, H.P., and Schleper, C. (2005). Novel genes for nitrite reductase and Amo-related proteins indicate a role of uncultivated mesophilic crenarchaeota in nitrogen cycling. Environ. Microbiol. *7*, 1985–1995.

Vancanneyt, M., Schut, F., Snauwaert, C., Goris, J., Swings, J., and Gottschal, J.C. (2001). *Sphingomonas alaskensis* sp. nov., a dominant bacterium from a marine oligotrophic environment. Int. J. Syst. Evol. Microbiol. *51*, 73–79.

Venter, J.C., Remington, K., Heidelberg, J.F., Halpern, A.L., Rusch, D., Eisen, J.A., Wu, D., Paulsen, I., Nelson, K.E., Nelson, W., *et al.* (2004). Environmental genome shotgun sequencing of the Sargasso Sea. Science *304*, 66–74.

Wagner, M., Smidt, H., Loy, A., and Zhou, J. (2007). Unravelling microbial communities with DNA-microarrays: challenges and future directions. Microb. Ecol. *53*, 498–506.

Wang, Y., Lau, P.C., and Button, D.K. (1996). A marine oligobacterium harboring genes known to be part of aromatic hydrocarbon degradation pathways of soil pseudomonads. Appl. Environ. Microbiol. *62*, 2169–2173.

Warming, E. (1875). Om nogle ved Danmarks Kyster levende Bakterien. Medd. naturhist. For. Kjöbenhavn *7*, 356–376.

Warming, E. (1876). Om nogle ved Danmarks Kyster levende Bakterien. Medd. naturhist. For. Kjöbenhavn *8*, 21–42.

Wright, T.D., Vergin, K.L., Boyd, P.W., and Giovannoni, S.J. (1997). A novel delta-subdivision proteobacterial lineage from the lower ocean surface layer. Appl. Environ. Microbiol. *63*, 1441–1448.

Wuchter, C., Schouten, S., Boschker, H.T., and Sinninghe Damste, J.S. (2003). Bicarbonate uptake by marine Crenarchaeota. FEMS Microbiol. Lett. *219*, 203–207.

Yooseph, S., Sutton, G., Rusch, D.B., Halpern, A.L., Williamson, S.J., Remington, K., Eisen, J.A., Heidelberg, K.B., Manning, G., Li, W., *et al.* (2007). The Sorcerer II Global Ocean Sampling expedition: expanding the universe of protein families. PLoS Biol. *5*, e16.

Yurkov, V.V., and Beatty, J.T. (1998). Aerobic anoxygenic phototrophic bacteria. Microbiol Mol Biol Rev *62*, 695–724.

Zengler, K., Toledo, G., Rappe, M., Elkins, J., Mathur, E.J., Short, J.M., and Keller, M. (2002). Cultivating the uncultured. Proc. Natl. Acad. Sci. U.S.A. *99*, 15681–15686.

ZoBell, C.E. (1946). Marine Microbiology: A Monograph on Hydrobacteriology. (Waltham, MA: Chronica Botanica Company.)

The Human Environment

9

Johan Dicksved, Liping Zhao and Janet K. Jansson

Abstract

Applications of recent advances in molecular methods have illuminated the previously hidden diversity of the microbial world that not only inhabits our bodies, but that also lives in a close symbiotic association with us. This human-associated microbiota, or human microbiome, is responsible for many key functions in our bodies. Increasing evidence suggests many important roles of individual members of the human microbiome and their respective influences towards ultimate health and disease of the host. This chapter highlights some of the important functions of the human microbiome, many of which were gleaned using different molecular approaches. The clinical field has thus greatly benefited from the molecular toolbox that was initially developed by microbial ecologists for investigation of other complex ecosystems, such as soil. As the field has progressively moved away from a dependence on cultivation-based approaches towards increasing reliance on molecular approaches, the amount of knowledge about the human microbiome composition and function has greatly expanded. Most recent studies using molecular tools, including various 'omics' approaches, have focused on the intestinal microbiota. Therefore, we have also primarily discussed the gut microbiota in this chapter. In addition, the influence of difference host-related factors, such as genetics, age, birth mode, diet and geographical location are discussed with respect to their impact on the composition and related function of the human microbiome. Some beneficial bacteria, such as probiotic strains, are discussed, in addition to those that are particularly detrimental to human health. Some of the latter include correlations of microbial compositions to intestinal diseases and cancer. The more information that we have about the key roles of specific members of the human microbiome, the more potential we have for manipulation of the composition of the microbiota to enhance the prevalence of beneficial species and to diminish the amounts of detrimental ones. This is a guiding vision for future research in this area.

Introduction

For a microbiologist it is particularly intriguing to know that one of the most densely populated microbial ecosystems on earth lies enclosed within our own bodies. It has been estimated that 1 gram of faeces contains 10^{11} bacteria (Franks *et al.*, 1998). However, it is far from as diversified as the soil ecosystem. Approximately 800–1000 different species and > 7000 different strains inhabit the GI tract (Bäckhed *et al.*, 2005) and the majority of these (> 80%) have never been cultured (Eckburg *et al.*, 2005). Therefore, the intestinal tract of humans represents an ecological niche that is largely unexplored. What we do know is that the colon is an anaerobic environment and anaerobic bacteria outnumber the aerobic by 1000:1. The bacteria in the human GI tract include commensal strains as well as potential pathogens. Many commensal members of the microbiota of the colon provide beneficial effects that are vital for maintenance of health of the host and this relationship is mutualistic, or beneficial to both partners. Some beneficial effects for the host include synthesis of vitamins and digestion of certain carbohydrates (dietary fibres) (Hooper *et al.*, 2002). For example, some members of the anaerobic community in the colon degrade indigestible polysaccharides like cellulose, pectin and starch that humans are otherwise unable to degrade. These polysaccharides are fermented to short-chain fatty acids (SCFAs) such as acetate, propionate and butyrate, and various gases (Topping and Clifton, 2001). Butyrate is the preferred substrate for the epithelial cells lining the gut, and it is thought to be important for maintaining a healthy intestinal cell wall. Additional important roles of the normal, commensal microbiota include their involvement in the development and maintenance of an effective immune system and in the prevention of colonization by pathogenic bacteria. Colonization resistance can be performed by various means, such as (1) competition for nutrients, (2) competition for attachment sites on the intestinal wall and thereby inhibition of adherence of pathogenic bacteria to the mucosa (Zhong *et al.*, 2004) and (3) production of SCFA which can prevent the overgrowth of pH-sensitive pathogenic bacteria (Topping and Clifton, 2001).

Other habitats that are colonized on the human include the rest of the GI tract from the mouth to the anus, as well as the skin, lungs and vagina. Basically, any human surface that is exposed to the external environment in some way is colonized by microorganisms. In this chapter we will focus on the intestinal tract that has been most intensively studied to date. However, it is important to keep in mind the growing number of studies of microbes associated with other sites of the body (see, for example, Andersson *et al.*, 2008; Bik *et al.*, 2006, Dethlefsen *et al.*, 2007).

Mammalian models for the human intestinal microbiota

To elucidate mechanisms involved in nutrition, inflammation and physiological dysfunction, it is sometimes necessary to access different compartments of the gastrointestinal tract. In humans this can be problematic due to difficulties and risks with invasive sampling and associated high costs. Therefore, animal models are commonly used as a replacement to address such questions. One must however be aware of the limitations that make extrapolation back to humans unreliable. Rodent models are commonly used when performing nutritional or inflammation studies. Compared with humans, rodents have significantly different food intakes and energy expenditures for their body size, a different life span and body proportions (Patterson *et al.*, 2008). In addition, there are differences in their intestinal morphology and composition of their enteric microbiota as well as other physiological differences. Even if the majority of the rodent microbial gut species are unique it is has been shown that the microbiota is similar to that of humans at the division level, with Firmicutes and Bacteroidetes predominating (Ley *et al.*, 2005). The species-level composition probably is more relevant with host metabolism and immunity. Attempts to establish human-flora associated (HFA) rodent models showed that some important members of the human gut microbiota, such as bifidobacteria, cannot colonize the rodent gut (Raibaud *et al.*, 1980). It was also found that HFA-rodents do not develop a full range of normal immune responses unless some important key members of their gut microbiota are reintroduced (Imaoka *et al.*, 2004).

Although no experimental animal model will perfectly mimic the human condition, the porcine model is superior compared to the other non-primate animal models. Despite some anatomic differences, the physiology of digestion and associated metabolic processes are very similar between pigs and humans. The most essential macroscopic difference between humans and pigs is the length of the intestine, which is far longer in pigs. In a pig the average small intestine length is 15–22 m and the large intestine is 4–6 m, whereas in humans the average small intestinal length varies between 5.5 and 7 m and large intestine 1.5 m (Patterson *et al.*, 2008). The spatial organization of the intestine differs considerably between humans and pigs, in particular that of the large intestine. In humans the large intestine is arranged in a square-like configuration, whereas in pigs the large intestine has a spiral shaped conformation. Anatomically the intestine of pig is similar to that of humans with an omnivorous diet, and nutritional requirements that are very similar to humans (Patterson *et al.*, 2008). In addition, the pig microbiota has a composition that resembles the human gut microbiota (Leser *et al.*, 2002). Recently, it was demonstrated that a HFA pig model could be established with the most predominant members of the human gut microbiota (Pang *et al.*, 2007). This HFA pig model has good potential for human physiology and gut ecology research. It can also be developed into a model for pharmaceutical research as the data obtained with this kind of model can be more relevant to the human situation.

Establishment of the gut microbiota

We are born sterile but immediately after birth a microbial community starts to establish (O'Hara and Shanahan, 2006). The first inoculum comes from the mother, by the birth canal, and/or a faecal-oral route, and from the surrounding area. For example, the same serotype of *Escherichia coli* was found in the mouths of newborn babies as in faeces from their mothers (Brook *et al.*, 1979). The microbial composition and colonization pattern varies widely from baby to baby (Palmer *et al.*, 2007). During the first months of life the community is highly variable and dominated with a facultative anaerobic microbiota (Fig. 9.1). When space and nutrients are not limiting, bacteria with high division rates are more prevalent. However, over time, when the system becomes more densely populated and nutrient supply is limited, a more specialized microbiota takes over (Hooper *et al.*, 1999). The microbiota of the children becomes progressively more complex and at the age of two the microbial community is stabilized and an adult-like flora established with a large predominance of anaerobic microbes (Favier *et al.*, 2002; Fig. 9.1). Without outer environmental stresses the dominant microbiota is considered to be remarkably stable over time (Ley *et al.*, 2006; Vanhoutte *et al.*, 2004; Zoetendal *et al.*, 1998). However, temporal variations have been observed in specific members of the microbiota, such as some lactobacilli and *Enterococcus* populations, whereas *Bacteroides* spp. and bifidobacteria remained relatively stable (Jernberg *et al.*, 2007; Vanhoutte *et al.*, 2004).

It has been postulated that the early establishment of the gut microbiota is of great importance for the postnatal development of the immune system (Hooper *et al.*, 2002). Interestingly, the birth delivery mode has been shown to influence the composition of the microbiota, with slower diversification and lack of anaerobic species, such as clostridia, in infants delivered by caesarean section (Grönlund *et al.*, 1999). Moreover, infants born by caesarean section were characterized by an absence of *Bifidobacterium* species (Biasucci *et al.*, 2008). In another study, a lower prevalence of clostridial species in the microbiota of caesarean section-delivered children was prolonged over a period of several years (Salminen *et al.*, 2004). However, it is unknown whether this difference in the microbial composition has any significant implications for host health.

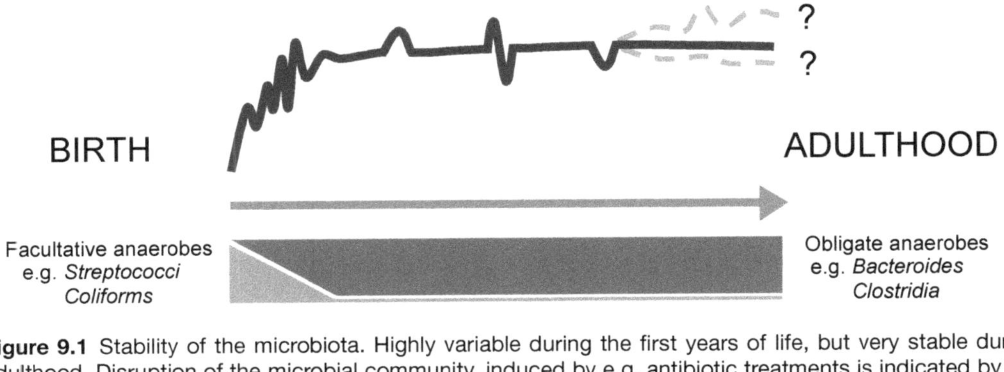

Figure 9.1 Stability of the microbiota. Highly variable during the first years of life, but very stable during adulthood. Disruption of the microbial community, induced by e.g. antibiotic treatments is indicated by the dashed grey lines followed by question marks and it is not clear how a disruption affects the community later in life. Infants have a microbiota dominated by facultative anaerobic bacteria (light grey) whereas obligate anaerobic bacteria outcompete aerobic bacteria in adults (dark grey).

The upper GI microbiota

A complex microbiota of more than 700 different species with representatives from six bacterial divisions, (Firmicutes, Bacteroidetes, Proteobacteria, Actinobacteria, Fusobacteria, and TM7), has been found in the oral cavity of humans (Aas *et al.*, 2005; Fig. 9.2). The mouth comprises many different types of surfaces, all with a distinct microbial profile. However, some genera are common to all sites of the mouth, and these include *Gemella*, *Granulicatella*, *Streptococcus*, and *Veillonella* (Aas *et al.*, 2005). Oral microbial communities are commonly found in biofilms and it is probable that a metabolic inter-dependency is important for shaping these communities (Jenkinson and Lamont, 2005). Bacteria are introduced into the oesophagus either by swallowing or by reflux from a colonized stomach (Pei *et al.*, 2004). Previously, it was generally thought that the oesophagus was sterile or contained transients from the oral cavity due to the short retention time and flushing of foods and fluids ingested by the host. However, in 2004, Pei and coworkers showed that a diverse microbiota colonizes the mucosal surfaces of the oesophagus. They found 95 phylotypes with representatives from six different divisions, (Firmicutes, Bacteroidetes, Proteobacteria, Actinobacteria, Fusobacteria, and TM7; Fig. 9.1). Many of these were novel but culturable. In addition, this microbiota shares many members that usually are found in the oral cavity (Pei *et al.*, 2004).

The stomach is the first line in the body's defence that microbes encounter as they enter the gastrointestinal tract. The acidic environment is thought to be an effective barrier, killing off most of the microbes in their passage through the stomach into the small bowel. In addition, a

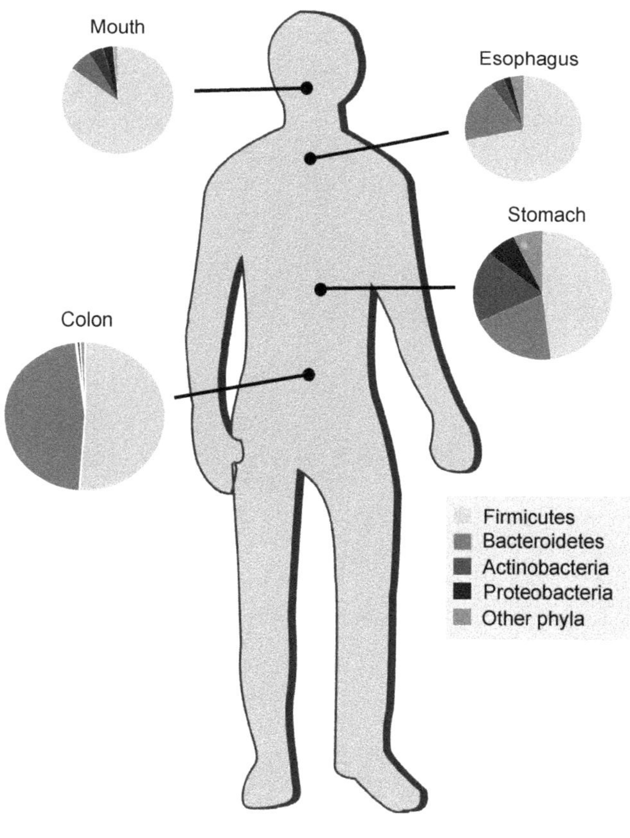

Figure 9.2 Distributions of bacterial phyla at different anatomical sites of the body. Modified from Dethlefsen *et al.* (2007). A colour version of this figure is located in the plate section at the back of the book.

short transit time contributes to the relatively low number of bacteria that are found in the stomach (Guarner, 2006). Previously the healthy stomach was considered to be otherwise sterile and not colonized by any microbes (Finegold *et al.*, 1983). However, the gastric mucosa was hiding a secret that was discovered more than two decades ago (Marshall and Warren, 1984); namely *Helicobacter pylori*. *Helicobacter pylori* is now known to be a colonizer of the gastric mucosa. Expression of the urease enzyme allows its survival in the acidic environment during its active passage into the mucus layer of the stomach lining where the pH is almost neutral. Additional bacteria have been found in the stomach, such as streptococci and lactobacilli, enterobacteria and other members of the oral cavity (Savage, 1977). However, there is no direct evidence that other bacteria than *H. pylori* can colonize the healthy stomach. Recent advances in molecular approaches to study the stomach associated microbiota have shown a more diversified gastric microbiota than previously thought. For example, in 2006 Bik and coworkers found 128 phylotypes in the gastric mucosa, with representatives from six bacterial divisions (Firmicutes, Bacteroidetes, Proteobacteria, Actinobacteria, Fusobacteria, and TM7; Fig. 9.2). Although these same divisions were represented in the stomach and the oesophagus (see above, Pei *et al.*, 2004); the set of organisms could be distinguished from what has been found in the mouth and oesophagus, suggesting that the stomach is home for a distinct microbial ecosystem (Bik *et al.*, 2006). Recently we characterized the gastric microbiota from ten patients with gastric cancer and compared to five dyspeptic controls using the molecular profiling approach, terminal-restriction fragment length polymorphism (T-RFLP), in combination with 16S rRNA gene cloning and sequencing (Dicksved *et al.*, 2009). T-RFLP analysis revealed a complex bacterial community in the cancer patients that was not significantly different from the controls. Representatives of several genera were detected with a predominance of Firmicutes, mainly of oral origin. The prevalence of *H. pylori* was very low and this bacterium was only found in samples where it had previously been isolated with culture-based techniques. Although we only examined a limited number of subjects, this study contributes the first DNA-based description of the bacterial community composition in the stomachs of individuals with gastric cancer and demonstrated a presence of a complex gastric microbiota with potential as a maintainer or bystander in the progression of gastric cancer. The respective role of these species in development of gastric cancer remains to be determined.

The microbiota of the bowel

The small bowel, comprising duodenum, jejunum and ileum, is the main site for digestion of foods and absorption of nutrients in the body. Bile, pancreatic secretions and flushing mechanisms restrict the bacterial density in the proximal parts of the small bowel (O'Hara and Shanahan, 2006) whereas a lower transit rate increases bacterial density dramatically in the distal regions (Booijink *et al.*, 2007). Due to technical difficulties in sampling, the residential microbiota is not well characterized. According to current knowledge, the duodenal and jejunal microbiota is predominated by facultative anaerobes, mainly acid tolerant lactobacilli and streptococci (Hayashi *et al.*, 2005; Wang *et al.*, 2005). The obligate anaerobic Firmicutes that dominate the microbiota of the large bowel are not commonly found in the upper parts of the small bowel (Hayashi *et al.*, 2005). However, the distal part of the ileum hosts a more diverse microbiota and it resembles that of the colonic microbiota.

The large bowel, comprising ascending, transverse, descending colon and rectum, is the primary site for microbial persistence in the human body. The number of bacteria is estimated to be 10^{11}–10^{12} bacteria/gram colonic content (O'Hara and Shanahan, 2006). The anaerobic condition promotes fermentation and obligate anaerobic bacteria outnumber facultative anaerobic bacteria. Even if the bacterial diversity is very high, the large bowel contains few phyla and is totally dominated by two, i.e. Bacteroidetes and Firmicutes, which account for over 90% of the bacteria found in the large bowel (Ley *et al.*, 2006; Fig. 9.2). Despite great physiological differences between the proximal and distal parts of the colon, bacteria are uniformly distributed along the length of the large bowel. There are, however, some differences in bacterial types found in biopsies compared with those found in faeces (Eckburg *et al.*, 2005; Lepage *et al.*, 2005; Nielsen *et al.*, 2003; Willing *et al.*, 2009; Zoetendal *et al.*, 2002).

The number of bacteria found in the luminal region of the intestinal lining outnumbers the bacteria found within the mucus layer. Approximately 10^5 to 10^6 bacteria have been detected per biopsy from the colonic wall (Zoetendal *et al.*, 2002). In contrast, several studies have failed to detect bacteria on the epithelial surface from healthy individuals (Kleessen *et al.*, 2002; Swidsinski *et al.*, 2007; van der Waaij *et al.*, 2005). It has been challenging to get correct estimates of the number and types of bacteria in the bowel using traditional cultivation-based approaches because the majority of the bacteria found in the large bowel have not yet been cultured. This challenge was exemplified in a large survey of the gut microbiota using molecular approaches; i.e. cloning and sequencing (Eckburg *et al.*, 2005), with the finding that 80% of the species represented by the clones had never previously been cultured.

Recently, there has been a surge of application of different 'omics' approaches for investigation of the composition of the human microbiome in various body sites and in different disease states. In one study, the metagenome of healthy human faecal samples was constructed using fosmid vectors and compared to that of a patient with Crohn's disease (CD). The CD microbiota had a lower diversity of especially Firmicutes, which was confirmed by fluorescence *in situ* hybridization (FISH) (Manichanh *et al.*, 2006). In another study a metagenomic library of faecal DNA was constructed and this library was screened for phylogenetic and functional genes. The information obtained from random sequencing of the metagenome was also used for the prediction of different metabolic functions (Gill *et al.*, 2006; see Chapter 2). In that study, no 16S rRNA gene sequences from the *Bacteroides* community could be detected using universal primers although this species is one of the most abundant microorganisms in the intestinal tract. This discrepancy was thought either to be due to the extraction method or to bias inherent in the PCR procedure such as choice of primers or the reaction conditions, but it could also reflect an unusually low number of *Bacteroides* in the sampled individuals. The sample size was too limited to enable these different options to be tested. A subsequent study also used a metagenomics sequencing approach to investigate the human microbiome composition and functional genes in 13 healthy Japanese individuals of various ages (Kurokawa *et al.*, 2007). They found the major number of assigned genes in the adults belonged to the *Bacteroides* group, followed by genera in the Firmicutes and *Bifidobacterium*, whereas bifidobacteria were more dominant in the infants. The samples from the Japanese study varied significantly in composition when compared to that from the previous American study by Gill *et al.* (2006). In addition to the differences in the *Bacteroides* already mentioned, the American samples had a higher number of archaeal sequences than the Japanese samples.

Recently, a shotgun metaproteomics approach was used to investigate the identities of thousands of proteins expressed by the gut microbiota in a pair of healthy Swedish individuals (Verberkmoes *et al.*, 2009; see Chapter 3 for a description of the approach used). It is interesting to compare the results from these various 'omics' approaches from different individuals and from different countries. The Swedish metaproteomes showed a lack of proteins for the methanogenesis pathway, but a high amount of proteins for acetogenesis. This could be a similar situation for the individuals in the Japanese study. One could speculate that different hydrogen disposal routes were active in the respective gut communities; i.e. a methanogenic route carried out by archaeal methanogens in the Americans, compared to a reductive acetogenic route in the Japanese and Swedish individuals. A high amount of the key enzyme in reductive acetogenesis was found in the Swedish gut proteomes; i.e. formyltetrahydrofolate synthetase, and this gene was also highly enriched in the Japanese metagenomes; although Kurokawa and coworkers speculated that the role of this enzyme was regulation of one carbon pools by formate (Kurokawa *et al.*, 2007). It is known that about 30–50% of people in Western countries dispose of hydrogen in their guts via a methanogenic route, whereas in the remainder hydrogen is consumed by sulphate reduction or reductive acetogenesis, and it is interesting that these different routes appear to be functioning in the small number of individuals investigated using these approaches to date. In addition, Kurokawa and coworkers found that genes involved in carbohydrate metabolism were highly enriched in the gut as also seen in the proteomics study by VerBerkmoes *et al.* (2009). Also, both the Japanese metagenomics study and the Swedish proteomics study found that genes and proteins, respectively, for flagella and chemotaxis were depleted, suggesting that motility is not an important requirement for residing in the human gut, but this hypothesis needs to be confirmed by additional studies.

Impact of antibiotics on the gut microbiota

Antibiotics can disrupt the complex microbial community in the human intestine since they not only affect the pathogens to which they are directed but may also impact other members of the intestinal microbiota. Another serious concern with antibiotic treatment is the emergence and spread of antibiotic resistance. How fast a bacterial population acquires resistance depends upon the selection pressure of the antibiotic and is thereby affected by the volume of drug use and the rate of development of resistant bacteria (Martinez and Baquero, 2000). Once resistance has been established in a bacterial community, resistance genes can be transferred between bacteria. For example *erm* genes, which confer resistance to clindamycin and macrolides like clarithromycin, can be transferred not only between *Bacteriodes* spp. but also to other species in the colon (Andremont, 2003). It is well acknowledged that the rate of resistance development is much faster than the rate in which resistance is lost within a bacterial population (Sjölund *et al.*, 2003).

The long-term impacts of antibiotics are currently being examined and discussed with contrasting results. It is likely that temporal impact of a given antibiotic is largely dependent on the mode of action of the antibiotic, the dose administered and the genetic and environmental influences experienced by the host. Dethlefsen and coworkers recently used 454 pyrosequencing to determine the impact of ciprofloxacin on the gut microbiota of three healthy volunteers. Ciprofloxacin treatment influenced the abundance of about a third of the bacterial taxa in the gut and resulted in a lower diversity of the intestinal microbiota. However, the magnitude of this effect varied among individuals and the disruption was only temporary. After 4 weeks the gut microbiota was re-established to pre-administration levels, although some taxa in some individuals were impacted for up to 6 months (Dethlefsen *et al.*, 2008). In another study, the long-term impact of another antibiotic, clindamycin, on the intestinal microbiota was studied (Jernberg *et al.*, 2007; Löfmark *et al.*, 2006). Clindamycin is excreted in bile and concentrations can be high in faeces. Compared to ciprofloxacin, administration of clindamycin often leads to major ecological disturbances in the intestine. For example, overgrowth of the pathogen *Clostridium difficile* is common, which in turn leads to diarrhoea. Cultivation data often show a decrease in the total number of anaerobes like the *Bacteroides* spp. after clindamycin or clarithromycin administration (Sullivan *et al.*, 2001). Although some *Bacteroides* spp. are suppressed by clindamycin (Sullivan *et al.*, 2003), others have been shown to be resistant to clindamycin and several other antibiotics (Hedberg and Nord, 2003; Löfmark *et al.*, 2006) and resistance to this antibiotic is increasing (Hedberg and Nord, 2003). In a recent study we used Rep-PCR (a high resolution fingerprinting technique) to monitor specific clones of *Bacteroides* species over time (Jernberg *et al.*, 2007). There was a large clonal variation between subjects, and almost all clone types were specific for respective individuals. Specific *Bacteroides* sp. clones were dramatically impacted upon clindamycin treatment and some clindamycin resistant clones dominated the community for long periods of time; even up to 2 years post treatment. In this study, T-RFLP also demonstrated that the *Bacteroides* communities were dramatically impacted by a 7-day clindamycin administration over the subsequent 2-year study period. In addition, clindamycin resistance genes were enriched in the faecal samples and persisted over the 2-year period as determined by quantitative real-time PCR with specific primers for *ermB* genes. These findings are particularly disturbing because they indicate long-term persistence of antibiotic resistance genes that could potentially be transferred to pathogenic strains in the intestinal environment, for example upon subsequent antibiotic administration that would select for resistant strains. These results and others support the use of caution in administration of some antibiotics.

Host genetics

The structure of the intestinal microbiota has been shown to be dependent to a certain extent on host genetics. In a recent study the gut microbiota of 60 mammalian species from 13 taxonomic orders was studied using cloning and sequencing. The results showed that the microbiota had a higher degree of similarity within species than between different species (Ley *et al.*, 2008). These authors suggested that the bacterial communities had co-diversified with their hosts and that the individual microbiome reflected variations in host diet and phylogeny. Studies among human hosts have also revealed similarities that correlate to host genetics. The faecal microbiota in healthy identical twins was examined using molecular approaches and a highly similar bacterial community within twin pairs was observed (Dicksved *et al.*, 2008). In addition, Zoetendal and coworkers (2001) studied the faecal microbiota of individuals with different degrees of genetic relatedness, ranging from monozygotic twins to unrelated individuals. Monozygotic twins that lived apart for years showed high similarities in their microbial profiles whereas marital partners that lived in the same environment and with comparable feeding habits did not (Zoetendal *et al.*, 2001). Van de Merwe and coworkers compared the culturable fraction of the faecal microbiota between mono and dizygotic twins and also found higher similarities within monozygotic pairs (Van de Merwe *et al.*, 1983). This difference was confirmed in young children, living in the same household (Stewart *et al.*, 2005). Animal studies have also supported the importance of genetic influence on the microbial composition. In 2005, Ley and coworkers, showed in a mouse model that three generations of offspring shared some similarities in their caecal microbiota, and these similarities were not seen when different families of mice were compared (Ley *et al.*, 2005). Taken together, these findings strongly argue that genetic factors provide a strong force in shaping the gut microbiota. However, colonization history and inoculum at birth from mother to child as discussed above, might also be contributing to the similar microbial communities found among twins.

Geographical impacts

There have also been reports that geographical region can have an impact on the gut microbial composition. For example, the gut microbiota has been shown to differ between Westernized developed countries, Asian and developing countries. *H. pylori* has been found to be more prevalent in developing countries than in developed countries (Genta *et al.*, 1995). A recent study, Li *et al.* (2008) showed that the species composition of a Chinese family cohort was significantly different from that of two American individuals (Gill *et al.*, 2006). In another study, the microbiota of Estonian infants was shown to differ from the microbiota of Swedish infants (Björksten *et al.*, 1999). However, it has not yet been properly investigated whether these region-based discordances are based on the local environment, climate, genetic, dietary, or other factors connected to different lifestyles. In one study, local variations in the microbiota of the Western European population were correlated to age, i.e. adult compared with elderly (Mueller *et al.*, 2006). For example, country and age interactions were observed for Germans and Italians, with inverse interactions for the predominating *Clostridium coccoides* and *Bacteroides-Prevotella* groups. Contrary to this observation, other studies have shown no significant differences in gut microbial compositions between individuals from different European countries (Dicksved *et al.*, 2007; Lay *et al.*, 2005).

Influence of ageing

Owing to an altered physiology in elderly, with a decreased intestinal motility, reduced secretion of gastric acid, and a change of dietary habits and lifestyle, the microbiota in elderly persons differs from younger adults, although with great individual variations. In elderly persons, it is common to have a reduction in numbers and diversity of *Bacteroides* and bifidobacteria, and a reduced production of SCFA and amylolytic activity. In addition, increased numbers of facultative anaerobes, fusobacteria, clostridia and eubacteria but decreased numbers of bifidobacteria have been reported in elderly people (Woodmansey, 2007). Although there is large disagreement in the findings between studies, probably due to individual differences and use of different methods, the abundance and diversity of bifidobacteria is consistently reported to be decreased in elderly individuals (Mueller *et al.*, 2006; Woodmansey, 2007).

Probiotics and the gut microbiota

Probiotics have been defined by FAO/WHO as 'live microorganisms which when administered in adequate amounts confer a health benefit on the host' (www.fao.org/es/ESN/Probio/reports.pdf). The most commonly used probiotic bacteria are members of the genera *Lactobacillus* and *Bifidobacterium* which are lactic-acid producers and also members of the normal microbiota. There are a number of probiotic foods and preparations available on the market and most common are the milk based products. The probable beneficial mechanisms of action of the probiotic strains in controlling intestinal pathogens are (1) competition for binding sites on the intestinal wall, (2) competition for nutrients, (3) modulation of immune responses, and (4) production of bacteriocins. These are the same positive benefits that the normal flora in the intestinal tract has in general. However, when the normal intestinal flora does not function well due to disease or drug treatment the idea is that probiotics could enhance these beneficial effects and maybe shorten the duration time of the imbalance in the flora. Some areas where the benefits of probiotics have been investigated are in a variety of disease conditions such as *H. pylori* infection, acute gastroenteritis, inflammatory bowel diseases and irritable bowel syndrome (Sullivan and Nord, 2005). In one study, the possible beneficial effects of a probiotic yoghurt during clindamycin treatment was investigated (Jernberg *et al.*, 2005). The T-RFLP analysis indicated that the disturbances of the intestinal microbiota were more prolonged in subjects not ingesting the probiotic. It was also shown that the probiotic lactobacilli strains dominated this more defined community during the ingestion of the probiotic. This was assessed by T-RFLP with a *Lactobacillus*-specific primer. Seven days after ingestion of the probiotic had stopped the probiotic strains could no longer be detected (Jernberg *et al.*, 2005). Previous reports have also shown that probiotic bacteria only transiently colonize the intestine (Gorbach, 2000; Tannock, 1999).

The microbiota and gut-related disorders

A balance between the host immune system and the commensal gut microbiota is crucial for maintaining health. When this balance is disturbed, the host–microbe relationship can progress towards a disease state. Although some members of the commensal gut microbiota protect our body from foreign infections, other members are partly responsible for several of the disorders within the gastrointestinal tract. For example, some microbes in the mouth are responsible for two of the most common bacterial infections in humans, namely caries and periodontitis. Moreover, the discovery of *H. pylori* and its role in the development of gastritis, duodenal ulcers and gastric cancer has previously been mentioned. In addition, the gut microbiota has been implicated to play an important role in several other disorders of the gastrointestinal tract, such as inflammatory bowel diseases (IBD), including Crohn's disease (CD) and ulcerative colitis (UC), irritable bowel syndrome (IBS), as well as colorectal cancer (CRC) and allergies. How, when and to which extent the microbiota is involved in the mechanisms causing these diseases remains an enigma. The mechanisms are thought to be multifactorial for most, if not all disorders of the gastrointestinal tract, including host susceptibility, microbial factors and environmental triggers.

In recent decades incidence rates of several of the immune dysfunctional disorders in the human gastrointestinal tract have increased dramatically (Guarner *et al.*, 2006). This increase in autoimmune disease incidence is, however, only observed in developed countries. A high level of hygiene has been identified as one risk factor for developing allergies, IBD, type I diabetes, and multiple sclerosis (Guarner *et al.*, 2006). The correlation between the modern hygienic lifestyle and the increase in allergies has given rise to the hygiene hypothesis, suggesting a link between decreased pathogen exposure and developing inflammatory disorders (Strachan, 1989). For example, there are supportive data to suggest that carriage of helminths, a multicellular worm parasite that is eliminated in Western societies, might protect the host from developing immunological disorders (Guarner *et al.*, 2006). However, an alternative interpretation suggests that disturbances of the intestinal microbiota occur due to an increased use of antibiotics and 'Westernized' dietary habits that disrupt the mechanisms involved in the development of immune tolerance (Noverr and Huffnagle, 2004).

The gut microbiota and IBD

Inflammatory bowel disease refers to chronic relapsing disorders of the intestinal tract with yet unknown aetiologies. It is the collective term for a group of heterogeneous tissue reactions, with similar clinical behaviour. The most familiar subtypes of IBD are CD and UC. CD can affect the whole intestinal tract and is characterized by discontinuous involvement of various portions of the gut, whereas UC is restricted to the colon and rectum and is generally characterized by a continuous inflammation of the large bowel. Heterogeneities in response to medication and clinical behaviour within CD and UC have further divided it into distinct disease phenotypes. For example, CD can be classified into a few main phenotypes, with respect to the disease location. The classical phenotype is located in the terminal ileum, i.e. ileal CD (ICD). Inflammation located in the colon is referred to as colonic CD (CCD) and inflammation involving both ileum and colon is addressed as ileocolonic disease (ICCD). Less common but still present is inflammation of the upper intestinal tract in some individuals (Satsangi *et al.*, 2006).

Genetics plays an important role for the disease prevalence, with higher concordance rates (both are sick) in monozygotic twins than in dizygotic twins with CD (approximately 50% versus < 10%). For UC the genetic factors are not as strong as for CD, shown by lower concordance rates in both mono- and dizygotic twins (approximately 15% and < 5% respectively) (Orholm *et al.*, 2000; Thompson *et al.*, 1996; Tysk *et al.*, 1988). Still, a large number of monozygotic twins are discordant for the disease (one is sick and one healthy), which emphasizes that other factors in addition to genetic predisposition play an important role in disease prevalence. There are several lines of evidence that argue for a microbial involvement in the disease aetiology:

1. Inflammation occurs in the regions with the highest bacterial densities.
2. Diversion of the faecal stream induces clinical remission of active CD; however, disease relapses shortly after the bowel continuity is restored (Harper *et al.*, 1983).
3. Germ-free animals do not develop colitis (Sartor, 2003).
4. A subset of CD patients have polymorphisms in genes that encode receptors that recognize bacterial antigens (Ahmad *et al.*, 2002).
5. Antibiotics and probiotics have shown therapeutic efficacy in IBD patients (Sartor, 2004).
6. The epithelial barrier is leaky with defects in mucus production in patients with IBD (Swidsinski *et al.*, 2007).
7. Decreased levels of antimicrobial peptides (alpha defensins) have been demonstrated in ileal tissues from a subset of CD patients (Wehkamp *et al.*, 2005).
8. A large number of studies have reported an altered bacterial composition and higher bacterial loads in IBD patients compared with healthy individuals (Baumgart *et al.*, 2007; Dicksved *et al.*, 2008; Frank *et al.*, 2007; Kleessen *et al.*, 2002; Seksik *et al.*, 2003; Swidsinski *et al.*, 2002, 2005; Willing *et al.*, 2009). In addition, the severity of the disease correlates with the bacterial density found in the intestinal mucosa (Swidsinski *et al.*, 2002).

Two strategies have been used to study the relationship between the intestinal bacteria and the pathogenesis of IBD. One focuses on 'candidate' microorganisms and the other is a 'global description strategy' (Marteau *et al.*, 2004).

The first strategy has been used to search for a specific microorganism as the 'infectious origin' of IBD. Intensive research investigations of *Mycobacterium avium* subsp. *paratuberculosis* as the cause of CD have been performed, because *M. avium* subsp. *paratuberculosis* could cause the related Johne's disease in ruminants (Chiodini, 1989; Thayer *et al.*, 1984). In 2003, Bull and coworkers detected *M. avium* subsp. *paratuberculosis* in fresh ileocolonic mucosal biopsy specimens from 92% and 26% individuals with and without CD respectively and the author thought that the significantly higher detection rate of *M. avium* subsp. *paratuberculosis* in CD patients implicated the pathogenic role of this bacterium (Bull *et al.*, 2003). However, many studies have subsequently tried to detect this bacterium using nested-PCR or measuring serum antibodies with conflicting or inconclusive results (Bernstein *et al.*, 2004; Kanazawa *et al.*, 1999). Therefore, it was postulated that MAP infection could be responsible for CD in a subset of patients that were susceptible or that this organism may selectively but accidentally colonize the ulcerated mucosa of CD patients but not initiate inflammation. Currently, most recent publications do not specifically point out MAP as a causal agent in CD, however its role in CD is neither excluded (Sartor, 2005).

Several studies have identified a higher prevalence of enterobacteria and in particular *E. coli* in mucosal tissues from IBD patients (Baumgart *et al.*, 2007; Darfeuille-Michaud *et al.*, 1998; Kotlowski *et al.*, 2007; Swidsinski *et al.*, 2007; Willing *et al.*, 2009). Most attention has been focused on adherent invasive *E. coli* (AIEC), that was isolated from and within ileal tissues of CD patients (Darfeuille-Michaud *et al.*, 1998). This bacterium has been found to persist and multiply within macrophages (Baumgart *et al.*, 2007). Furthermore, it has been reported almost exclusively localized to ileal tissues from CD patients with the ileal disease phenotype. Moreover, a recent study by Mpofu and coworkers, showed that supernatants from MAP cultures impaired the ability of adherent monocytes and monocyte-derived macrophages to kill phagocytosed *E. coli* and that could explain the increased prevalence of *E. coli* attached to and within the epithelia (Mpofu *et al.*, 2007).

A possible link between sulphate-reducing bacteria (SRB) and IBD has also been evoked because their metabolic end product, hydrogen sulphide can have deleterious effects (Gibson *et al.*, 1993). Studies have indicated that SRB were ubiquitous in the bowel (Fite *et al.*, 2004; Zinkevich and Beech, 2000) but the exact role of these bacteria in the aetiology of IBD has not been identified (Fite *et al.*, 2004; Pitcher *et al.*, 2000). Other bacteria, such as *Helicobacter* sp., *Bacteroides vulgatus, Listeria monocytogenes* and many more have been proposed as playing a potential aetiological role, but none of them has been confirmed as the cause of IBD (Chen *et al.*, 2000; Fujita *et al.*, 2002; Oliveira *et al.*, 2004; Streutker *et al.*, 2004). Virus have also been proposed as the cause of IBD, especially for CD (Ekbom *et al.*, 1996; Wakefield *et al.*, 1993), but the association has also failed to be confirmed (Fisher *et al.*, 1997).

Intestinal ecologists have used a combination of culture based and molecular techniques to make significant progress towards comparing the bacterial composition in IBD patients and controls (Marteau *et al.*, 2004). Because faecal samples are easy to collect, most efforts have concentrated on analysis of the bacterial composition in faecal samples. It has been reported that the dominant bacteria in individuals with IBD are unusual when compared to healthy subjects (Seksik *et al.*, 2003; Sokol *et al.*, 2006). In addition, bifidobacteria, which were thought to be potentially protective against IBD, were found to be decreased in CD (Favier *et al.*, 1997). In our twin study, where the microbiota of healthy twins pairs were compared with twin pairs that were discordant (one is healthy and one is sick) or concordant (both are sick) for CD, a highly similar microbiota within any given healthy pair were observed, whereas within the discordant and the concordant pairs, large differences in the microbial compositions were recorded (Dicksved *et al.*, 2008; Fig. 9.3). For example, we found that individuals with CD had higher relative abundances of *Bacteroides ovatus* and *B. vulgatus* and a lower relative abundance of *B. uniformis* using *Bacteroides* group specific primers. Furthermore the CD patients with the ileal disease phenotype showed a considerably different microbial composition compared with both healthy individuals and those with the colonic disease phenotype (Fig. 9.4). The microbial diversity also differed between the healthy and the sick individuals and in particular among the discordant twin pairs all diseased individuals had the lowest microbial diversity (Dicksved *et al.*, 2008). A reduced diversity of the bacterial phylum of Firmicutes in the faecal microbiota of CD patients was also revealed using a metagenomic approach (Manichanh *et al.*, 2006).

One interesting difference in a specific bacterial population in individuals with ileal CD compared to healthy individuals or those with

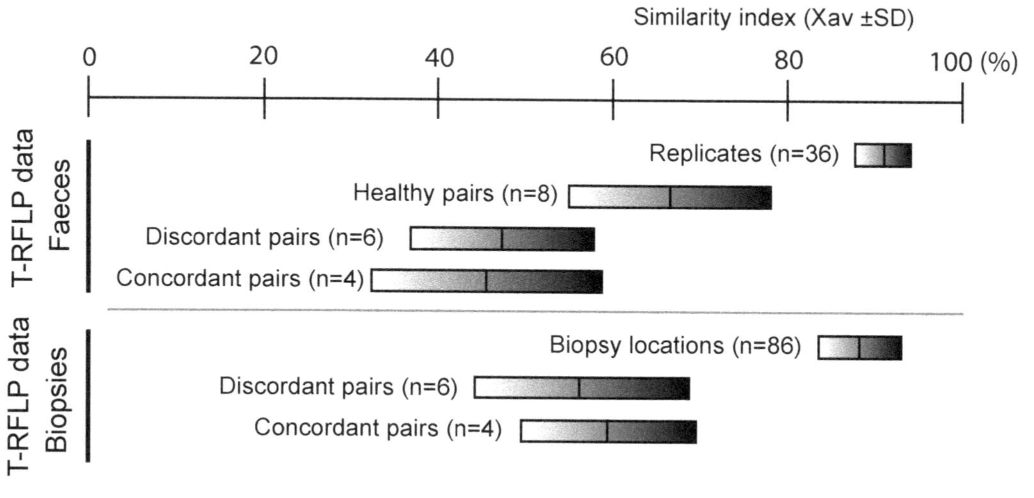

Figure 9.3 Similarity scores for pairwise comparisons of microbial community profiles obtained from faecal samples and biopsies from healthy twin pairs and twin pairs that are either concordant or discordant for Crohn's disease. Reproducibility of the T-RFLP method (from duplicated faecal samples), is indicated in the upper box 'replicates'. Similarities of T-RFLP profiles derived from different biopsy locations are indicated 'biopsy location'. The data originates from pairwise comparisons of Manhattan distances. The boxes represent the mean and standard deviations for all comparisons made within a group. n; represents the number of pairwise comparisons performed (data derived from Dicksved *et al.*, 2008 and Willing *et al.*, 2009).

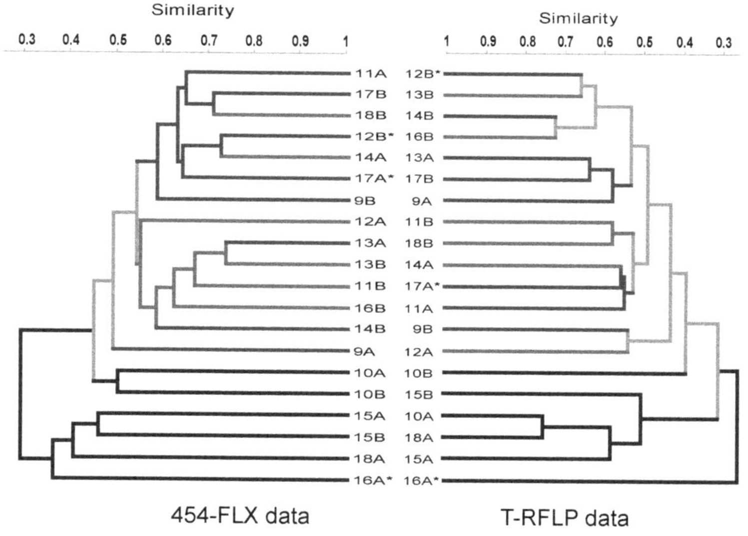

Figure 9.4 Clustering of bacterial community profiles in faecal samples originated from twin pairs that were either healthy, condordant or discordant for Crohn's disease (samples derived from Dicksved *et al.*, 2008; Willing *et al.*, 2009; see Fig. 9.3, which represents same sample cohort). Bray Curtis index was used as a distance matrix and the dendrogram was based on UPGMA clustering representing data from T-RFLP analysis and 454-FLX pyrosequencing of 16S rRNA genes. The numbers represent the twin pairs (Dicksved *et al.*, 2008) and the letters (A or B) represent the individuals within a pair. Shading of branches refers to the disease phenotype: dark grey, Ileal Crohn's disease (ICD); light grey, Colonic Crohn's disease (CCD); and black, healthy individuals. *Represent individuals that originally were classified as ileocolonic disease phenotype, but were reclassified as predominantly having ICD or CCD (Willing *et al.*, 2009). A colour version of this figure is located in the plate section at the back of the book.

colonic CD is a reduced number of bacteria in the *Clostridium leptum* group, and in particular *Faecalibacterium prausnitzii* (Baumgart *et al.*, 2007; Manichanh *et al.*, 2006; Sokol *et al.*, 2006; Willing *et al.*, 2009) (see Fig. 9.5). Recently *F. prausnitzii* was proposed as a probiotic for Crohn's disease based on the reduction of disease symptoms in mice after inoculation (Sokol *et al.*, 2008). However, this remains to be tested in humans. Some members of the *C. leptum* group (such as *F. prausnitzii*) are thought to play a beneficial role in the gut, for example as major producers of SCFA, particularly butyrate, which has been shown to act as a fuel and inflammation repressor for the colonic epithelial cells (Pryde *et al.*, 2002). This observation has been supported by the observation of decreased faecal levels of acetate and butyrate in CD patients (Marchesi *et al.*, 2007; Takaishi *et al.*, 2008). The study by Sokol and coworkers (2008) showed that reduction of *F. prausnitzii* increased the risk of developing postoperative recurrence of ileal CD and that live *F. prausnitzii* or its supernatant markedly reduced the severity of trinitrobenzenesulphonic acid (TNBS) colitis partly though blocking NF-κB activation and IL8 production.

Since the mucosa-associated bacteria could induce a local immune response (Poxton *et al.*, 1997) and the mucosa-associated bacterial composition is significantly different from that in faeces (Zoetendal *et al.*, 2002), some recent studies have focused on the mucosa-associated bacteria in IBD patients. Microbial population levels in washed biopsies were studied in 305 IBD patients and 40 healthy controls. The results indicated that the concentration of mucosal bacteria increased in accordance with the disease severity (Swidsinski *et al.*, 2002). In 2004, Ott and coworkers, found that the bacterial diversity was reduced in IBD patients and this result was proven to be associated with the loss of normal bacteria, such as *Bacteroides, Eubacterium* and *Lactobacillus* (Ott *et al.*, 2004). In another study a decreased population level of anaerobic bacteria,

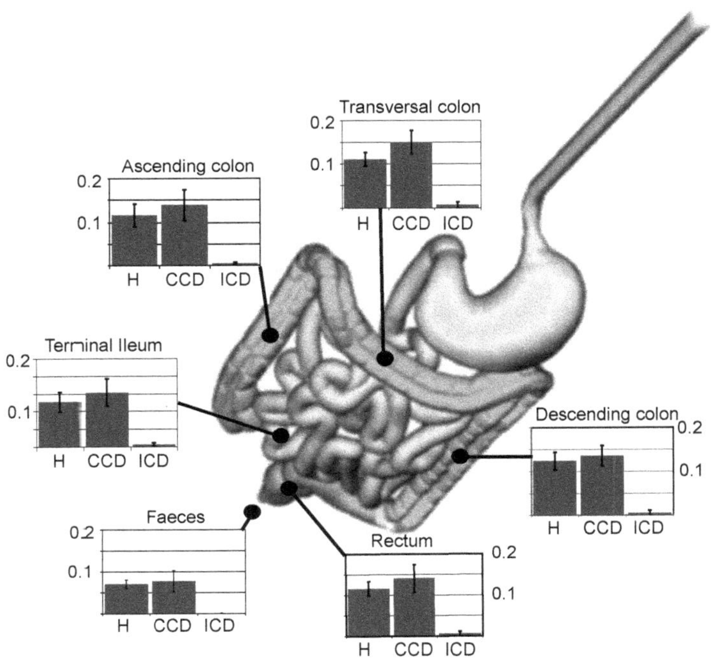

Figure 9.5 Relative abundance of *Faecalibacterium prausnitzii* along the length of the bowel and in faecal samples based on T-RFLP data. Bar charts illustrate the relative abundance of the representative phyotype in the different individuals with the following disease phenotypes: H; healthy individuals, CCD; colonic Crohn's disease, ICD; ileal Crohn's disease (derived from Dicksved *et al.*, 2008; Willing *et al.*, 2009).

especially *B. vulgatus*, in paediatric patients was found (Conte *et al.*, 2006), but some studies have instead reported an increase of *Bacteroides* spp. population levels in IBD using both culture-dependent and culture-independent methods (Giaffer *et al.*, 1991; Lucke *et al.*, 2006; Ruseler-van Embden and Both-Patoir, 1983; Swidsinski *et al.*, 2002) and a *Bacteroides fragilis* biofilm was found to be the main feature of IBD (Swidsinski *et al.*, 2005). In 2004, Macfarlane and coworkers detected 72 bacterial taxa (18 genera) in nine UC patients and 10 patients without IBD. The results indicated that *Peptostreptococcus* was only detected in UC patients and the population level of bifidobacteria in UC decreased significantly compared with that in controls (Macfarlane *et al.*, 2004). In addition, increased numbers of *Clostridia* and *E. coli* and decreased numbers of bifidobacteria were found in IBD patients (Mylonaki *et al.*, 2005).

Since the mucosa-associated bacterial composition varies from person to person (Zoetendal *et al.*, 2002), it may be more relevant to compare the intestinal bacteria within the same individual over time or in different regions of the intestine (Lepage *et al.*, 2005). No significant difference in the amount of bacteria between inflamed and non-inflamed biopsies have been found in patients with IBD (Swidsinski *et al.*, 2002). In another study, the dominant microbiota did not differ qualitatively between the ulcerated biopsy and non-ulcerated biopsy in Crohn's disease patients (Seksik *et al.*, 2005). In addition, a disruption of the microbial structure within an individual was not localized to a specific region but stable along the length of the large bowel (Willing *et al.*, 2009). Understanding the role that the intestinal microbiota play in the pathogenesis of IBD will be very helpful to decrease the incidence rate of this disease and to eventually develop the most effective clinical therapy.

Diet and the gut microbiota

Although the composition of the intestinal microbiota is generally perceived to be relatively stable in healthy subjects, it can fluctuate due to external influences, such as diet (Gibson and Roberfroid, 1995) and lifestyle (Dicksved *et al.*, 2007). Dietary habits are considered to impact the microbial composition in the gut, in particular during the first years of life. For example, the faecal microbiota differs between breast-fed and formula-fed infants, with more lactic acid bacteria and bifidobacteria in the breast-fed infants (Balmer and Wharton, 1991; Harmsen *et al.*, 2000). However broadly defined diets such as a 'Western diet', comprising a high intake of fat and animal proteins but low fibre, and a 'Japanese diet', with less fat and more vegetables, have revealed only moderate differences in the gut microbiota involving a few bacterial genera (Finegold and Sutter, 1978). Similar observations were made by Aries and coworkers in 1971, that only observed small differences in the bacterial composition when persons on a normal mixed diet were compared with strict vegetarians (Aries *et al.*, 1971). In another study a shift in the bacterial SCFA profile was observed between individuals that had an uncooked extreme vegan diet for 1 month compared with individuals that had a Westernized diet. However, this shift was not reflected in the bacterial community composition (Peltonen *et al.*, 1992). Diet is likely to have larger effects on microbial communities in the small intestine due to a more diversified substrate supply in that location compared to the colon. The dietary effects of the small bowel microbiota could explain the large temporal variations of the microbiota in small intestinal digesta (Booijink *et al.*, 2007) but not in faeces (Zoetendal *et al.*, 1998). However, recently it was shown that diet could promote a division wide change of the microbiota. Obese people that were randomly assigned a fat-restricted or carbohydrate-restricted low-calorie diet over a year, showed a pronounced increase in the abundance of Bacteroidetes with a corresponding decrease of Firmicutes (Ley *et al.*, 2006).

Other lifestyle factors in addition to diet can have an impact on the gut microbiota. We found that there were significant differences in the diversity of the gut microbiota of children raised according to an anthroposophic lifestyle, characterized by a restricted use of antibiotics and ingestion of fermented vegetables, compared with children raised on a farm, characterized by higher intake of fresh milk (Dicksved *et al.*, 2007). The children raised according to the anthroposophic lifestyle had a higher gut microbial diversity than the farm children, suggesting that diet or some other uncharacterized lifestyle factors had an input on the gut microbial composition. Normally a higher diversity is considered advantageous with respect to functional resiliency of a community. However, our study also showed that there were differences in the lactobacillus communities in the children that were dominated by one or another phylotype. The functional capabilities of dominant members of the microbiota can theoretically outweigh the contribution of a high diversity of bacteria with redundant functions and it is not yet known what the importance of a high gut microbial diversity is to human health.

Diet, intestinal bacteria and colorectal cancer

The development of colorectal cancer is not only related to genetic changes but also associated with some external environmental factors such as food and internal environmental factors, including the intestinal microbiota (Kanazawa *et al.*, 1996). Studies of colorectal cancer and diet indicate human and animals with high-fat and low-fibre diets have a high risk of colorectal cancer. In 1995, Moore and Moore, compared the faecal microbial composition of polyp patients in Japanese-Hawaiians, who had a Western-type diet and were considered to have high risk of colorectal cancer, and the rural native Japanese and Africans who had very little red meat in their diets and were considered to have low risk of developing colorectal cancer. They found significant differences of the bacterial composition between the two groups: i.e. *Bacteroides vulgatus* and *Bacteroides stercoris* showed significant association with high risk individuals, whereas some *Lactobacillus* species and *Eubacterium aerofaciens* related to individuals in low risk of colon cancer (Moore and Moore, 1995). Furthermore, rats with a high-fat diet were more likely to develop colorectal cancer and had higher DNA damage compared to a control group with a normal diet when challenged by 1,2-dimethylhydrazine (DMH) (Hambly *et al.*, 1997a; Reddy *et al.*, 1975). A high-beef diet in male F344 rats resulted in increased levels of nitroreductase, azoreductase, and beta-glucuronidase produced by the colon microbiota and caecal beta-glucuronidase activity also increased in human flora-associated rats fed with a high-risk diet (Goldin and Gorbach, 1976; Hambly *et al.*, 1997b). These enzymes have been implicated in the conversion of procarcinogens into carcinogens (Gorbach and Goldin, 1990). The patients with colorectal cancer had high concentrations of faecal bile acids and cholesterol metabolites compared with the controls (Reddy and Wynder, 1977). Faecal bile acids and cholesterol metabolites which resulted from high a fat diet were then modified by the intestinal bacteria and may play an important role in large bowel carcinogenesis (Reddy *et al.*, 1975). Moreover, the tryptophan concentration and tryptophanase activity, which produced indole from tryptophan in faeces of rats fed with all meat diet, were significantly higher than rats on a normal diet. This implies that the large intestine with bacterial populations under a meat rich diet had more chances for contacting with carcinogens such as indole (Chung *et al.*, 1975). All of these suggest that high fat intake might be related to the development of colorectal cancer in humans and animals.

Culture dependent and biochemical methods for structural analysis of the gut microbiota in patients and animal models indicated that some members of the intestinal bacteria were correlated to colorectal cancer. Distinctly different microbiota compositions were found between populations with high and low risks of colorectal cancer and populations with colorectal cancer and control subjects (Legakis *et al.*, 1981; Moore and Moore, 1995; Scanlan *et al.*, 2008; Seksik *et al.*, 2003).

Clinical studies suggest that some patients with chronic intestinal inflammatory diseases such as ulcerative colitis and Crohn's disease were associated with high risk of colorectal cancer as well (Maykel *et al.*, 2006). The changes in the composition of the gut microbiota in individuals with IBD that were discussed above may also contribute to colorectal cancer development. In rats treated with DMH, *Escherichia coli*, *Enterococcus faecium*, and several strains of *Bacteroides* and *Clostridium* species aggravated the precancerous lesions of colorectal cancer, while rats that were inoculated with bacteria from a healthy donor had less lesions (Onoue *et al.*, 1997). Moreover, some herbs such as caraway, *Piper nigrum* L. with anti-microbial activity could modify some intestinal bacterial enzyme activities, including beta-glucuronidase and mucinase, and significantly suppressed the progress of preneoplastic lesions (Deeptha *et al.*, 2006, Kamaleeswari *et al.*, 2006; Nalini *et al.*, 1998). One particular microbe, namely *Streptococcus bovis,* has been implicated in promotion of colonic carcinogenesis, and antigens extracted from its cell wall have provoked proinflammatory responses associated with carcinogenesis in animal models (Biarc *et al.*, 2004; Ellmerich *et al.*, 2000).

Different members of the gut microbiota presumably play different roles in progression of colorectal cancer. Some species in the gut have been considered to produce free radicals and H_2S which could damage the colon epithelium (McGarr *et al.*, 2005; Stone and Papas, 1997). In addition, some species such as *Clostridium scindens,* have bile acid 7a-dehydroxylation activity which produces a dehydroxylation enzyme and could transform primary bile acids to secondary bile acids (Kitahara *et al.*, 2004), and these have been considered to promote tumours in the colon. Additionally, beta-glucuronidase activity of some intestinal bacteria could release activated carcinogens from an inactive glucuronide conjugate in the colon and increase the DNA adducts and ACF formation in the rat colon (Fiala, 1975).

On the other hand, some bacteria have preventive and therapeutic effects on colorectal cancer. For example, some probiotics, such as *Bifidobacterium* and *Lactobacillus* species, could inhibit DNA damage induced by carcinogens (Bolognani *et al.*, 2001; Le Leu *et al.*, 2005). Some non-digestible disaccharides could decrease the number of bile acid 7a-dehydroxylating bacteria and decrease secondary bile acid. Therefore, milk fermented by some lactobacilli could be used as colorectal cancer chemoprevention agents (Balansky *et al.*, 1999; Minamida *et al.*, 2005). The antitumor mechanisms of probiotic bacteria might be involved in decreasing the activity of some specific intestinal bacterial enzymes implicated in converting the procarcinogens to carcinogens (Goldin and Gorbach 1984; Naidu *et al.*, 1999), regulating immune response (Matsuzaki *et al.*, 2004), or producing short chain fatty acids (SCFA), such as butyrate that has been implicated in protection against colitis and colorectal cancer (McIntyre *et al.*, 1993). Several *Clostridium* spp., including some *Ruminococcus* species, have been shown to produce butyrate or decrease bile acid 7a-dehydroxylation and could thus play potential roles in protecting against colorectal cancer (Hold *et al.*, 2003; Minamida *et al.*, 2005; Pryde *et al.*, 2002).

Metabolic profiling

Metabolic profiling (metabonomics/metabolomics) is a top-down, discovery approach which can retrieve molecular composition data of samples containing mixed metabolites excreted from all living cells of an organism (Nicholson and Wilson, 2003). As urine samples may contain various metabolic wastes excreted from all types of cells (human and symbiotic bacteria) within a host body, they become one of the best 'windows of health' accessible with metabolic profiling technologies (Nicholson *et al.*, 2005). Metabolites detected in urine samples are of three physiological origins: (1) produced by host metabolic pathways, (2) excreted from gut microbiota metabolic activities and (3) co-processed by enzymes from both host and symbiotic bacteria. A new functional metagenomics method has recently been developed to understand the possible roles of members of the gut microbiota to host physiology using a systems based approach (Li *et al.*, 2008). This method is based on the assumption that if a particular species or strain of gut bacteria has an important interaction with the host, the variation of its population level in the gut will somehow affect related pathways of the host whose change can be monitored by metabolic profiling of urine metabolites. Thus, covariation analysis between molecular profiles of gut microbiota and host urine metabolites of a cohort of people over a period of time can lead to identification of associations between a particular bacterial population and host metabolites. These links can be employed as a basis for developing new hypotheses for further molecular studies to both verify the links and understand mechanisms. In the first example of this kind of analysis, urine and faecal samples were collected from a seven-member, four generation Chinese family over one month intervals. Inter-and intra-individual variations of gut microbiota and urine metabolites were monitored with PCR-DGGE analysis and NMR-based metabolic profiling respectively. A multivariate statistical analysis was applied to the two matrices of data for identifying all possible correlations between a particular gut bacterial species and urine metabolites. Ten phylotypes of gut bacteria were identified to have a linkage with at least one metabolite in the urine. A *Faecalibacterium prausnitzii*-like bacterium was found to be associated with eight urine metabolites, indicating that a wide range of host physiology may be affected by this bacterium (Li *et al.*, 2008). Inoculation of isolates of this species into germ-free pigs can be used to study the molecular mechanisms of interactions between the host and this potentially important gut bacterium (Pang *et al.*, 2007).

Method considerations

For studies of the human gut, it is valuable to have knowledge about external or host related factors that might affect the gut microbial composition and to use this information when designing studies to reduce the number of confounding factors when possible. Sample origins and types should be carefully considered depending on the scientific question to be addressed. For example, some studies have shown differences in the luminal and the mucosa-associated microbiota in the gut (Eckburg *et al.*, 2005; Lepage *et al.*, 2005; Sanduleanu *et al.*, 2001; Zoetendal *et al.*, 2002). As mentioned above, in some studies the faecal microbiota has been shown to differ from the colonic microbiota found on mucosal biopsies. When studying intestinal diseases, it is often stated that the mucosal microbiota are more relevant to study due to their closer interaction and ability to crosstalk with the host epithelial cell layer. However, mucosal samples are harder to obtain than faecal samples and sampling is invasive to the human host. By contrast, faecal samples are non- invasive, easier to collect, and thereby more feasible for use in dynamic studies of the microbiota. Another consideration to take into account is that biopsy collection is usually preceded by a washout of the bowel. However, this procedure might introduce a biased view of the mucosa-associated community when comparing with a faecal sample or an unwashed biopsy. In one study, it was shown that the faecal microbiota differed from its original composition after colonoscopy in three out of five patients. This effect was observed 6–8 weeks after colonoscopy and it is not known how long this disruption of the faecal microbiota lasts (Mai *et al.*, 2006).

The use of molecular methods for analysing the human gut microbiota has increased extensively during the past decade. These methods provide large sets of data that are commonly interpreted with multivariate statistical methods. There is no single standard method for examining differences in microbial community structure. In general, data from fingerprinting approaches are analysed by ordination approaches, such as principal component analysis (PCA; Dicksved *et al.*, 2007; Dicksved *et al.*, 2008; Jernberg *et al.*, 2005; Wang *et al.*, 2004). Cluster analysis is also commonly used for data generated from molecular fingerprints (Blackwood *et al.*, 2003; Dicksved *et al.*, 2008; Kitts, 2001; Sakata *et al.*, 2005). In cluster analysis, pairwise comparisons of samples are calculated as distances and sorted according to the calculated value of the distance (Fig. 9.4). The use of multivariate tools for interpretation of large sets of data has become invaluable to identify gradients or subgroups within a sample cohort. However, the complex nature of biological data and differences in sampling designs should be taken into account prior to selection of the analysis method. There is a danger in only relying on the observation from a multivariate statistical approach without further confirmation. Different methods sometimes give inconsistent results because they are used incorrectly.

Future perspectives

The key question of importance for inflammatory diseases of the gut remains unanswered. Is the inflammation itself causing a shift in the composition of the gut microbiota or is the shift in the microbiota the cause of inflammation? To be able to answer that question the first critical issue is to understand the relative contributions of the host and environmental factors to the dynamics of the microbiota. Current improvement in technologies, such as 454 pyrosequencing (see Chapter 2), enables a more in depth analysis of the structural composition of the microbial community than was previously possible. Future advances in sequencing technologies might lower sequencing costs and make this approach more feasible for screening large numbers of samples. This will provide important insights about microbial diversity and dynamics in large sets of individuals. In parallel, the functional repertoire of genes represented in the human microbiome and the expression of these genes as well as the metabolic products of active microbial populations in the gut (in a healthy compared to a diseased state) should be explored. This information would be useful to correlate structural shifts in the microbiota to an impaired metabolic function and will provide important insights into the beneficial or detrimental roles of specific microorganisms as an initiator/ maintainer, or guardian against intestinal diseases. These studies require good model systems to detect significant differences in human populations, and one excellent example is to continue to study identical twins.

References

Aas, J.A., Paster, B.J., Stokes, L.N., Olsen, I., and Dewhirst, F.E. (2005). Defining the normal bacterial flora of the oral cavity. J. Clin. Microbiol. *43*, 5721–5732.

Ahmad, T., Armuzzi, A., Bunce, M., Mulcahy-Hawes, K., Marshall, S.E., Orchard, T.R., Crawshaw, J., Lange, O., de Silva, A., Cook, J.T., *et al.* (2002). The molecular classification of the clinical manifestations of Crohn's disease. Gastroenterology *122*, 854–866.

Andersson, A., F., Lindberg, M., Jakobsson, H., Bäckhed, F., and Nyren, P. (2008). Comparative analysis of human gut microbiota by barcoded pyrosequencing. PLoS One *e2836*, 1–8.

Andremont, A. (2003). Commensal flora may play key role in spreading antibiotic resistance. ASM News *69*, 601–607.

Aries, V.C., Crowther, J.S., Drasar, B.S., Hill, M.J., and Ellis, F.R. (1971). The effect of a strict vegetarian diet on the faecal flora and faecal steroid concentration. J. Pathol. *103*, 54–56.

Bäckhed, F., Ley, R.E., Sonnenburg, J.L., Peterson, D.A., and Gordon, J.I. (2005). Host-bacterial mutualism in the human intestine. Science *307*, 1915–1920.

Balansky, R., Gyosheva, B., Ganchev, G., Mircheva, Z., Minkova, S., and Georgiev, G. (1999). Inhibitory effects of freeze-dried milk fermented by selected *Lactobacillus bulgaricus* strains on carcinogenesis induced by 1,2-dimethylhydrazine in rats and by diethylnitrosamine in hamsters. Cancer Lett. *147*, 125–137.

Balmer, S.E., and Wharton, B.A. (1991). Diet and faecal flora in the newborn: iron. Arch. Dis. Child. *66*, 1390–1394.

Baumgart, M., Dogan, B., Rishniw, M., Weitzman, G., Bosworth, B., Yantiss, R., Orsi, R.H., Weidmann, M., McDonough, P., Kim, S.G., *et al.* (2007). Culture independent analysis of ileal mucosa reveals a selective increase in invasive *Escherichia coli* of novel phylogeny relative to depletion of *Clostridiales* in Crohn's disease involving the ileum. ISME J. *1*, 403–418.

Bernstein, C.N., Blanchard, J.F., Rawsthorne, P., and Collins, M.T. (2004). Population-based case control study of seroprevalence of *Mycobacterium paratuberculosis* in patients with Crohn's disease and ulcerative colitis. J. Clin. Microbiol. *42*, 1129–1135.

Biarc, J., Nguyen, I.S., Pini, A., Gosse, F., Richert, S., Thierse, D., Van Dorssealer, A., Leize-Wagner, E., Raul, F., Klein, J.P., *et al.* (2004). Carcinogenic properties of proteins with pro-inflammatory activity from *Streptococcus infantarius* (formerly *S.bovis*). Carcinogenesis *25*, 1477–1484.

Biasucci, G., Benenati, B., Morelli, L., Bessi, E., and Boehm, G. (2008). Cesarean delivery may affect the early biodiversity of intestinal bacteria. J. Nutr. *138*, 1796S-1800S.

Bik, E.M., Eckburg, P.B., Gill, S.R., Nelson, K.E., Purdom, E.A., Francois, F., Perez–Perez, G., Blaser, M.J., and Relman, D.A. (2006). Molecular analysis of the bacterial microbiota in the human stomach. Proc. Natl. Acad. Sci. U.S.A. *103*, 732–737.

Björksten, B., Naaber, P., Sepp, E., and Mikelsaar, M. (1999). The intestinal microflora in allergic Estonian and Swedish 2-year-old children. Clin. Exp. Allergy *29*, 342–346.

Blackwood, C.B., Marsh, T., Kim, S.H., and Paul, E.A. (2003). Terminal restriction fragment length polymorphism data analysis for quantitative comparison of microbial communities. Appl. Environ. Microbiol. *69*, 926–932.

Bolognani, F., Rumney, C.J., Pool-Zobel, B.L., and Rowland, I.R. (2001). Effect of lactobacilli, bifidobacteria and inulin on the formation of aberrant crypt foci in rats. Eur. J. Nutr. *40*, 293–300.

Booijink, C.C., Zoetendal, Kleerebezem, E.G., M., and de Vos, W.M. (2007). Microbial communities in the human small intestine: coupling diversity to metagenomics. Future Microbiol. *2*, 285–295.

Brook, I., Barrett, C.T., Brinkman, C.R., 3rd, Martin, W.J., and Finegold, S.M. (1979). Aerobic and anaerobic bacterial flora of the maternal cervix and newborn gastric fluid and conjunctiva: a prospective study. Pediatrics *63*, 451–455.

Bull, T.J., McMinn, E.J., Sidi-Boumedine, K., Skull, A., Durkin, D., Neild, P., Rhodes, G., Pickup, R., and Hermon-Taylor, J. (2003). Detection and verification of *Mycobacterium avium* subsp. *paratuberculosis* in fresh ileocolonic mucosal biopsy specimens from individuals with and without Crohn's disease. J. Clin. Microbiol. *41*, 2915–2923.

Chen, W., Li, D., Paulus, B., Wilson, I., and Chadwick, V.S. (2000). Detection of *Listeria monocytogenes* by polymerase chain reaction in intestinal mucosal biopsies from patients with inflammatory bowel disease and controls. J. Gastroenterol. Hepatol. *15*, 1145–1150.

Chiodini, R.J. (1989). Crohn's disease and the mycobacterioses: a review and comparison of two disease entities. Clin. Microbiol. Rev. *2*, 90–117.

Chung, K.T., Fulk, G.E., and Slein, M.W. (1975). Tryptophanase of fecal flora as a possible factor in the etiology of colon cancer. J. Natl. Cancer Inst. *54*, 1073–1078.

Conte, M.P., Schippa, S., Zamboni, I., Penta, M., Chiarini, F., Seganti, L., Osborn, J., Falconieri, P., Borelli, O., and Cucchiara, S. (2006). Gut-associated bacterial microbiota in paediatric patients with inflammatory bowel disease. Gut *55*, 1760–1767.

Darfeuille-Michaud, A., Neut, C., Barnich, N., Lederman, E., Di Martino, P., Desreumaux, P., Gambiez, L., Joly, B., Cortot, A., and Colombel, J.F. (1998). Presence of adherent *Escherichia coli* strains in ileal mucosa of patients with Crohn's disease. Gastroenterology *115*, 1405–1413.

Deeptha, K., Kamaleeswari, M., Sengottuvelan, M., and Nalini, N. (2006). Dose dependent inhibitory effect of dietary caraway on 1,2-dimethylhydrazine induced colonic aberrant crypt foci and bacterial enzyme activity in rats. Invest. New Drugs *24*, 479–488.

Dethlefsen, L., McFall-Ngai, M., and Relman, D.A. (2007). An ecological and evolutionary perspective on human-microbe mutualism and disease. Nature *449*, 811–818.

Dethlefsen, L., Huse, S., Sogin, M.L., and Relman, D.A. (2008). The pervasive effects of an antibiotic on the human gut microbiota, as revealed by deep 16S rRNA sequencing. PLoS Biol. *6*, e280.

Dicksved, J., Flöistrup, H., Bergström, A., Rosenquist, M., Pershagen, G., Scheynius, A., Roos, S., Alm, J.S., Engstrand, L., Braun-Fahrländer, C., *et al.* (2007). Molecular fingerprinting of the fecal microbiota of children raised according to different lifestyles. Appl. Environ. Microbiol. *73*, 2284–2289.

Dicksved, J., Halfvarson, J., Rosenquist, M., Järnerot, G., Tysk, C., Apajalahti, J., Engstrand, L., and Jansson, J.K. (2008). Molecular analysis of the gut microbiota of identical twins with Crohn's disease. ISME J. *2*, 716–727.

Dicksved, J., Lindberg, M., Rosenquist, M., Enroth, H., Jansson, J.K., and Engstrand, L. (2009). Molecular characterization of the stomach microbiota in patients with gastric cancer and controls. J. Med. Microbiol. *58*, 509–516.

Eckburg, P.B., Bik, E.M., Bernstein, C.N., Purdom, E., Dethlefsen, L., Sargent, M., Gill, S.R., Nelson, K.E., and Relman, D.A. (2005). Diversity of the human intestinal microbial flora. Science *308*, 1635–1638.

Ekbom, A., Daszak, P., Kraaz, W., and Wakefield, A.J. (1996). Crohn's disease after in-utero measles virus exposure. Lancet *348*, 515–517.

Ellmerich, S., Schöller, M., Duranton, B., Gosse, F., Galluser, M., Klein, J.P., and Raul, F. (2000). Promotion of intestinal carcinogenesis by *Streptococcus bovis*. Carcinogenesis *21*, 753–756.

Favier, C., Neut, C., Mizon, C., Cortot, A., Colombel, J.F., and Mizon, J. (1997). Fecal beta-D-galactosidase production and bifidobacteria are decreased in Crohn's disease. Dig. Dis. Sci. *42*, 817–822.

Favier, C.F., Vaughan, E.E., De Vos, W.M., and Akkermans, A.D. (2002). Molecular monitoring of succession of bacterial communities in human neonates. Appl. Environ. Microbiol. *68*, 219–226.

Fiala, E. (1975). Investigations into the metabolism and mode of action of the colon carcinogen 1, 2-dimethylhydrazine. Cancer *36*, 2407–2412.

Finegold, S.M., and Sutter, V.L. (1978). Fecal flora in different populations, with special reference to diet. Am. J. Clin. Nutr. *31*, S116-S122.

Finegold, S.M., Sutter, V.L., and Mathisen, G.E. (1983). Normal indigenous intestinal flora. In Human intestinal microflora in health and disease, Hentges, D.J., ed. (New York: Academic Press), pp. 3–29.

Fisher, N.C., Yee, L., Nightingale, P., McEwan, R., and Gibson, J.A. (1997). Measles virus serology in Crohn's disease. Gut *41*, 66–69.

Fite, A., Macfarlane, G.T., Cummings, J.H., Hopkins, M.J., Kong, S.C., Furrie, E., and Macfarlane, S. (2004). Identification and quantitation of mucosal and faecal desulfovibrios using real time polymerase chain reaction. Gut *53*, 523–529.

Frank, D.N., St Amand, A.L., Feldman, R.A., Boedeker, E.C., Harpaz, N., and Pace, N.R. (2007). Molecular-phylogenetic characterization of microbial community imbalances in human inflammatory bowel diseases. Proc. Natl. Acad. Sci. U.S.A. *104*, 13780–13785.

Franks, A.H., Harmsen, H.J., Raangs, G.C., Jansen, G.J., Schut, F., and Welling, G.W. (1998). Variations of bacterial populations in human feces measured by fluorescent *in situ* hybridization with group-specific 16S rRNA-targeted oligonucleotide probes. Appl. Environ. Microbiol. *64*, 3336–3345.

Fujita, H., Eishi, Y., Ishige, I., Saitoh, K., Takizawa, T., Arima, T., and Koike, M. (2002). Quantitative analysis of bacterial DNA from Mycobacteria spp., *Bacteroides vulgatus*, and *Escherichia coli* in tissue samples from patients with inflammatory bowel diseases. J. Gastroenterol. *37*, 509–516.

Genta, R.M., Gurer, I.E., and Graham, D.Y. (1995). Geographical pathology of *Helicobacter pylori* infection: is there more than one gastritis? Ann. Med. *27*, 595–599.

Giaffer, M.H., Holdsworth, C.D., and Duerden, B.I. (1991). The assessment of faecal flora in patients with inflammatory bowel disease by a simplified bacteriological technique. J. Med. Microbiol. *35*, 238–243.

Gibson, G.R., and Roberfroid, M.B. (1995). Dietary modulation of the human colonic microbiota: introducing the concept of prebiotics. J. Nutr. *125*, 1401–1412.

Gibson, G.R., Macfarlane, G.T., and Cummings, J.H. (1993). Sulphate reducing bacteria and hydrogen metabolism in the human large intestine. Gut *34*, 437–439

Gill, S.R., Pop, M., Deboy, R.T., Eckburg, P.B., Turnbaugh, P.J., Samuel, B.S., Gordon, J.I., Relman, D.A., Fraser-Liggett, C.M., and Nelson, K.E. (2006). Metagenomic analysis of the human distal gut microbiome. Science *312*, 1355–1359.

Goldin, B.R., and Gorbach, S.L. (1976). The relationship between diet and rat fecal bacterial enzymes implicated in colon cancer. J. Natl. Cancer Inst. *57*, 371–375.

Goldin, B.R., and Gorbach, S.L. (1984). The effect of milk and *Lactobacillus* feeding on human intestinal bacterial enzyme activity. Am. J. Clin. Nutr. *39*, 756–761

Gorbach, S.L., and Goldin, B.R. (1990). The intestinal microflora and the colon cancer connection. Rev. Infect Dis. *12* Suppl. *2*. S252-S261.

Gorbach, S.L. (2000). Probiotics and gastrointestinal health. Am. J. Gastroenterol. *95*, S2–4.

Grönlund, M.M., Lehtonen, O.P., Eerola, E., and Kero, P. (1999). Fecal microflora in healthy infants born by different methods of delivery: permanent changes in intestinal flora after cesarean delivery. J. Pediatr. Gastroenterol. Nutr. *28*, 19–25.

Guarner, F. (2006). Enteric flora in health and disease. Digestion 73 *Suppl 1*, 5–12.

Guarner, F., Bourdet-Sicard, R., Brandtzaeg, P., Gill, H.S., McGuirk, P., van Eden, W., Versalovic, J., Weinstock, J.V., and Rook, G.A. (2006). Mechanisms of disease: the hygiene hypothesis revisited. Nat. Clin. Pract. Gastroenterol. Hepatol. *3*, 275–284.

Hambly, R.J., Rumney, C.J., Cunninghame, M., Fletcher, J.M., Rijken, P.J., and Rowland, I.R. (1997a). Influence of diets containing high and low risk factors for colon cancer on early stages of carcinogenesis in human flora-associated (HFA) rats. Carcinogenesis *18*, 1535–1539.

Hambly, R.J., Rumney, C.J., Fletcher, J.M., Rijken, P.J., and Rowland, I.R. (1997b). Effects of high- and low-risk diets on gut microflora-associated biomarkers of colon cancer in human flora-associated rats. Nutr. Cancer *27*, 250–255.

Harmsen, H.J., Wildeboer-Veloo, A.C., Raangs, G.C., Wagendorp, A.A., Klijn, N., Bindels, J.G., and Welling, G.W. (2000). Analysis of intestinal flora development in breast-fed and formula-fed infants by using molecular identification and detection methods. J. Pediatr. Gastroenterol. Nutr. *30*, 61–67.

Harper, P.H., Truelove, S.C., Lee, E.C., Kettlewell, M.G., and Jewell, D.P. (1983). Split ileostomy and ileocolostomy for Crohn's disease of the colon and ulcerative colitis: a 20 year survey. Gut *24*, 106–113.

Hayashi, H., Takahashi, R., Nishi, T., Sakamoto, M., and Benno, Y. (2005). Molecular analysis of jejunal, ileal, caecal and recto-sigmoidal human colonic microbiota using 16S rRNA gene libraries and terminal restriction fragment length polymorphism. Journal of Medical Microbiology *54*, 1093–1101.

Hedberg, M., and Nord, C.E. (2003). Antimicrobial susceptibility of *Bacteroides fragilis* group isolates in Europe. Clin. Microbiol. Infect. *9*, 475–488.

Hold, G.L., Schwiertz, A., Aminov, R.I., Blaut, M., and Flint, H.J. (2003). Oligonucleotide probes that detect quantitatively significant groups of butyrate-producing bacteria in human feces. Appl. Environ. Microbiol. *69*, 4320–4324.

Hooper, L.V., Xu, J., Falk, P.G., Midtvedt, T., and Gordon, J.I. (1999). A molecular sensor that allows a gut commensal to control its nutrient foundation in a competitive ecosystem. Proc. Natl. Acad. Sci. U.S.A. *96*, 9833–9838.

Hooper, L.V., Midtvedt, T., and Gordon, J.I. (2002). How host–microbial interactions shape the nutrient environment of the mammalian intestine. Annu. Rev. Nutr. *22*, 283–307.

Imaoka, A., Setoyama, H., Takagi, A., Matsumoto, S., and Umesaki, Y. (2004). Improvement of human faecal flora-associated mouse model for evaluation of the functional foods. J. Appl. Microbiol. *96*, 656–663.

Jenkinson, H.F., and Lamont, R.J. (2005). Oral microbial communities in sickness and in health. Trends in Microbiology *13*, 589–595.

Jernberg, C., Sullivan, Å., Edlund, C., and Jansson, J.K. (2005). Monitoring of antibiotic-induced alterations in the human intestinal microflora and detection of probiotic strains by use of terminal restriction fragment length polymorphism. Appl. Environ. Microbiol. *71*, 501–506.

Jernberg, C., Löfmark, S., Edlund, C., and Jansson, J.K. (2007). Long-term ecological impacts of antibiotic administration on the human intestinal microbiota. ISME J. *1*, 56–66.

Kamaleeswari, M., Deeptha, K., Sengottuvelan, M., and Nalini, N. (2006). Effect of dietary caraway (*Carum carvi* L.) on aberrant crypt foci development, fecal steroids, and intestinal alkaline phosphatase activities in 1,2-dimethylhydrazine-induced colon carcinogenesis. Toxicol. Appl. Pharmacol. *214*, 290–296.

Kanazawa, K., Konishi, F., Mitsuoka, T., Terada, A., Itoh, K., Narushima, S., Kumemura, M., and Kimura, H. (1996). Factors influencing the development of sigmoid colon cancer. Bacteriologic and biochemical studies. Cancer *77*, 1701–1706.

Kanazawa, K., Haga, Y., Funakoshi, O., Nakajima, H., Munakata, A., and Yoshida, Y. (1999). Absence of *Mycobacterium paratuberculosis* DNA in intestinal tissues from Crohn's disease by nested polymerase chain reaction. J. Gastroenterol. *34*, 200–206.

Kitahara, M., Sakata, S., Sakamoto, M., and Benno, Y. (2004). Comparison among fecal secondary bile acid levels, fecal microbiota and *Clostridium scindens* cell numbers in Japanese. Microbiol. Immunol. *48*, 367–375.

Kitts, C.L. (2001). Terminal restriction fragment patterns: a tool for comparing microbial communities and assessing community dynamics. Curr. Issues Intest. Microbiol. *2*, 17–25.

Kleessen, B., Kroesen, A.J., Buhr, H.J., and Blaut, M. (2002). Mucosal and invading bacteria in patients with inflammatory bowel disease compared with controls. Scand. J. Gastroenterol. *37*, 1034–1041.

Kotlowski, R., Bernstein, C.N., Sepehri, S., and Krause, D.O. (2007). High prevalence of *Escherichia coli* belonging to the B2+D phylogenetic group in inflammatory bowel disease. Gut *56*, 669–675.

Kurokawa, K., Itoh, T., Kuwahara, T., Oshima, K., Toh, H., Toyoda, A., Takami, H., Morita, H., Sharma, V.K., Srivastava, T.P., *et al.* (2007). Comparative metagenomics revealed commonly enriched gene sets in human gut microbiomes. DNA Res. *14*, 169–181.

Lay, C., Rigottier-Gois, L., Holmström, Rajilic, K., M., Vaughan, E.E., de Vos, W.M., Collins, M.D., Thiel, R., Namsolleck, P., Blaut, M., *et al.* (2005). Colonic microbiota signatures across five northern European countries. Appl. Environ. Microbiol. *71*, 4153–4155.

Le Leu, R.K., Brown, I.L., Hu, Y., Bird, A.R., Jackson, M., Esterman, A., and Young, G.P. (2005). A synbiotic combination of resistant starch and *Bifidobacterium lactis* facilitates apoptotic deletion of carcinogen-damaged cells in rat colon. J. Nutr. *135*, 996–1001.

Legakis, N.J., Ioannides, H., Tzannetis, S., Golematis, B., and Papavassiliou, J. (1981). Faecal bacterial flora in patients with colon cancer and control subjects. Zentralbl. Bakteriol. Mikrobiol. Hyg. [A] *251*, 54–61.

Lepage, P., Seksik, P., Sutren, M., de la Cochetiere, M.F., Jian, R., Marteau, P., and Dore, J. (2005). Biodiversity of the mucosa-associated microbiota is stable along the distal digestive tract in healthy individuals and patients with IBD. Inflamm. Bowel Dis. *11*, 473–480.

Leser, T.D., Amenuvor, J.Z., Jensen, T.K., Lindecrona, R.H., Boye, M.,and Möller, K. (2002). Culture-independent analysis of gut bacteria: the pig gastrointestinal tract microbiota revisited. Appl. Environ. Microbiol. *68*, 673–690.

Ley, R.E., Bäckhed, F., Turnbaugh, P., Lozupone, C.A., Knight, R.D., and Gordon, J.I. (2005). Obesity alters gut microbial ecology. Proc. Natl. Acad. Sci. U.S.A. *102*, 11070–11075.

Ley, R.E., Turnbaugh, P.J., Klein, S., and Gordon, J.I. (2006). Microbial ecology: human gut microbes associated with obesity. Nature *444*, 1022–1023.

Ley, R.E., Hamady, M., Lozupone, C., Turnbaugh, P.J., Ramey, R.R., Bircher, J.S., Schlegel, M.L., Tucker, T.A., Schrenzel, M.D., Knight, R., *et al.* (2008). Evolution of mammals and their gut microbes. Science *320*, 1647–1651.

Li, M., Wang, B., Zhang, M., Rantalainen, M., Wang, S., Zhou, H., Zhang, Y., Shen, J., Pang, X., Zhang, M., *et al.* (2008). Symbiotic gut microbes modulate human metabolic phenotypes. Proc. Natl. Acad. Sci. U.S.A. *105*, 2117–2122.

Löfmark, S., Jernberg, C., Jansson, J.K., and Edlund, C. (2006). Clindamycin-induced enrichment and long-term persistence of resistant *Bacteroides* spp. and resistance genes. J. Antimicrob. Chemother. *58*, 1160–1167.

Lucke, K., Miehlke, S., Jacobs, E., and Schuppler, M. (2006). Prevalence of *Bacteroides* and *Prevotella* spp. in ulcerative colitis. J. Med. Microbiol. *55*, 617–624.

Macfarlane, S., Furrie, E., Cummings, J.H., and Macfarlane, G.T. (2004). Chemotaxonomic analysis of bacterial populations colonizing the rectal mucosa in patients with ulcerative colitis. Clin. Infect. Dis. *38*, 1690–1699.

Mai, V., Greenwald, B., Morris, J.G., Jr, Raufman, J.P., and Stine, O.C. (2006). Effect of bowel preparation and colonoscopy on post-procedure intestinal microbiota composition. Gut *55*, 1822–1823.

Manichanh, C., Rigottier-Gois, L., Bonnaud, E., Gloux, K., Pelletier, E., Frangeul, L., Nalin, R., Jarrin, C., Chardon, P., Marteau, P., *et al.* (2006). Reduced diversity of faecal microbiota in Crohn's disease revealed by a metagenomic approach. Gut *55*, 205–211.

Marchesi, J.R., Holmes, E., Khan, F., Kochhar, S., Scanlan, P., Shanahan, F., Wilson, I.D., and Wang, Y. (2007). Rapid and noninvasive metabonomic characterization of inflammatory bowel disease. J. Proteome Res. *6*, 546–551.

Marshall, B.J., and Warren, J.R. (1984). Unidentified curved bacilli in the stomach of patients with gastritis and peptic ulceration. Lancet *1*, 1311–1315.

Marteau, P., Lepage, P., Mangin, I., Suau, A., Dore, J., Pochart, P., and Seksik, P. (2004). Review article: gut flora and inflammatory bowel disease. Aliment. Pharmacol. Ther. *20 Suppl 4*, 18–23.

Martinez, J.L., and Baquero, F. (2000). Mutation frequencies and antibiotic resistance. Antimicrob. Agents Chemother. *44*, 1771–1777.

Matsuzaki, T.A., Takagi, A., Ikemura, H., Matsuguchi, T., and Yokokura, T. (2004). Antitumor activity and action mechanisms of *Lactobacillus casei* through regulation of immune responses. Biofactors *22*, 63–66.

Maykel, J.A., Hagerman, G., Mellgren, A.F., Li, S.Y., Alavi, K., Baxter, N.N., and Madoff, R.D. (2006). Crohn's colitis: the incidence of dysplasia and adenocarcinoma in surgical patients. Dis. Colon Rectum *49*, 950–957.

McGarr, S.E., Ridlon, J.M., and Hylemon, P.B. (2005). Diet, anaerobic bacterial metabolism, and colon cancer: a review of the literature. J. Clin. Gastroenterol. *39*, 98–109.

McIntyre, A., Gibson, P.R., and Young, G.P. (1993). Butyrate production from dietary fibre and protection against large bowel cancer in a rat model. Gut *34*, 386–391.

Minamida, K., Kaneko, M., Ohashi, M., Sujaya, I.N., Sone, T., Wada, M., Yokota, A., Hara, H., Asano, K.,and Tomita, F. (2005). Effects of difructose anhydride III (DFA III) administration on bile acids and growth of DFA III-assimilating bacterium *Ruminococcus productus* on rat intestine. J. Biosci. Bioeng. *99*, 548–554.

Moore, W.E., and Moore, L.H. (1995). Intestinal floras of populations that have a high risk of colon cancer. Appl. Environ. Microbiol. *61*, 3202–3207.

Mpofu, C.M., Campbell, B.J., Subramanian, S., Marshall-Clarke, S., Hart, C.A., Cross, A., Roberts, C.L., McGoldrick, A., Edwards, S.W., and Rhodes, J.M. (2007). Microbial mannan inhibits bacterial killing by macrophages: a possible pathogenic mechanism for Crohn's disease. Gastroenterology *133*, 1487–1498.

Mueller, S., Saunier, K., Hanisch, C., Norin, E., Alm, L., Midtvedt, T., Cresci, A., Silvi, S., Orpianesi, C., Verdenelli, M.C., *et al.* (2006). Differences in fecal microbiota in different European study populations in relation to age, gender, and country: a cross-sectional study. Appl. Environ. Microbiol. *72*, 1027–1033.

Mylonaki, M., Rayment, N.B., Rampton, D.S., Hudspith, B.N., and Brostoff, J. (2005). Molecular characterization of rectal mucosa-associated bacterial flora in inflammatory bowel disease. Inflamm. Bowel Dis. *11*, 481–487.

Naidu, A.S., Bidlack, W.R., and Clemens, R.A. (1999). Probiotic spectra of lactic acid bacteria (LAB). Crit. Rev. Food Sci. Nutr. *39*, 13–126

Nalini, N., Sabitha, K., Viswanathan, P., and Menon, V.P. (1998). Influence of spices on the bacterial (enzyme) activity in experimental colon cancer. J. Ethnopharmacol. *62*, 15–24

Nicholson, J.K., and Wilson, I.D. (2003). Opinion: understanding 'global' systems biology: metabonomics and the continuum of metabolism. Nat. Rev. Drug Discov. *2*, 668–676.

Nicholson, J.K., Holmes, E., and Wilson, I.D. (2005). Gut microorganisms, mammalian metabolism and personalized health care. Nat. Rev. Microbiol. *3*, 431–438.

Nielsen, D.S., Möller, P.L., Rosenfeldt, V., Paerregaard, A., Michaelsen, K.F., and Jakobsen, M. (2003). Case study of the distribution of mucosa-associated *Bifidobacterium* species, *Lactobacillus* species, and other lactic acid bacteria in the human colon. Appl. Environ. Microbiol. *69*, 7545–7548.

Noverr, M.C., and Huffnagle, G.B. (2004). Does the microbiota regulate immune responses outside the gut? Trends Microbiol. *12*, 562–568.

O'Hara, A.M., and Shanahan, F. (2006). The gut flora as a forgotten organ. EMBO Rep. *7*, 688–693.

Oliveira, A.G., das Gracas Pimenta Sanna, M., Rocha, G.A., Rocha, A.M., Santos, A., Dani, R., Marinho, F.P., Moreira, L.S., de Lourdes Abreu Ferrari, M., Moura, S.B., *et al.* (2004). *Helicobacter* species in the intestinal mucosa of patients with ulcerative colitis. J. Clin. Microbiol. *42*, 384–386.

Onoue, M., Kado, S., Sakaitani, Y., Uchida, K., and Morotomi, M. (1997). Specific species of intestinal bacteria influence the induction of aberrant crypt foci by 1,2-dimethylhydrazine in rats. Cancer Lett. *113*, 179–186.

Orholm, M., Binder, V., Sorensen, T.I., Rasmussen, L.P., and Kyvik, K.O. (2000). Concordance of inflammatory bowel disease among Danish twins. Results of a nationwide study. Scand. J. Gastroenterol. *35*, 1075–1081.

Ott, S.J., Musfeldt, M., Wenderoth, D.F., Hampe, J., Brant, O., Fölsch, U.R., Timmis, K.N., and Schreiber, S. (2004). Reduction in diversity of the colonic mucosa associated microflora in patients with active inflammatory bowel disease. Gut *53*, 685–693.

Palmer, C., Bik, E.M., Digiulio, D.B., Relman, D.A., and Brown, P.O. (2007). Development of the human infant intestinal microbiota. PLoS Biol. *5*, 1556–1573.

Pang, X., Hua, X., Yang, Q., Ding, D., Che, C., Cui, L., Jia, W., Bucheli, P., and Zhao, L. (2007). Inter-species transplantation of gut microbiota from human to pigs. ISME J. *1*, 156–162.

Patterson, J.K., Lei, X.G., and Miller, D.D. (2008). The pig as an experimental model for elucidating the mechanisms governing dietary influence on mineral absorption. Exp. Biol. Med. (Maywood) *233*, 651–664.

Pei, Z., Bini, E.J., Yang, L., Zhou, M., Francois, F., and Blaser, M.J. (2004). Bacterial biota in the human distal esophagus. Proc. Natl. Acad. Sci. U.S.A. *101*, 4250–4255.

Peltonen, R., Ling, W.H., Hanninen, O., and Eerola, E. (1992). An uncooked vegan diet shifts the profile of human fecal microflora: computerized analysis of direct stool sample gas-liquid chromatography profiles of bacterial cellular fatty acids. Appl. Environ. Microbiol. *58*, 3660–3666.

Pitcher, M.C., Beatty, E.R., and Cummings, J.H. (2000). The contribution of sulphate reducing bacteria and 5-aminosalicylic acid to faecal sulphide in patients with ulcerative colitis. Gut *46*, 64–72.

Poxton, I.R., Brown, R., Sawyerr, A., and Ferguson, A. (1997). Mucosa-associated bacterial flora of the human colon. J. Med. Microbiol. *46*, 85–91.

Pryde, S.E., Duncan, S.H., Hold, G.L., Stewart, C.S., and Flint, H.J. (2002). The microbiology of butyrate formation in the human colon. FEMS Microbiol. Lett. *217*, 133–139.

Raibaud, P., Ducluzeau, R., Dubos, F., Hudault, S., Bewa, H., and Muller, M.C. (1980). Implantation of bacteria from the digestive tract of man and various animals into gnotobiotic mice. Am. J. Clin. Nutr. *33*, 2440–2447.

Reddy, B.S. (1999). Possible mechanisms by which pro- and prebiotics influence colon carcinogenesis and tumor growth. J. Nutr. *129*, 1478S-1482S.

Reddy, B.S., and Wynder, E.L. (1977). Metabolic epidemiology of colon cancer. Fecal bile acids and neutral sterols in colon cancer patients and patients with adenomatous polyps. Cancer *39*, 2533–2539.

Reddy, B.S., Narisawa, T., Maronpot, R., Weisburger, J.H., and Wynder, E.L. (1975). Animal models for the study of dietary factors and cancer of the large bowel. Cancer Res. *35*, 3421–3426.

Ruseler-van Embden, J.G.,and Both-Patoir, H.C. (1983). Anaerobic gram-negative faecal flora in patients with Crohn's disease and healthy subjects. Antonie Van Leeuwenhoek *49*, 125–132.

Sakata, S., Tonooka, T., Ishizeki, S., Takada, M., Sakamoto, M., Fukuyama, M., and Benno, Y. (2005). Culture-independent analysis of fecal microbiota in infants, with special reference to *Bifidobacterium* species. FEMS Microbiol. Lett. *243*, 417–423.

Salminen, S., Gibson, G.R., McCartney, A.L., and Isolauri, E. (2004). Influence of mode of delivery on gut microbiota composition in seven year old children. Gut *53*, 1388–1389.

Sanduleanu, S., Jonkers, D., De Bruine, A., Hameeteman, W., and Stockbrugger, R.W. (2001). Non-*Helicobacter pylori* bacterial flora during acid-suppressive therapy: differential findings in gastric juice and gastric mucosa. Aliment. Pharmacol. Ther. *15*, 379–388.

Sartor, R.B. (2003). Targeting enteric bacteria in treatment of inflammatory bowel diseases: why, how, and when. Curr. Opin. Gastroenterol. *19*, 358–365.

Sartor, R.B. (2004). Therapeutic manipulation of the enteric microflora in inflammatory bowel diseases: antibiotics, probiotics, and prebiotics. Gastroenterology *126*, 1620–1633.

Sartor, R.B. (2005). Does *Mycobacterium avium* subspecies *paratuberculosis* cause Crohn's disease? Gut *54*, 896–898.

Satsangi, J., Silverberg, M.S., Vermeire, S., and Colombel, J.F. (2006). The Montreal classification of inflammatory bowel disease: controversies, consensus, and implications. Gut *55*, 749–753.

Savage, D.C. (1977). Microbial ecology of the gastrointestinal tract. Ann. Rev. Microbiol. *31*, 107–133.

Scanlan, P.D., Shanahan, F., Clune, Y., Collins, J.K., O'Sullivan, G.C., O'Riordan, M., Holmes, E., Wang, Y., and Marchesi, J.R. (2008). Culture-independent analysis of the gut microbiota in colorectal cancer and polyposis. Environ. Microbiol. *10*, 789–798.

Seksik, P., Rigottier-Gois, L., Gramet, G., Sutren, M., Pochart, P., Marteau, P., Jian, R., and Dore, J. (2003). Alterations of the dominant faecal bacterial groups in patients with Crohn's disease of the colon. Gut *52*, 237–242.

Seksik, P., Lepage, P., de la Cochetiere, M.F., Bourreille, A., Sutren, M., Galmiche, J.P., Dore, J., and Marteau, P. (2005). Search for localized dysbiosis in Crohn's disease ulcerations by temporal temperature gradient gel electrophoresis of 16S rRNA. J. Clin. Microbiol. *43*, 4654–4658.

Sjölund, M., Wreiber, K., Andersson, D.I., Blaser, M.J., and Engstrand, L. (2003). Long-term persistence of resistant *Enterococcus* species after antibiotics to eradicate *Helicobacter pylori*. Ann. Intern. Med. *139*, 483–487.

Sokol, H., Seksik, P., Rigottier-Gois, L., Lay, C., Lepage, P., Podglajen, I., Marteau, P.,and Dore, J. (2006).

Specificities of the fecal microbiota in inflammatory bowel disease. Inflamm. Bowel Dis. *12*, 106–111.

Sokol, H., Pigneur, B., Watterlot, L., Lakhdari, O., Bermudez-Humaran, L.G., Gratadoux, J.J., Blugeon, S., Bridonneau, C., Furet, J.P., Corthier, G., *et al.* (2008). *Faecalibacterium prausnitzii* is an anti-inflammatory commensal bacterium identified by gut microbiota analysis of Crohn's disease patients. Proc. Natl. Acad. Sci. U.S.A. *105*, 16731–16736.

Stewart, J.A., Chadwick, V.S., and Murray, A. (2005). Investigations into the influence of host genetics on the predominant eubacteria in the faecal microflora of children. J. Med. Microbiol. *54*, 1239–1242.

Stone, W.L., and Papas, A.M. (1997). Tocopherols and the etiology of colon cancer. J. Natl. Cancer Inst. *89*, 1006–1014.

Strachan, D.P. (1989). Hay fever, hygiene, and household size. BMJ *299*, 1259–1260.

Streutker, C.J., Bernstein, C.N., Chan, V.L., Riddell, R.H., and Croitoru, K. (2004). Detection of species-specific helicobacter ribosomal DNA in intestinal biopsy samples from a population-based cohort of patients with ulcerative colitis. J. Clin. Microbiol. *42*, 660–664.

Sullivan, A., Edlund, C., and Nord, C.E. (2001). Effect of antimicrobial agents on the ecological balance of human microflora. Lancet Infect Dis. *1*, 101–114.

Sullivan, A., Barkholt, L., and Nord, C.E. (2003). *Lactobacillus acidophilus, Bifidobacterium lactis* and *Lactobacillus* F19 prevent antibiotic-associated ecological disturbances of Bacteroides fragilis in the intestine. J. Antimicrob. Chemother. *52*, 308–311.

Sullivan, A., and Nord, C.E. (2005). Probiotics and gastrointestinal diseases. J. Intern. Med. *257*, 78–92.

Swidsinski, A., Ladhoff, A., Pernthaler, A., Swidsinski, S., Loening-Baucke, V., Ortner, M., Weber, J., Hoffman, U., Schreiber, S., Dietel, M., *et al.* (2002). Mucosal flora in inflammatory bowel disease. Gastroenterology *122*, 44–54.

Swidsinski, A., Weber, J., Loening-Baucke, V., Hale, L.P., and Lochs, H. (2005). Spatial organization and composition of the mucosal flora in patients with inflammatory bowel disease. J. Clin. Microbiol. *43*, 3380–3389.

Swidsinski, A., Loening-Baucke, V., Theissig, F., Engelhardt, H., Bengmark, S., Koch, S., Lochs, H., and Dorffel, Y. (2007). Comparative study of the intestinal mucus barrier in normal and inflamed colon. Gut *56*, 343–350.

Takaishi, H., Matsuki, T., Nakazawa, A., Takada, T., Kado, S., Asahara, T., Kamada, N., Sakuraba, A., Yaijma, T., Higuchi, H., *et al.* (2008). Imbalance in intestinal microflora constitution could be involved in the pathogenesis of inflammatory bowel disease. Int. J. Med. Microbiol. *298*, 463–472.

Tannock, G.W. (1999). Analysis of the intestinal microflora: a renaissance. Antonie Van Leeuwenhoek *76*, 265–278.

Thayer, WR., Jr, Coutu, J.A., Chiodini, R.J., Van Kruiningen, H.J., and Merkal, R.S. (1984). Possible role of mycobacteria in inflammatory bowel disease. II. Mycobacterial antibodies in Crohn's disease. Dig. Dis. Sci. *29*, 1080–1085.

Thompson, N.P., Driscoll, R., Pounder, R.E., and Wakefield, A.J. (1996). Genetics versus environment in inflammatory bowel disease: results of a British twin study. BMJ *312*, 95–96.

Topping, D.L., and Clifton, P.M. (2001). Short-chain fatty acids and human colonic function: roles of resistant starch and nonstarch polysaccharides. Physiol. Rev. *81*, 1031–1064.

Tysk, C., Lindberg, E., Järnerot, G., and Floderus-Myrhed, B. (1988). Ulcerative colitis and Crohn's disease in an unselected population of monozygotic and dizygotic twins. A study of heritability and the influence of smoking. Gut *29*, 990–996.

Van de Merwe, J.P., Stegeman, J.H.,and Hazenberg, M.P. (1983). The resident faecal flora is determined by genetic characteristics of the host. Implications for Crohn's disease? Antonie van Leeuwenhoek *49*, 119–124.

van der Waaij, L.A., Harmsen, H.J., Madjipour, M., Kroese, F.G., Zwiers, M., van Dullemen, H.M., de Boer, N.K., Welling, G.W.,and Jansen, P.L. (2005). Bacterial population analysis of human colon and terminal ileum biopsies with 16S rRNA-based fluorescent probes: commensal bacteria live in suspension and have no direct contact with epithelial cells. Inflamm. Bowel Dis. *11*, 865–871.

Vanhoutte, T., Huys, G., De Brandt, E., and Swings, J. (2004). Temporal stability analysis of the microbiota in human feces by denaturing gradient gel electrophoresis using universal and group-specific 16S rRNA gene primers. FEMS Microbiol. Ecol. *48*, 437–446.

Verberkmoes, N.C., Russell, A.L., Shah, M., Godzik, A., Rosenquist, M., Halfvarson, J., Lefsrud, M.G., Apajalahti, J., Tysk, C., Hettich, R.L., *et al.* (2009). Shotgun metaproteomics of the human distal gut microbiota. ISME J. 3, 179–189.

Wakefield, A.J., Pittilo, R.M., Sim, R., Cosby, S.L., Stephenson, J.R., Dhillon, A.P., and Pounder, R.E. (1993). Evidence of persistent measles virus infection in Crohn's disease. J. Med. Virol. 39, 345–353.

Wang, M., Ahrne, S., Antonsson, M., and Molin, G. (2004). T-RFLP combined with principal component analysis and 16S rRNA gene sequencing: an effective strategy for comparison of fecal microbiota in infants of different ages. J. Microbiol. Methods *59*, 53–69.

Wang, M., Ahrne, S., Jeppsson, B., and Molin, G. (2005). Comparison of bacterial diversity along the human intestinal tract by direct cloning and sequencing of 16S rRNA genes. FEMS Microbiol. Ecol. *54*, 219–231.

Wehkamp, J., Salzman, N.H., Porter, E., Nuding, S., Weichenthal, M., Petras, R.E., Shen, B., Schaeffeler, E., Schwab, M., Linzmeier, R., *et al.* (2005). Reduced Paneth cell alpha-defensins in ileal Crohn's disease. Proc. Natl. Acad. Sci. U.S.A. *102*, 18129–18134.

Willing, B., Halfvarson, J., Dicksved, J., Rosenquist, M., Järnerot, G., Engstrand, L., Tysk, C., and Jansson, J.K. (2009). Twin studies reveal specific imbalances in the mucosa-associated microbiota of patients with ileal Crohn's disease. Inflamm. Bowel Dis. *15*, 653–660.

Woodmansey, E.J. (2007). Intestinal bacteria and ageing. J. Appl. Microbiol. *102*, 1178–1186.

Zhong, S.S., Zhang, Z.S., Wang, J.D., Lai, Z.S., Wang, Q.Y., Pan, L.J., and Ren, Y.X. (2004). Competitive inhibition of adherence of enterotoxigenic *Escherichia coli*, enteropathogenic *Escherichia coli* and *Clostridium difficile* to intestinal epithelial cell line Lovo by purified adhesin of *Bifidobacterium adolescentis* 1027. World J. Gastroenterol. *10*, 1630–1633.

Zinkevich, V.V., and Beech, I.B. (2000). Screening of sulfate-reducing bacteria in colonoscopy samples from healthy and colitic human gut mucosa. FEMS Microbiol. Ecol. *34*, 147–155.

Zoetendal, E.G., Akkermans, A.D., and de Vos, W.M. (1998). Temperature gradient gel electrophoresis analysis of 16S rRNA from human fecal samples reveals stable and host-specific communities of active bacteria. Appl. Environ. Microbiol. *64*, 3854–3859.

Zoetendal, Akkermans, E.G., A.D.L., Akkermans-van Vliet, W.M., de Visser, J.A.G.M., and de Vos, W.M. (2001). The host genotype affects the bacterial community in the human gastrointestinal tract. Microb. Ecol. Health Dis. *13*, 129–134.

Zoetendal, E.G. von Wright, A., Vilpponen-Salmela, T., Ben-Amor, K., Akkermans, A.D., and de Vos, W.M. (2002). Mucosa-associated bacteria in the human gastrointestinal tract are uniformly distributed along the colon and differ from the community recovered from feces. Appl. Environ. Microbiol. *68*, 3401–3407.

Wastewater Treatment

10

Satoshi Okabe and Yoichi Kamagata

Abstract

Of Earth's diverse microbial habitats, wastewater treatment processes are one of the most elaborate anthropogenic niches geared towards one purpose: cleaning up water. Recent application of molecular techniques is unveiling the microbial composition and architecture of the complex communities involved in the treatment processes. It is now recognized that wastewater processes harbour a vast variety of microorganisms most of which are yet-to-be cultured, hence uncharacterized. In this chapter, the latest knowledge on diversity, structure and functions of microbial communities in nitrifying processes, anaerobic ammonia oxidation processes and methane fermenting processes are summarized.

Introduction

Biological wastewater treatment is undoubtedly one of the most important biotechnological processes, which have been used for over a century to treat municipal and industrial wastewaters. Wastewater treatment processes, however, have long been regarded as 'a black box' in practice primarily due to the lack of analytical techniques to determine the identities of microorganisms involved in the processes and the limitation of their resolutions, as represented by traditional culture-dependent techniques. In the 1990s, culture-independent molecular techniques emerged and since then, they have also been used to study the diversity and physiology of ecologically important microorganisms in wastewater treatment processes. A number of new exciting insights into structure, function, and dynamics of complex microbial communities in wastewater treatment processes have since been gained, which have been significantly expanding our understanding of process design, operation and control. This chapter aims to highlight recent findings about the role of microbes in cycling of nitrogen in wastewater treatment processes (including anaerobic ammonia oxidation processes) and methane fermentation processes, based on recent molecular and physicochemical techniques.

Bacterial biofilms in wastewater treatment

Most microbes in wastewater treatment processes, including nitrifying bacteria, prefer an immobilized life-style; attached to surfaces (i.e. biofilms; see Chapter 11) and aggregated together (for example in activated sludge and granules). In addition, multiple electron donors and electron acceptors are present in wastewaters. As a result, wastewater biofilms are very complex multispecies consortia, displaying considerable heterogeneity, with respect to both the microorganisms present and their physicochemical microenvironments. High metabolic rates and molecular diffusion limitations create steep microgradients of different physicochemical parameters, resulting in a vertical zonation of predominant respiratory processes within the biofilms with a typical thickness of only a few millimetres. Such microbial communities have their own particular activities that cannot be achieved by individual microorganisms. Without knowing the microenvironments in the biofilms, it would be impossible to understand relationships between the microbial community structures and their *in situ* functions. Analytical techniques and tools with a high spatial and temporal resolution are therefore required to directly link the microbial community structure and *in situ* functions.

Polyphasic approach

Understanding the structure and function of complex microbial communities is a central theme in wastewater microbiology. However, canonical cultivation-dependent techniques are inadequate to fulfil this task because most members of microbial communities in natural and engineered ecosystems cannot be cultured (Amann *et al.*, 1995). New molecular biological techniques (virtually based on the 16S ribosomal RNA gene sequences (see Chapter 1) provide comprehensive information on molecular diversity of microbial communities, including heretofore-unknown microorganisms, without the need for cultivation. Here we will focus on molecular approaches that have proven to be particularly useful for study of microorganisms in wastewater treatment systems. For example, the fluorescence *in situ* hybridization (FISH) technique allows *in situ* detection, localization, and quantification of target microorganisms in complex heterogeneous microbial communities such as biofilms; a key feature of wastewater treatment (see Chapter 11 for a discussion of biofilms). Several reviews provide technical details, applications, and limitation of FISH and the rRNA approach (Amann and Ludwig, 2000; Head *et al.*, 1998).

Linking community structure with community function (activity) is one of the most intriguing topics for microbial ecologists. In biofilms, substrate transfer often limits microbial conversion rates. Thus, the biofilms develop various complex microenvironments due to substrate transfer constraints. One way to overcome this limitation is to use microelectrodes to study processes in biofilms. Microelectrode measurement is a reliable and direct way to determine microenvironments within biofilms with high spatial and temporal resolution (Amann and Kühl, 1998; Schramm, 2003, Okabe *et al.*, 2004a). One advantage of the use of microelectrodes is their ability to detect, directly and with minimum disturbance, *in situ* microbial activities, which are derived from the substrate profiles if the transport process (usually molecular diffusion) is known. Microelectrodes, however, only measure net chemical profiles. The spatial resolution of microelectrodes is at the 'community level', which is, however, adequate to characterize the concentration gradients across biofilms and to calculate the net rates of production and consumption at a certain depth within the biofilm. However, the spatial resolution is not sufficient to determine the activities of individual microbial cells. In addition, when the substrates used by unidentified microorganisms are not known, or the abundance of the targeted microorganisms is low, the chemical profiles and fluxes cannot be correlated with the abundance of specific bacterial populations. FISH cannot provide sufficient information on metabolic capabilities, because phylogeny and phenotypes are rarely congruent.

To overcome these technical problems, microautoradiography (MAR) combined with FISH (MAR-FISH) has been successfully developed (Lee *et al.*, 1999; Ouverney and Fuhrman, 1999) and applied in a limited, but rapidly increasing, number of studies to evaluate uptake of substrates by different phylogenetic groups of microbes in natural and engineered environments (Okabe *et al.*, 2004b; Wagner *et al.*, 2006). The MAR-FISH technique provides a direct link between 16S rRNA phylogeny and *in situ* physiology (i.e. substrate uptake patterns) of both cultivated and uncultivated bacterial species on a single-cell level, providing further insights into the ecophysiological importance of key microorganisms in various environments. Recently, the scope of MAR-FISH was extended by combining it with microelectrode measurements (Gieseke *et al.*, 2005), and other techniques (Wagner *et al.*, 2006).

The experimental procedure of MAR-FISH has been summarized from principle to application (Lee *et al.*, 1999; Nielsen *et al.*, 1999; Nielsen and Nielsen, 2005; Okabe *et al.*, 2004b). The typical MAR-FISH procedure is composed of the following steps: (i) incubation with radiolabelled substrates, (ii) fixation and handling, (iii) staining (FISH, DAPI, or gram-staining), (iv) microautoradiography (MAR), and (v) microscopic observation. When MAR-FISH is applied to complex microbial communities, several important methodological options have been described (Nielsen *et al.*, 1999). MAR-FISH is very sensitive compared to DNA- and RNA-stable isotope probing (SIP) (see Chapter 4 for a description of these techniques). In addition, MAR-FISH requires relatively short incubation times (generally a few hours) and thus minimizes substrate cross-feeding by other bacteria that are not primary substrate consumers (Okabe *et al.*, 2004; Wagner *et al.*, 2006). Although MAR-FISH analysis is generally qualitative, a more quantitative MAR-FISH has been recently developed which allows more precise determination of substrate uptake rates by target bacteria (Nielsen *et al.*, 2003).

Nitrifying microbial communities

Nitrogen cycling in wastewater biofilms

One of the key goals of wastewater treatment is to reduce the amounts of nitrogen as part of the clean up process and this is largely dependent on microbial enzymatic activities. Microbial nitrification, the oxidative part of the nitrogen cycle (i.e. oxidation of ammonia to nitrate) is a sequential two-step process and is carried out by two phylogenetically unrelated groups of lithoautotrophic bacteria, the ammonia-oxidizing bacteria (AOB) and nitrite-oxidizing bacteria (NOB). A better understanding of ecophysiology (i.e. diversity, abundance, spatial distribution, and *in situ* activity) of these nitrifying bacteria in wastewater biofilms is essential for maintaining or improving process performance and control.

Diversity among AOB in wastewater biofilms

Ammonia-oxidizing bacteria (AOB) comprise the following two monophyletic lineages, based on 16S rRNA gene sequences: the genera *Nitrosomonas* and *Nitrosospira* (containing the former genera *Nitrosovibrio* and *Nitrosolobus*) within the *Betaproteobacteria* and the genus *Nitrosococcus* within the *Gammaproteobacteria* (Head *et al.*, 1993; Purkhold *et al.*, 2000). The *Nitrosomonas europaea/eutropha* lineage has been traditionally considered as the most important AOB in various wastewater treatment systems (Gieseke *et al.*, 2001; Mobarry *et al.*, 1996; Okabe *et al.*, 1999; Satoh *et al.*, 2006; Schramm *et al.*, 1996). FISH analyses in activated sludge and biofilms, however, showed that other AOB, like *Nitrosococcus mobilis* which was originally isolated from brackish water (Koops *et al.*, 1976), were the dominant AOB in several wastewater treatment plants (Daims *et al.*, 2001; Juretschko *et al.*, 1998; Juretschko *et al.*, 2002; Okabe *et al.*, 2004a). In addition, *Nitrosospira* and the well-known *Nitrosomonas* have also been detected (Okabe *et al.*, 2004a; Siripong and Rittmann, 2007). AOB within the Gammaproteobacteria (i.e. *Nitrosococcus*) have not been detected by FISH, or by *amoA*-analyses in nitrifying wastewater treatment plants. For nitrification, since phylogenetic affiliation corresponds well with metabolic function, bacteria that are identified as members of one of the aforementioned six groups by 16S rRNA gene sequence analysis or FISH are consequently considered as ammonia-oxidizers.

Diversity among NOB in wastewater biofilms

Nitrite-oxidizing bacteria (NOB) are more phylogenetically diverse than AOB. All known NOB belong to one of four different genera: the genus *Nitrobacter* within the *Alphaproteobacteria*, the genus *Nitrococcus* within the *Deltaproteobacteria*, the genus *Nitrospina* within the *Betaproteobacteria*, and the genus *Nitrospira* within a separate phylum *Nitrospirae* (Daims *et al.*, 2000; Ehrich *et al.*, 1995; Teske *et al.*, 1994). Although *Nitrobacter* has long been considered as the most important NOB in wastewater treatment plants, recent observation using molecular techniques revealed that yet-to-be cultured *Nitrospira*-like NOB were more frequently dominant NOB in wastewater biofilms (Schramm *et al.*, 1998; Okabe *et al.*, 1999; Okabe *et al.*, 2004a; Kindaichi *et al.*, 2006) and in activated sludge (Juretschko *et al.*, 1998; Burrell *et al.*, 1999; Dionisi *et al.*, 2002; Harms *et al.*, 2003; Siripong and Rittmann, 2007). The predominance of *Nitrospira*-like bacteria over *Nitrobacter* in most wastewater treatment plants is generally explained by their different survival strategies (Schramm *et al.*, 1999). It has been speculated that *Nitrospira* is a slow-growing *k*-strategist with high affinities to nitrite and oxygen, while *Nitrobacter* has a higher μ_{max} and low affinities (*r*-strategist) (Andrews and Harris, 1986; Schramm *et al.*, 1999; Okabe *et al.*, 1999). Thus, *Nitrospira* would outcompete *Nitrobacter* because nitrite and oxygen concentrations are usually low in most wastewater treatment systems and in wastewater biofilms (Kindaichi *et al.*, 2006). The coexistence of *Nitrospira* and *Nitrobacter* was, however, observed in a sequence batch biofilm reactor with relatively high nitrite concentrations (Daims *et al.*, 2001).

Several AOB and NOB who perform the same task frequently coexist in some wastewater treatment plants (Lydmark *et al.*, 2006; Siripong and Rittmann, 2007). Lydmark *et al.* (2006) found that thin biofilms in a full-scale nitrifying trickling filter housed six coexisting populations of AOB, NOB and anaerobic ammonia-oxidizing bacteria (anammox bacteria; see discussion below). Among AOB populations, the two *Nitrosomonas oligotropha* populations showed different vertical distribution patterns within the biofilm, indicating different ecophysiologies, although they belong to the same AOB lineage. The coexistence of different AOB and NOB in wastewater treatment plants is the evidence of functional redundancy, which may allow communities to maintain physiological capabilities when conditions change, thus leading to the stability of reactor performance.

In situ *spatial distribution of AOB and NOB in biofilms*

To investigate the spatial distributions of AOB and NOB within biofilms, *in situ* hybridization with specific oligonucleotide probes can be carried out for entire vertical sections of biofilms. The community composition and architecture of nitrifying biofilms in full-scale wastewater treatment plants have been studied using molecular methods (Egli *et al.*, 2003; Lydmark *et al.*, 2006; Okabe *et al.*, 1999; Persson *et al.*, 2002; Rowan *et al.*, 2003; Schramm *et al.*, 1996). Schramm *et al.* (1996) were the first to use both FISH and microelectrodes to study the spatial distribution and local activities of AOB and NOB in trickling filter biofilms. In their study, both AOB and NOB formed dense cell clusters, which were not restricted to the oxic zone of the biofilm but were also detected in the anoxic layers. Okabe *et al.*, (1999) found at least three different AOB to be present throughout a wastewater biofilm, while *Nitrospira*-like NOB inhabited only the deeper part of the biofilms (Fig. 10.1A and B). As a result, microelectrode measurements showed that active ammonia and nitrite oxidation occurred in two distinct zones, in the outer and inner part of the biofilms, respectively (Fig. 10.1C and D). The separated oxidation of NH_4^+ and NO_2^- found by microelectrode measurements coincides with locations where higher abundances of AOB and NOB were detected. This is a good example showing a correlation between the distribution of microbial species and accompanying expected activity in the biofilm.

Spatial distributions and related local activities of AOB and NOB have also been studied in simple experimental biofilms grown in artificial medium. *In situ* hybridization clearly indicated that the majority of stained AOB formed dense spherical clusters consisting of rod-shaped cells. The spherical clusters of AOB were detected throughout the oxic biofilm strata, indicating more or less a homogeneous spatial distribution (Fig. 10.2A). In contrast, the clusters of stained *Nitrospira*-like cells formed irregularly shaped aggregates, which were primarily detected in the deeper part of the biofilm. Some AOB clusters closely associated with NO_2^--oxidizing bacteria, demonstrating the sequential metabolism of ammonia via nitrite to nitrate. No hybridization signal was observed when *Nitrobacter*-specific probes NIT2 and NIT3 were used with any of the samples, indicating the dominant NOB in this biofilm were *Nitrospira*-like NOB as previously reported (Kindaichi *et al.*, 2004, Okabe *et al.*, 1999, 2002; Satoh *et al.*, 2003). The microelectrode measurements also showed that the vertically separated oxidation of NH_4^+ and NO_2^- occurred in the outer and inner part of the biofilms, respectively (Fig. 10.2B and C). The layered structure of AOB and NOB has been reported in several studies (Okabe *et al.*, 1999; Kindaichi *et al.*, 2006; Schramm *et al.*, 2000; Gieseke *et al.*, 2003). The spatial segregation of AOB and NOB could be explained as a result of the lower growth rate of *Nitrospira* and its adaptation to, or persistence in lower oxygen and nitrite concentrations (Kindaichi *et al.*, 2006; Okabe *et al.*, 1999; Gieseke *et al.*, 2003). It should be noted that visualization of these micro-scale spatial organizations of two phylogenetically unrelated species was only possible by use of the FISH technique.

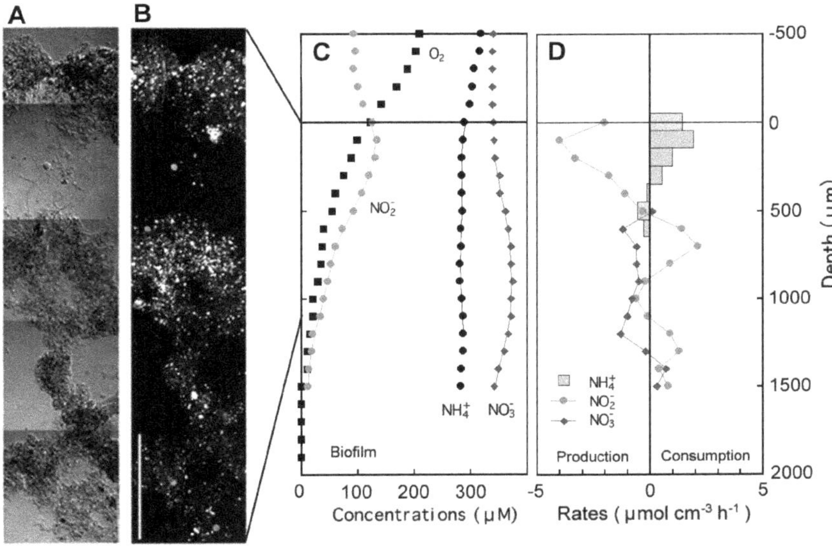

Figure 10.1 Simultaneous *in situ* hybridization of vertical sections of a domestic wastewater biofilm with TRITC-labelled Nso190 probe and FITC-labelled Ntsp454 probe. A composite DIC image (A) and corresponding CSLM projection image (B) of the entire biofilm vertical section (scale bar = 200 μm). The biofilm thickness was about 950 μm. Steady-state concentration profiles of O_2, NH_4^+, NO_2^-, and NO_3^- in a domestic wastewater biofilm incubated in air-saturated medium with 0 μM acetate, 300 μM NH_4^+, 300 μM NO_3^-, and 100 μM NO_2^- (C) and the spatial distribution of the estimated volumetric consumption and production rates of NH_4^+, NO_2^-, and NO_3^- (D). The points are measured mean values ± standard deviations, the lines are calculated profiles. The biofilm surface is at a depth of 0 μm. Figures are reproduced from Okabe *et al.* (1999). A colour version of this figure is located in the plate section at the back of the book.

Microbial community interactions

Microbial interactions, such as commensal relationships, are important for wastewater treatment processes. For example, nitrification is a two-step process, mediated by the sequential action of AOB and NOB. In addition, the coexistence of a high level of heterotrophic bacteria with nitrifying bacteria has often been found in autotrophic nitrifying biofilms cultured without an external organic carbon supply as determined using a 16S rRNA gene fingerprinting approach (Kindaichi *et al.*, 2004; Okabe *et al.*, 1999; Okabe *et al.*, 2002). It has been hypothesized that heterotrophic bacteria utilize soluble microbial products produced by nitrifying bacteria. This hypothesis was recently tested by applying microautoradiography combined with FISH; i.e. the MAR-FISH technique described earlier (Kindaichi *et al.*, 2004; Okabe *et al.*, 2005). These studies revealed that [^{14}C] carbon, that was fixed primarily by nitrifying bacteria (AOB and NOB), was utilized mainly by *Bacteroides* and *Chloroflexi* under nitrifying (cultured with NH_4^+) and non-nitrifying (cultured without NH_4^+) conditions, respectively. These results indicated that the members of *Chloroflexi* preferentially utilized microbial products derived mainly from decay of nitrifying bacteria, whereas the members of *Bacteroides* utilized metabolites of nitrifying bacteria. The *Chloroflexi* detected by FISH with different specific probes (Björnsson *et al.*, 2002; Gich *et al.*, 2001) were affiliated with the *Chloroflexi* subphylum 1. The subphylum 1 contains the most diverse environmental clones and thus their *in situ* metabolic functions are presently not known. Application of the MAR-FISH technique can, however, provide a direct link between 16S rRNA phylogeny and their *in situ* physiology (i.e. substrate uptake patterns) of uncultivated *Chloroflexi*. Based on the resulting experimental results, it could be speculated that the coexisting heterotrophic bacteria (e.g. *Bacteroides* and *Chloroflexi*) efficiently clean up

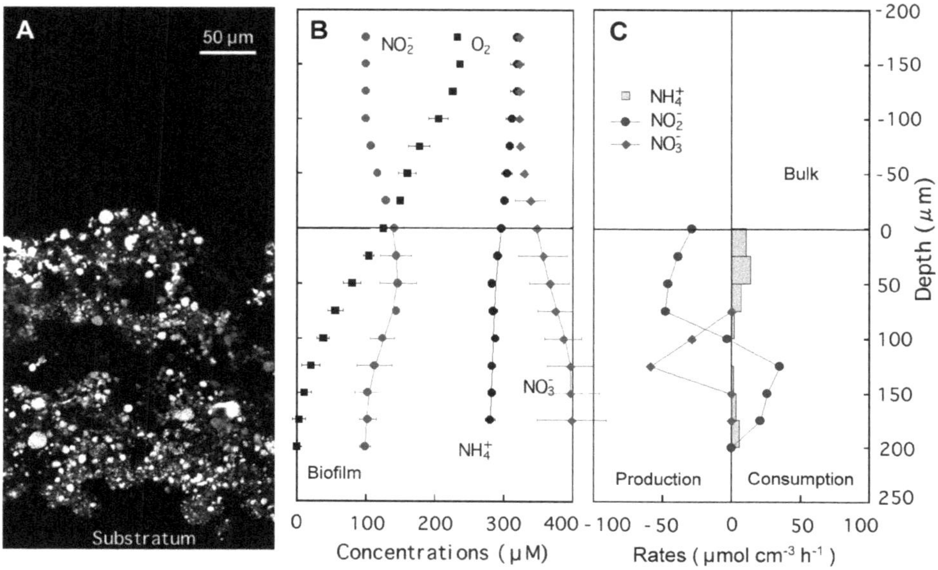

Figure 10.2 Simultaneous *in situ* hybridization of vertical sections of an autotrophic nitrifying biofilm with TRITC-labelled NEU probe and FITC-labelled Ntsp454 probe. A CSLM projection image of the entire biofilm vertical section (scale bar = 100 μm) (A). The biofilm thickness was about 350 μm. Steady-state concentration profiles of O_2, NH_4^+, NO_2^-, and NO_3^- in an autotrophic nitrifying biofilm incubated in air-saturated medium with 0 M acetate, 300 μM NH_4^+, 300 μM NO_3^-, and 100 μM NO_2^- (B) and the spatial distribution of the estimated volumetric consumption and production rates of NH_4^+, NO_2^-, and NO_3^- (C). The points are measured mean values ± standard deviations, the lines are calculated profiles. The biofilm surface is at a depth of 0 μm. Figures are reproduced from Okabe *et al.*, (1999). A colour version of this figure is located in the plate section at the back of the book.

and prevent accumulation of waste products of nitrifying bacteria, which consequently creates a stable biofilm ecosystem.

In the biofilm systems, the competitive interaction between heterotrophs and nitrifiers for dissolved oxygen and space is also well known (Okabe *et al.*, 1996, 2004; Satoh *et al.*, 2000). In the presence of organic carbon, nitrifiers are usually outcompeted by heterotrophic bacteria due to the slow growth rate and low growth yield. Satoh *et al.*, (2000) and Okabe *et al.*, (2004) demonstrated that the substrate C/N ratio was a regulating factor for distribution and activity of nitrifying bacteria in the biofilm and that abundance and activity of AOB decreased at the biofilm surface with increasing substrate C/N ratios.

Anaerobic ammonium oxidation (anammox)

Anaerobic ammonium oxidation (anammox) is a biological process in which ammonium is directly converted to dinitrogen gas with nitrite as the electron acceptor under anoxic conditions (Jetten *et al.*, 1999). Anammox is carried out by chemolithoautotrophic bacteria belonging to the order *Planctomycetales*. Based on a full-cycle 16S rRNA approach (from clone library analysis to FISH with specific oligonucleotide probes (Schmid *et al.*, 2005), anammox bacteria were detected in many freshwater and marine ecosystems and engineered systems. To date, the anammox bacteria belonging to three genera: *Brocadia* including *B. anammoxidans* (Strous *et al.*, 1999) and *B. fulgida* (Kartal *et al.*, 2004), *Kuenenia* including *K. stuttgartiensis* (Schmid *et al.*, 2000) and *Scalindua* including *S. sorokinii* (Kuyper *et al.*, 2003), *S. brodae* (Schmid *et al.*, 2003), and *S. wagneri* (Schmid *et al.*, 2003) have been proposed. Recently, a mixotrophic anammox bacterium Candidatus *Anammoxglobus propionicus* was also described (Kartal *et al.*, 2006). Since these bacteria have not been isolated in pure culture yet, the current information about their physiology has been obtained from enrichment culture studies (Egli *et al.*, 2001, Strous *et al.*, 1998 and Toh *et al.*, 2002).

Application of the anammox process

The anammox process is a new and promising alternative to the conventional nitrogen removal processes. The application of anammox to nitrogen removal would lead to a significant reduction of costs for aeration and exogenous electron donor (up to 90% reduction of operational costs) as compared to the conventional nitrification-denitrification process (van Dongen *et al.*, 2001). Anammox bacteria were detected and enriched from various types of wastewater treatment plants, indicating that they are indigenous and can be implemented in nitrogen removal processes. However, one of the main drawbacks common to application of the anammox process is requirement of a long start-up period due to the extremely slow growth of anammox bacteria (Egli *et al.*, 2001 and van Dongen *et al.*, 2001) The doubling time of anammox bacteria was first reported to be approximately 11 days by Strous *et al.*, (1998), and a shorter doubling time of 3.6–5.4 days was recently reported using molecular techniques (Tsushima *et al.*, 2007a). Additionally, since anammox bacteria are strictly anaerobes and autotrophs, they are very difficult to culture. This suggests that a rapid and efficient enrichment and start-up of the anammox process is the key for optimization of wastewater treatment. For example, appropriate seeding sludge or starter cultures should be used to promptly establish the anammox process. To screen for appropriate seeding sludge with high anammox potential, a real-time quantitative PCR assay has been developed (Tsushima *et al.*, 2007a). An appropriate seeding sludge with a high abundance and activity of anammox bacteria was selected by using the real-time PCR assay and then inoculated into an up-flow anaerobic fixed-bed biofilm column reactor (Tsushima *et al.*, 2007b). The anammox reaction was observed within 50 days and the highest nitrogen removal rate of 26.0 kg-N m^{-3} day^{-1} was attained after 247 days. This rate was the highest nitrogen removal rate by anammox bacteria that has been reported so far.

Community structure and *in situ* activity of anammox biofilms

For future optimization and efficient designing of the anammox process, a better understanding of ecophysiology (i.e. microbial community structure and *in situ* activity) of anammox bacteria in complex biofilms must be obtained and related to the reactor performance. However, only a few studies have been carried out to achieve this objective so far (Kindaichi *et al.*, 2007; Nielsen *et al.*, 2005; Tsushima *et al.*, 2007b). Tsushima *et al.* (2007) have recently analysed the community structure, spatial organization and *in situ* activity of the anammox biofilms developed in an anaerobic fixed-bed column reactor by the combined use of a 16S rRNA based approach and microelectrodes (Fig. 10.3). *In situ* hybridization results revealed that '*Brocadia*'-like anaerobic anammox bacteria were distributed throughout the biofilm (accounting for more than 80% of total bacteria) (Fig. 10.3A and B). They coexisted with

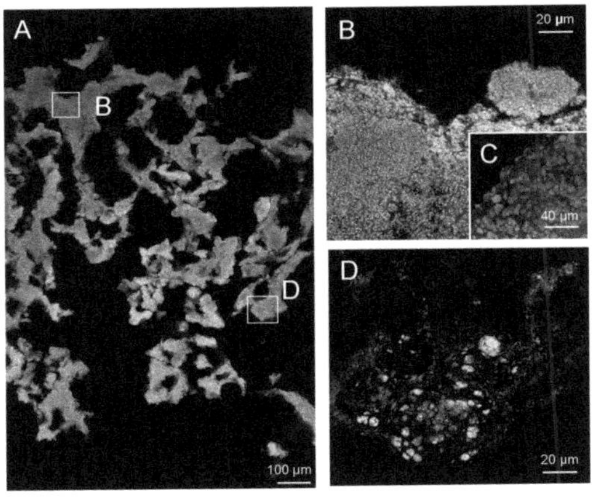

Figure 10.3 CLSM images of vertical sections (30 µm thick) of the anammox biofilm. FISH with TRITC-labelled EUB338 probe mixture (green) and FITC-labelled Amx820 probe (red) (specific to anammox bacteria; Candidatus *Brocadia anammoxidans* and Candidatus *Kuenenia stuttgartiensis*). Yellow signals result from binding both probes to one cell (**A**, **B** and **D**). FISH with FITC-labelled Amx820 probe (green) and TRITC-labelled Nse1472 probe (red) (specific to *Nitrosomonas eutropha*, *N. europaea* and *N. Halophila*) (**C**). The biofilm surfaces are at the top of all images. Figures are reproduced from Tsushima *et al*. (2007). A colour version of this figure is located in the plate section at the back of the book.

Nitrosomonas-like aerobic ammonia-oxidizing bacteria (AAOB) in the surface biofilm (Fig. 10.3C). To determine the affiliation of the anammox bacteria detected in this biofilm, a 16S rRNA gene clone library was constructed using a *Planctomycetals*-specific PCR primer set; pla46f (Neef *et al.*, 1998) and 1492r (Weisburg *et al.*, 1991). All of the 16S rRNA genes analysed had the same sequence and had at most 95% sequence similarity to the known Anammox species, the Candidatus *Brocadia anammoxidans* (Fig. 10.4) This indicated that the dominant anammox-like bacterium detected in this study was most likely a member of a novel species of the genus '*Brocadia*'. It should be noted that different anammox species are rarely found in the same habitat. This suggests that each anammox species has a well-defined, but yet unknown ecological niche (Kartal *et al.*, 2006). For the aerobic AOB case, they also occupy unique niches, but the niche differentiation is not well understood as well.

In the above study, microelectrodes were used to determine the concentration profiles of O_2, NH_4^+, NO_2^-, NO_3^-, and pH in the anammox biofilm when the average nitrogen removal rate of the reactor was 3.0 kg-N m^{-3} day^{-1} (Fig. 10.5). The dissolved oxygen concentration was under the detection limit throughout the biofilm. NH_4^+ and NO_2^- concentration profiles indicated that NO_2^- and NH_4^+ concentrations simultaneously decreased in the upper 800 μm of the biofilm, indicating high anammox reaction rates [4.45 g-N of (NH_4^+ + NO_2^-) m^{-2} day^{-1}]. The NH_4^+ concentration, however, gradually increased below a depth of 800 μm, most likely due to anaerobic mineralization of organic compounds, whereas the NO_3^- concentration steadily increased in the upper 800 μm. In addition, the pH increased from 7.3 to 7.8 in the upper 700 μm of the biofilm. Based on the microprofiles, the total consumption rates of NH_4^+, NO_2^- and NO_3^- were calculated to be 2.04, 2.41 and –0.67 g-N m^{-2} day^{-1}, showing a ratio of about 1:1.2:–0.33 in the anammox zone, similar to that previously reported (Strous *et al.*, 1998).

Coexistence of anammox bacteria and aerobic AOB

In wastewater treatment systems, *B. anammoxidans*-like anammox bacteria and *Nitrosomonas*-like AOB may be exposed to high NO_2^- concentrations, because NO_2^- is an obligatory intermediate in the oxidation of ammonia (Schmidt *et al.*, 2001). In the presence of NO_2^-,

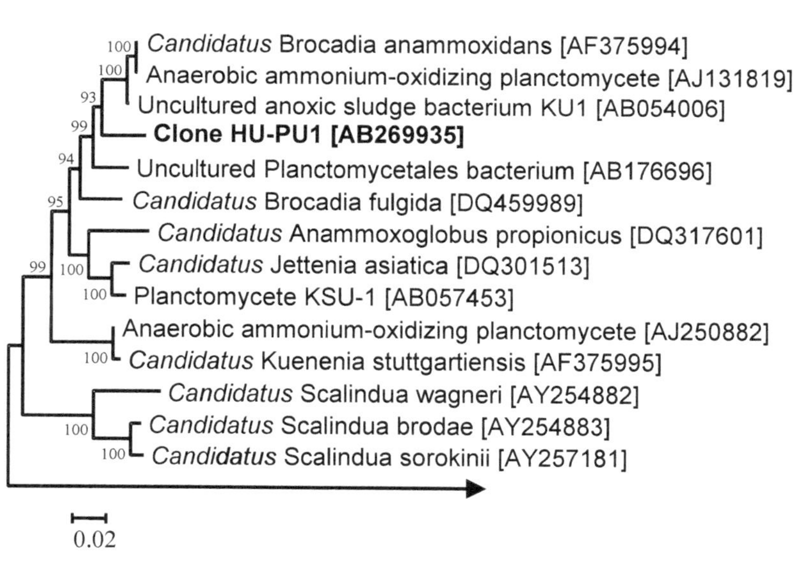

Figure 10.4 Phylogenetic trees showing the positions of the clone retrieved (OTU HU-PU1) from our anammox biofilm in relation to previously published anammox-like bacterial sequences (Tsushima *et al.*, 2007). The tree was generated by using 1,429 bp of the 16S rRNA genes and the neighbor-joining method. Scale bars = 2% sequence divergence. The values at the nodes are bootstrap values (1000 times resampling analysis). Figure is reproduced from Tsushima *et al.*, (2007).

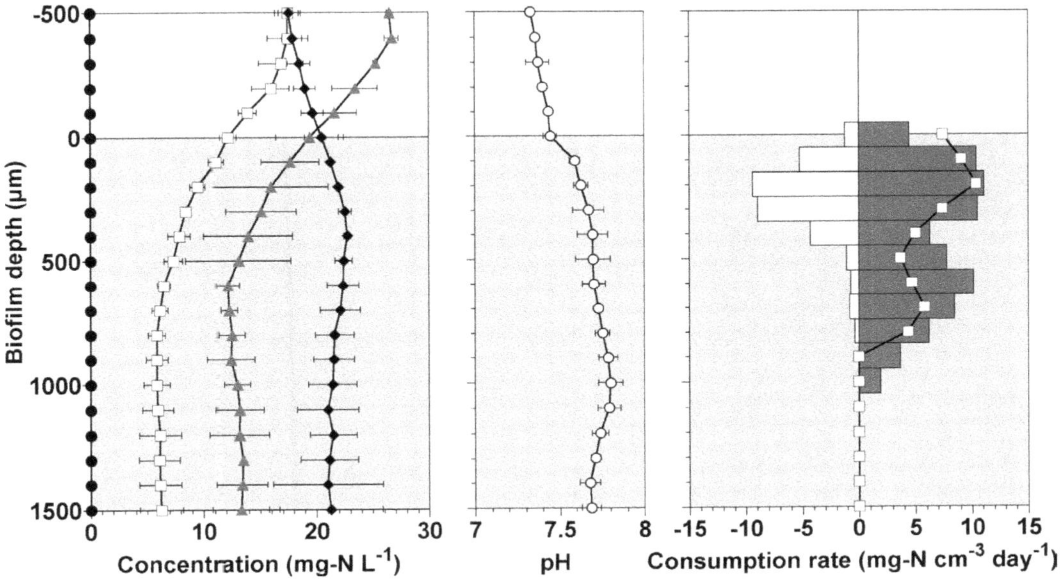

Figure 10.5 Steady-state concentration profiles of O_2, NH_4^+, NO_2^-, NO_3^- and pH in the anammox biofilm (A and B). The points are measured mean values ± standard deviations. The spatial distribution of the estimated volumetric consumption and production rates of NH_4^+, NO_2^-, and NO_3^- (C). Negative values indicate production rates. The biofilm surface is at a depth of 0 μm. Figures are reproduced from Tsushima *et al.*, (2007).

the specific ammonia oxidation activity of *B. anammoxidans* increased, and *Nitrosomonas*-like AOB could carry out NO_2^--dependent anaerobic ammonia oxidation (Schmidt *et al.*, 2002). Although the two groups of ammonia-oxidizing bacteria compete for ammonia, cooperation is also possible because *Nitrosomonas* produces NO_2^- during anaerobic ammonia oxidation, which can be consumed by *B. anammoxidans*. Thus, the anammox bacteria (*B. anammoxidans*) and aerobic AOB (*Nitrosomonas*) are able to coexist under anoxic conditions (Fig. 10.3C), leading to higher nitrogen removal activities. This is of relevance in natural environments and for technical applications. The NO_2^-/NH_4^+ ratio would decrease to 1.32 when active ammonia-oxidizing *Nitrosomonas* and high NO_2^- were present in the biofilm.

Ammonia-oxidizing archaea

Recently it was discovered that autotrophic oxidation of ammonia is not restricted to the domain *Bacteria* (Konneke *et al.*, 2005). An ammonia-oxidizing archaeon (AOA), *Nitrosopumilus maritimus*, was isolated from the rocky substratum of a tropical marine aquarium tank (Konneke *et al.*, 2005). *N. maritimus* grows chemolithoautotrophically by aerobically oxidizing ammonia to nitrite as described by the following stoichiometry:

$$NH_3 + 1.5\ O_2 \rightarrow NO_2^- + H_2O + H^+ \ (\Delta G^{0'} = -235\,kJ\,mol^{-1})$$

In addition, AOA occurred in some activated sludge bioreactors operating at low dissolved oxygen levels and long retention times, suggesting AOA may also be widespread in nitrifying wastewater treatment plants (Park *et al.*, 2006). The wide distribution of AOA in so many different habitats, their abundance, and autotrophic metabolism indicate that they must contribute significantly to global nitrogen cycle and nitrogen removal in wastewater treatment systems (Francis *et al.*, 2005; Ingalls *et al.*, 2006; Konneke *et al.*, 2005; Treusch *et al.*, 2005). However, their physiology and ecological impact remained unclear. Future studies will be imperative for better understanding their contributions to the global nitrogen cycle.

Microbial communities in methane fermenting processes

Methane fermenting processes

In anaerobic ecosystems, organic substances are degraded into methane and carbon dioxide unless efficient terminal electron acceptors such as nitrate, sulphate and oxidized forms of metals (for instance Fe^{3+}) are present for respiration. As most of wastewater does not contain sufficient amounts of those electron acceptors, methanogenesis is usually the primary pathway for degradation of organic substances.

Methane fermentation for wastewater treatment has over 100 years of history. However, it had not been a common and popular system until a totally new fermenting process emerged in the 1980s (Lettinga 1995). The wastewater treatment system currently known as up-flow anaerobic sludge blanket reactor (UASB reactor) has completely changed our view of what the methane fermentation process is. UASB process is now a mature technology for treating medium and high strength organic wastewaters and over 1000 full-scale reactors are now in use worldwide.

One of the characteristic features of the process is self-granulation of organisms, in other words, self-aggregation or self-immobilization of organisms, that results in retaining a much denser biomass than any other methane fermenting processes ever had, thus leading to high speed treatment of organic wastewater at high loading rates (Lettinga 1995; Schmidt and Ahring, 1996; Sekiguchi and Kamagata, 2004). In the process, anaerobic degradation of organic matter, methane formation and subsequent separation is performed in a single vessel. The wastewater is introduced from the bottom of the reactor and organic matters are digested as the wastewater moves up to the top. When digestion is occurring, methane gas bubbles are produced and carry the sludge upwards resulting in the formation of dense microbial aggregates ('granules') that are readily settled. Those granules occur naturally inside the vessel by virtue of the structure of the reactor, the path of wastewater supply and gas bubble formation. Those factors select granule-forming microorganisms, including all microbes necessary for anaerobic digestion and methanogenesis.

Molecular ecology of microbial communities in methanogenic granular sludge – overview

One pellet of granular sludge, several mm in diameter, from a UASB reactor is probably the best specimen to understand who are present in an anaerobic environment, how they make a living, and how organic substances are degraded into methane, as it is a functional unit of all of the different trophic groups of bacteria and archaea. It represents a range of competitive and syntrophic associations between organisms that together form a characteristic community architecture to make entire reactions energetically and physiologically feasible.

As shown in Fig. 10.6, methanogenic degradation of organic matter is performed by at least four different trophic groups, namely (1) fermentative heterotrophs (2) proton-reducing H_2-producing substrate-oxidizing bacteria which are mainly involved in degradation of relatively small molecules such as fatty acids, alcohols and simple aromatic compounds, (3) H_2-utilizing methanogenic archaea and (4) acetate-consuming (aceticlastic) methanogenic archaea. Each trophic group contains a wide array of organisms with different catabolic traits. One of the most important features that distinguish aerobic microorganisms from anaerobic ones is that those three

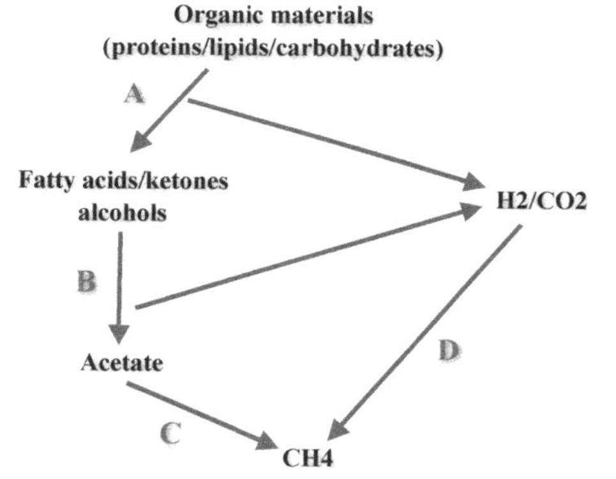

Figure 10.6 Methanogenic degradation of organic wastes. (A) heterotrophic organisms decomposing high molecules of organic materials; (B) heterotrophic proton-reducing H_2-producing organisms (syntroph); (C) aceticlastic methanogens; (D) H_2-consuming methanogens.

trophic groups share substrates, thus share energy (Schink 1997). Of the four trophic groups, proton-reducing H_2-producing substrate-oxidizing bacteria are tightly associated with methanogens in terms of substrate oxidation and energy conservation. For instance, fatty acids, such as butyrate, are oxidized by proton-reducing, H_2-producing substrate-oxidizing bacteria to acetate and H_2/CO_2 and acetate and H_2/CO_2 are subsequently converted to methane by aceticlastic methanogens and H_2-utilizing methanogens, respectively. In particular, H_2 is the key intermediate and it has to be consumed by H_2-utilizing methanogens since fatty acid oxidation with H_2 production is an endergonic reaction whereas H_2 consumption by methanogens is exergonic. Hence H_2-utilizing methanogens play a crucial role to keep the H_2 partial pressure sufficiently low to make whole reactions energetically feasible. In other words, proton-reducing H_2-producing substrate-oxidizing bacteria are very much dependent on H_2-utilizing methanogens as a H_2-scavenging partner. This is the primary reason why proton-reducing H_2-producing substrate-oxidizing bacteria are referred to as 'syntrophs'. From this viewpoint, H_2-utilzing methanogens make an essential contribution not only to methane production, but also in driving the initial step of oxidation of organic substrates to be degraded by syntrophs.

The syntrophic association between organisms and overall structure within UASB granules were successfully visualized in thin-sections of granules using 16S rRNA-targeted FISH (Sekiguchi *et al.*, 1999; Rocheleau *et al.*, 1999; Wu *et al.*, 2001; Gonzalez-Gil *et al.*, 2001; Saiki *et al.*, 2002; Liu *et al.*, 2002; Díaz *et al.*, 2006). FISH combined with confocal laser scanning microscopy (CSLM) revealed the spatial distribution of bacteria and archaea (Fig. 10.7). When hybridized with bacterial universal and archaeal universal probes, the granular biofilm was found to be a layered structure. The outer layer was dominated by bacterial cells, whereas the inner layer was occupied mainly by archaeal cells. As

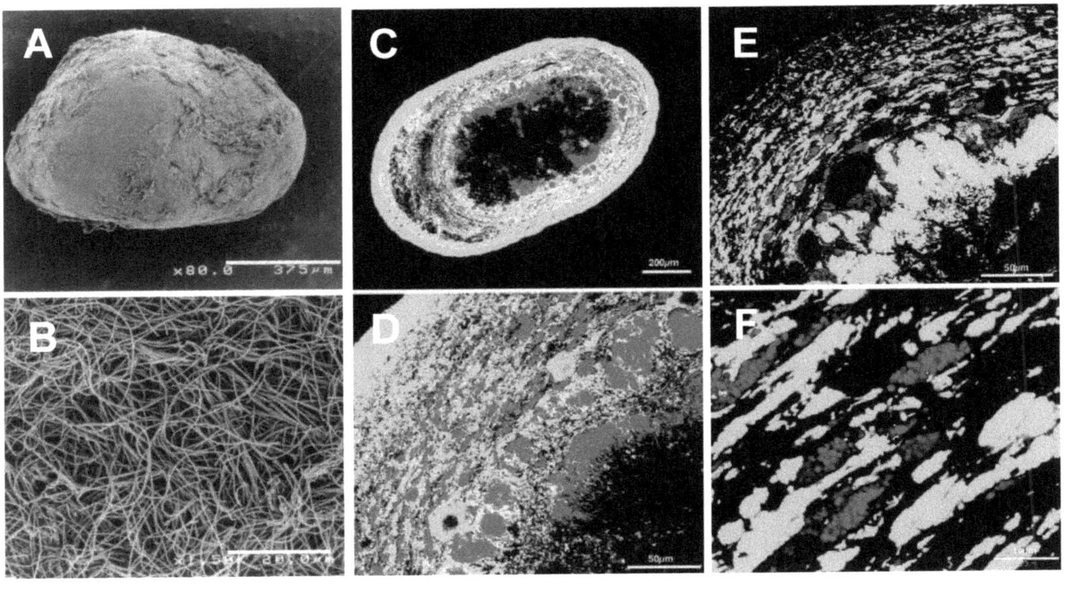

Figure 10.7 Micrographs of mesophilic methanogenic granular sludge. Scanning electron micrographs at **A** high and **B** low magnifications. **C–F** *In situ* hybridization of sections from a mesophilic granule viewed by confocal laser scanning microscopy. **C** and **D** Sections simultaneously hybridized with Cy-5-labelled bacterial-domain probe (EUB338) (green) and rhodamine-labelled archaeal-domain probe (ARC915) (red). **C** and **D** are at low and high magnifications, respectively. **E** and **F** Sections simultaneously hybridized with Cy-5-labelled archaeal-domain probe (ARC915) (green) and rhodamine-labelled *Syntrophobacter* specific probe (SYB701) (red). **E** and **F** are at low and high magnifications, respectively. Figures are reproduced from Sekiguchi *et al.* (1999). A colour version of this figure is located in the plate section at the back of the book.

the granules become larger, bacterial and archaeal cells formed an annual growth ring structure. Also, the non-staining center was always observed in larger granules indicating that the inner part of granules are inert due to the depletion of available substrates. This unique architecture is considered to be a result of substrate profiles formed within the granule. Given that organic materials in the wastewater are present on the surface of a granule, anaerobes in the outer layer of granules ferment them to produce intermediates such as fatty acids, alcohols and H_2. The organisms present in the middle layer including syntrophs oxidize these intermediates to form acetate and H_2. As mentioned above, the H_2 partial pressure must be kept sufficiently low for syntrophic degradation to proceed, hence H_2-consuming methanogens are spatially associated with syntrophs in their immediate proximity (Fig. 10.8).

Molecular ecology of methanogens in methanogenic granular sludge

FISH analysis also revealed the spatial distribution and abundance of methanogens, syntrophic bacteria and several interesting heterotrophs in granular sludge. Based on a 16S rRNA gene cloning analysis (Sekiguchi *et al.*, 1998), it was obvious that *Methanosaeta* was the predominant aceticlastic methanogen, though another aceticlastic methanogen, *Methanosarcina*, was also present at a lower abundance. The *Methanosaeta* cells were visualized using a *Methanosaeta* specific probe. They formed a number of micro-colonies inside the granules, and together with bacterial cells, they formed multiple layers. Considering that acetate is the most dominant low molecular weight intermediate and a precursor for methane that accounts for, in general, 70% of total methane produced, the dominance of *Methanosaeta* and juxtaposition of *Methanosaeta* and bacterial cells should be a consequence of interspecies acetate transfer between the two trophic groups.

In addition to *Methanosaeta*, several H_2-utilizing methanogens were visualized. *Methanobrevibacter*, *Methanobacterium*, and *Methanospirillum* all of which are H_2-utilizing methanogens were detected. In a high-temperature (55°C) lab-scale UASB reactor, *Methanobacterium* (currently recognized as *Methanothermobacter*) was abundant. When preliminary hybridization was attempted to detect *Methanobacterium* cells (Sekiguchi *et al.*, 1999), only very weak or no signals were obtained using a conventional *in situ* hybridization protocol. However, freeze and thaw cycles enhanced penetration of the probes into the cells and allowed the cells to hybridize with the FISH probes. Indeed, Nakamura *et al* (2006) reported that the *Methanothermobacter* cells grown syntrophically with an acetate-oxidizing H_2-producing syntroph thicken their cell walls, whereas the cells in pure culture of the same strain have much thinner ones, suggesting that the difference in cell wall thickness is the crucial factor influencing the hybridization efficiency. Interestingly, they also reported that the hybridizing signal intensity of *Methanothermobacter thermautotrophicus* cells in co-culture and other related methanogens in microbial communities was significantly improved by pseudomurein endoisopeptidase (Pei), a lytic enzyme that specifically acts on their cell walls. These results suggest that pretreatment is an attractive option for visualization of cells that are relatively impermeable to FISH probes.

Molecular ecology of syntrophs in methanogenic granular sludge

The abundance and distribution of syntrophs in the UASB granules were visualized in the same granular sludge as mentioned above. It was found that a number of micro-colonies of syntrophs and methanogens are present inside the granule forming a well-organized architecture to perform interspecies H_2 transfer between micro-colonies. One of the fascinating findings was that a propionate-oxidizing H_2-producing syntroph (later named *Pelotomaculum thermopropionicum* strain SI) formed micro-colonies that are surrounded by H_2-utilizing methanogens and their proximity is close enough to make propionate oxidation and subsequent methanogenesis via interspecies H_2 transfer energetically feasible (Imachi *et al.*, 2000, 2002). Interestingly, recent reports based on defined co-culture experiments showed that the propionate-oxidizing syntroph (strain SI) forms large aggregates with *Methanothermobacter thermautotrophicus* delta H cells (Ishii *et al.*, 2005). Moreover, field emission-scanning electron microscopy (FE-SEM) revealed that SI and delta H cells were connected with the

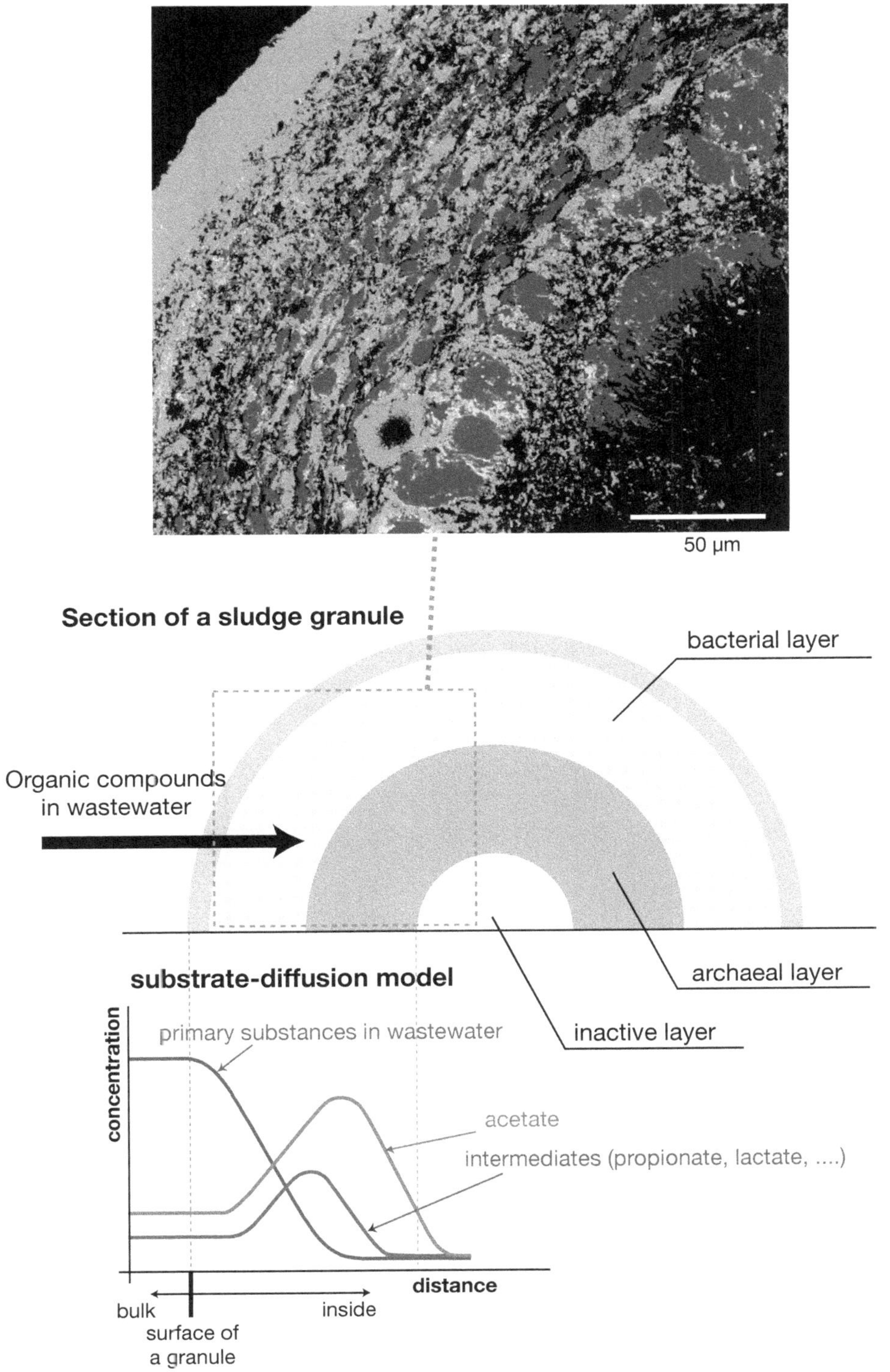

Figure 10.8 A *In situ* detection of bacteria (shown as green) and archael (indicated as red) cells cells within a section of a mesophilic methanogenic granule viewed by CSLM (see Fig. 10.7). **B** Schematic diagram of the typical internal structure of a sludge granule, indicating a microbial layered structure and flux profile of organic substrates and intermediates. Figures are reproduced from Sekiguchi (2006). A colour version of this figure is located in the plate section at the back of the book.

flagellum-like filaments of SI cells (Fig. 10.9). It is likely that this is the initial step for SI cells to make contact with delta H cells. The two organisms eventually intertwine with each other. These results suggest that flagella-like filaments of strain SI play a role in coaggregation of SI and delta H cells for their close contact. Very recently, Gorby *et al* (2006) reported that the flagellum-like appendages of SI cells connecting with delta H cells were electrically conductive nanowires based on scanning tunneling microscopy and tunneling spectroscopy indicating that electron transfer is occurring through the wire. Studies on syntrophs are now progressing to nano-scale physiology and electrochemistry to understand interspecies H_2 (electron) transfer.

Molecular ecology of heterotrophs in methanogenic granular sludge

A combination of 16S rRNA gene cloning analysis and other molecular technologies have now enabled microbiologists to access a huge variety of microbes that were previously not detected using conventional approaches such as isolation. In was late 1990s cloning based phylogeny of microbial constituents in UASB granules began to be studied. Sekiguchi *et al* (1998) investigated microbial diversity in mesophilic and thermophilic UASB processes fed with exactly the same substrates using same methanogenic sludge as a common inoculum. After the reactors became stable, they performed a 16S rRNA gene cloning analysis. Regardless of the common inoculum and same substrates, the two reactors were found to harbour a distinct array of organisms. Later, several other investigators also surveyed the bacterial diversity of various methanogenic reactors including classical methane fermenting processes (Godon *et al.*, 1997; Wintzingerode *et al.*, 1999; Wu *et al.*, 2001; Sekiguchi and Kamagata, 2004; Sekiguchi 2006). These results clearly showed that, without any exceptions, not only phyla that contain cultured representatives such as *Proteobacteria, Firmicutes, Spirochaetes, Actinobacteria, Choloroflexi, Synergistes, Planctomycetes, Chlorobi, Verrucomicrobia, Acidobacteria* and *Nitrospira*, but also the phyla comprised solely of environmental clones including OP10, OP9, OP8, OP3, TM6, TM7, EM3, BA024, and the termite group reside in methanogenic communities. These observations strongly indicate that methanogenic ecosystems virtually contain most of the bacterial phyla except for some specialized phyla found only in some particular extreme environments.16S rRNA gene cloning analysis and subsequent design of probes were used to visualize several yet-to-be cultured populations in the granular sludge and to speculate their *in situ* function. For instance, based on the cloning analysis, some particular bacterial clones belonging to subphylum I within the phylum Chloroflexi were found to be abundant accounting for 16% of the total clone library constructed from thermophilic methanogenic granules (Sekiguchi *et al* 1998, 1999). The phylum Chloroflexi had been divided into four subphyla and subphylum III contained cultured organisms belonging to several genera such *Chloroflexus, Sphaerobacter,* and *Oscillochloris.* Of

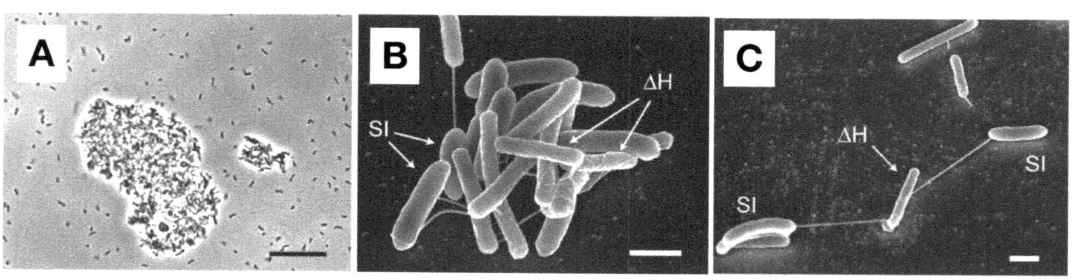

Figure 10.9 Micrographs of a coculture of *Pelotomaculum thermopropionicum* strain SI (propionate-oxidizing H_2-producing syntroph) and *Methanothermobacter thermautotrophicus* delta H (H_2-consuming methanogen) grown on propionate. (A) Aggregate of the coculture by phase contrast microscopy, bar = 25 μm. (B and C) Representative field emission-scanning electron microscopy (FE-SEM) images of the coculture. Bar = 1 μm. Figures are reproduced from Ishii *et al.* (2005).

the other three subphyla, subphylum I is a well-known group that is composed solely of diverse environmental clones (Sekiguchi and Kamagata, 2004). Importantly, members of subphylum I are widespread in anaerobic ecosystems and are frequently retrieved from various anoxic environments. Therefore, a specific probe was designed based on the clone sequences retrieved from the granular sludge and applied to the thin sections of granules. Surprisingly, the organisms that lit up with the probe were thin filaments and predominated at the outer most layers of the granules (Fig. 10.7) (Sekiguchi *et al.*, 1999, 2001). Considering the fact that the reactor was fed with sucrose and several fatty acids, and fatty acids could be consumed by other organisms inside the granule, it was very likely that the filamentous organism was a carbohydrate fermenter. In fact, the organism was later isolated from the granule and this trait was confirmed by pure culture (Sekiguchi *et al.*, 2001, 2003).

Isolation of key organisms in methane fermentation

It should be noted that molecular techniques also open the door to strategic isolation of yet-to-be cultured microorganisms. If cloned sequences that may represent an uncultured organism are obtained at a high frequency, it would be possible to detect or visualize the organisms by designing a primer set for PCR-based detection or a specific probe for microscopic observation, and applying them to the sample. One could also speculate the physiology, function, and plausible substrates to be used if the cloned sequence is similar to that of organisms already cultivated.

As described earlier, a huge variety of microorganisms still remain uncultivated, but by taking advantage of recently developed molecular techniques, several methanogens, syntrophs and heterotrophic organisms were successfully isolated. One of the earliest examples was isolation of a syntrophic propionate degrader from thermophilic UASB granules (Imachi *et al.*, 2000). Syntrophs are among the most difficult organisms to cultivate because: (1) they require partner methanogens or other H_2-scavenging organisms such as hydrogenotrophic sulphate reducers. This strict dependency frequently hinders isolation in pure culture; (2) most syntrophs are slow-growing organisms as the energy they could gain is generally very low. Despite these difficulties, the syntrophic propionate degrader mentioned above was isolated by the following procedure. The thermophilic granule was fed with propionate, thus the entire community was able to oxidize propionate. The isolation was initiated with a simple primary enrichment culture using propionate as the sole substrate. After one month of incubation, propionate was degraded together with methane formation. The culture was successively transferred into fresh medium every two months. Over ten successive transfers, the microbial component seemed to be relatively simplified. However, attempts to cultivate the syntroph using a roll tube method with propionate containing medium in the presence of an H_2-utilizing methanogen were unsuccessful. Therefore, the community in the propionate enrichment was analysed by 16rRNA gene cloning and sequencing and the dominant population was found to be affiliated with the *Desulphotomaculum* group. Based on the information that the cultured organisms within this group utilize various substrates with and without sulphate, the strategy for isolation of the target organism was altered: i.e. several substrates other than propionate were tested to determine whether or not the target microorganism could grow by monitoring the culture using a newly designed probe that specifically detected the target organism. Of the tested substrates, pyruvate was found to allow the growth of the target organism in pure culture, and the organism was eventually isolated. Interestingly, it was able to grow in pure culture depending upon the substrates used. The organism was finally inoculated into the medium containing propionate and H_2-utilizing methanogens and the ability to oxidize propionate was confirmed. The isolate was later designated *Pelotomaculum thermopropionicum* and its in-depth ecology, biochemistry, physiology and genome structure were extensively studied (Imachi *et al.*, 2000, 2002; Ishii *et al.*, 2005; Kosaka *et al.*, 2006; Gorby *et al.*, 2006, Imachi *et al.*, 2006). This example demonstrates that molecular techniques combined with classical isolation can facilitate isolation of uncultured organisms and shed light on those currently hidden from microbiologists.

Concluding remarks and future trends

Biological wastewater treatment processes are functional units of active microbial communities. Over the last decade, a number of studies have elucidated the microbial diversities of nitrifying, denitrifying, phosphorous removing, and methane fermenting processes. Recently, anammox has been drawing attention as a new nitrogen-removing system hence the communities responsible for this idiosyncratic reaction have been intensively studied. Molecular techniques have unveiled the entity of microbial constituents in wastewater treatment systems and the spatial distribution of each member to address the fundamental questions about microbial identities, functions and spatial organization. These approaches have also revealed that a vast variety of microorganisms in wastewater treatment systems are yet-to-be cultured and currently uncharacterized.

Molecular techniques are now directing microbiologists towards further ways to address the *in situ* function of each organism in wastewater treatment systems, the material transfer and flux between organisms and the changes in communities over time, depending upon the changes in operation conditions or perturbations. Ideally, these analyses have to be done in a short period of time, as rapid and ample monitoring would facilitate maintenance and alternation of processes as needed. These technical developments are important for practical contribution of knowledge gained about microbial ecology that can contribute to process engineering.

One of the typical needs in process engineering is the prediction of bulking in treatment plants. If the outbreak of bulking could be predicted in advance, for example using molecular techniques, on-site prevention or remediation would be possible to minimize the risk of disasters. There are a number of different types of causative organisms that are detrimental to treatment systems, including *Sphaerotilus natans* and *Thiothrix* spp., that have long been referred to as Type 021N filamentous bacteria. FISH probes specific for each group of bacteria have been designed and applied to activated sludges (Kanagawa *et al.*, 2000). Bulking in methane-fermenting UASB processes can also occur and several different types of filamentous bacteria belonging to the phylum Chloroflexi and the candidate phylum KSB3 have been found to be causative agents of that process (Sekiguchi *et al.*, 2001; Yamada *et al.*, 2007).

Together with better understanding of the physiology and ecophysiology of bulking organisms, establishing quick and accurate ways to detect and enumerate the target organisms in a reasonably economical way would be very helpful for monitoring of wastewater treatment plants. Thus, molecular based microbial ecology is now expected to have a large impact in engineering fields.

References

Amann, R., and Kühl, M. (1998). *In situ* methods for assessment of microorganisms and their activities. Curr. Opin. Microbiol. *1*, 352–358.

Amann, R., and Ludwig, W. (2000). Ribosomal RNA-targeted nucleic acid probes for studies in microbial ecology. FEMS Microbiol. Rev. *24*, 555–565.

Amann, R.I., Stomley, J., Devereux, R.K., and Stahl, D.A. (1992). Molecular and microscopic identification of sulfate-reducing bacteria in multispecies biofilms. Appl. Environ. Microbiol. *58*, 614–623.

Andrews, J.H., and Harris, R.F. (1986). *r*- and *k*-selection and microbial ecology. Adv. Microl. Ecol. *9*, 99–147.

Björnsson, L., Hugenholtz, P., Tyson, G.W., and Blackall, L.L. (2002). Filamentous *Chloroflexi* (green non-sulfur bacteria) are abundant in wastewater treatment processes with biological nutrient removal. Microbiology *148*, 2309–2318.

Burrell, P., Keller, J., and Blackall, L.L. (1999). Characterization of the bacterial consortium involved in nitrite oxidation in activated sludge. Water Sci. Technol. *39*, 45–52.

Daims, H., Purkhold, U., Bjerrum, L., Arnold, E., Wilderer, P.A., and Wagner, M. (2001). Nitrification in sequencing biofilm batch reactors: lessons from molecular approaches. Water Sci. Technol. *43*, 9–18.

Daims, H., Nielsen, P.H., Nielsen, J., Juretschko, S., and Wagner, M. (2000). Novel *Nitrospira*-like bacteria as dominant nitrite-oxidizers in biofilms from wastewater treatment plants: diversity and *in situ* physiology. Water Sci. Technol. *41*, 85–90.

Díaz, E.E., Stams, A.J.M., Amils, R., and Sanz, J.L. (2006). Phenotypic properties and microbial diversity of methanogenic granules from a full-scale upflow anaerobic sludge bed reactor treating brewery wastewater. Appl. Environ. Microbiol. *72*, 4942–4949.

Dionisi, H.M., Layton, A.C., Harms, G., Gregory, I.R., Robinson, K.G., and Sayler, G.S. (2002). Quantification of *Nitrosomonas oligotropha*-like ammonia- oxidizing bacteria and *Nitrospira* spp. from full-scale wastewater treatment plants by competitive PCR. Appl. Environ. Microbiol. *68*, 245–253.

Egli, K., Fanger, U., Alvarez, P.J.J., Siegrist, H., van der Meer, J.R., and Zehnder, A.J.B. (2001). Enrichment and characterization of an anammox bacterium from a rotating biological contactor treating ammonium-rich leachate. Arch. Microbiol. *175*, 198–207.

Egli, K., Bosshard, F. Werlen, C., Lais, P., Siegrist, H., Zehnder, A.J., and van der Meer, J.R. (2003). Microbial composition and structure of a rotating biological contactor biofilm treating ammonium-rich wastewater without organic carbon. Microb. Ecol. *45*, 419–432.

Ehrich, S., Behrens, D., Lebedeva, E., Ludwig, W., and Bock, E. (1995). A new obligately chemolithoautotrophic nitrite-oxidizing bacterium, Nitrospira moscoviensis sp. nov. and its phylogenetic relationship. Arch. Microbiol. *164*, 16–23.

Francis, C.A., Roberts, K.J., Beman, J.M., Santoro, A.E., and Okley, B.B. (2005). Ubiquity and diversity of ammonia-oxidizing archaea in water columns and sediments of the ocean. Proc. Natl. Acad. Sci. U.S.A. *102*, 14683–14688.

Gich, F., Garcia-Gil, J., and Overmann, J. (2001). Previously unknown and phylogenetically diverse members of the green nonsulfur bacteria are indigenous to freshwater lakes. Arch. Microbiol. *177*, 1–10.

Gieseke, A., Purkhold, U., Wagner, M., Amann, R., and Schramm, A. (2001). Community structure and activity dynamics of nitrifying bacteria in a phosphate-removing biofilm. Appl. Environ. Microbiol. *67*, 1351–1362.

Gieseke, A., Bjerrum, L., Wagner, M., and Amann, R. (2003). Structure and activity of multiple nitrifying bacterial populations co-existing in a biofilm. Environ. Microbiol. *5*, 355–369.

Gieseke, A., Nielsen, J.L., Amann, R., Nielsen, P.H., and de Beer, D. (2005). *In situ* substrate conversion and assimilation by nitrifying bacteria in a model biofilm. Environ. Microbiol. 7, 1392–1404.

Godon, J.J., Zumstein, E., Dabert, P., Habouzit, F., and Moletta, R. (1997). Molecular microbial diversity of an anaerobic digestor as determined by small-subunit rDNA sequence analysis. Appl. Environ. Microbiol. *63*, 2802–2813.

Gonzalez-Gil, G, Lens, P.N.L., Van Aelst, A., Van As, H., Versprille, A.I., and Lettinga, G. (2001). Cluster structure of anaerobic aggregates of an expanded granular sludge bed reactor. Appl. Environ. Microbiol. *67*, 3683–3692.

Gorby, Y.A., Yanina, S., McLean, J.S., Rosso, K.M., Moyles, D., Dohnalkova, A., Beveridge, T.J., Chang, Kim, I.S., B.H., Kim, K.S., Culley, D.E., Reed, S.B., Romine, M.F., Saffarini, D.A., Hill, E.A., Shi, L., Elias, D.A., Kennedy, D.W., Pinchuk, G., Watanabe, K., Ishii, S., Logan, B., Nealson, K.H., and Fredrickson, J.K. (2006). Electrically conductive bacterial nanowires produced by *Shewanella oneidensis* strain MR-1 and other microorganisms. Proc. Natl. Acad. Sci. U.S.A. *103*, 11358–11363.

Harms, G., Layton, A.C., Dionisi, H.M., Gregory, I.R., Garrett, V.M., Hawkins, S.A., Robinson, K.G., and Sayler, G.S. (2003). Real-time PCR quantification of nitrifying bacteria in a municipal wastewater treatment plant. Environ. Sci. Technol. 37, 343 -351.

Head, I.M., Hiorns, W.D., Embley, T.M., McCarthy, A.J., and Sunders, J.R. (1993). The phylogeny of autotrophic ammonia-oxidizing bacteria as determined by analysis of 16S ribosomal RNA gene sequences. J. Gen. Microbiol. *139*, 1147–1153.

Head, I.M., Saunders, J.R., and Pickup, R.W. (1998). Microbial evolution, diversity and ecology: a decade of ribosomal RNA analysis of uncultured microorganisms. Microb. Ecol. *35*, 1–21.

Imachi, H., Sekiguchi, Y., Kamagata, Y., Ohashi, A., and Harada, H. (2000). Cultivation and *in situ* detection of a thermophilic bacterium capable of oxidizing propionate in syntrophic association with hydrogenotrophic methanogens in a thermophilic methanogenic granular sludge. Appl. Environ. Microbiol. *66*, 3608–3615.

Imachi, H., Sekiguchi, Y., Kamagata, Y., Hanada, S., Ohashi, A., and Harada, H. (2002). *Pelotomaculum thermopropionicum* gen. nov., sp. nov., a novel anaerobic, thermophilic syntrophic propionate-oxidizing bacterium. Int. J. Syst. Evol. Microbiol., *52*, 1729–1735.

Imachi, H., Sekiguchi, Y., Kamagata, Y., Loy, A., Qiu, Y-L., Hugenholtz, P., Kimura, N., Wagner, M., Ohashi, A., and Harada, H. (2006). Non-sulfate-reducing, syntrophic bacteria affiliated with the *Desulfotomaculum* cluster I are widely distributed in methanogenic environments. Appl. Environ. Microbiol. *72*, 2080–2091.

Ingalls, A.E., Shah, S.R., Hansman, R.L., Aluwihare, L.I., Sanyos, G.M., Druffel, E.R.M., and Pearson, A. (2006). Quantifying archaeal community autotrophy in the mesopelagic ocean using natural radiocarbon. Proc. Natl. Acad. Sci. U.S.A. *103*, 6442–6447.

Ishii, S., Kosaka,T., Hori, K., Hotta, Y., and Watanabe, K. (2005). Coaggregation facilitates interspecies hydrogen transfer between *Pelotomaculum thermopropionicum* and *Methanothermobacter thermautotrophicus*. Appl. Environ. Microbiol. *71*, 7838–7845.

Jetten, M.S.M., Strous, M., van de Pas-Schoonen, K.T., Schalk, J., van Dongen, U.G.J.M., van de Graaf, A.A., Logemann, S., Muyzer, G., van Loosdrecht, M.C.M., and Kuenen, J.G. (1999). The anaerobic oxidation of ammonium. FEMS Microbiol. Rev. *22*, 421–437.

Juretschko, S., Timmermann, G., Schmid, M., Schleifer, K.-H., Pommerening-Roser, A., Koops, H.-P., and Wagner, M. (1998). Combined molecular and conventional analyses of nitrifying bacterium diversity in activated sludge: *Nitrosococcus mobilis* and *Nitrospira*-like bacteria as dominant populations. Appl. Environ. Microbiol. *64*, 3042–3051.

Juretschko, S., Loy, A., Lehner, A., and Wagner, M. (2002). The microbial community composition of a nitrifying-denitrifying activated sludge from an industrial sewage treatment plant analyzed by the full-cycle rRNA approach. Syst. Appl. Microbiol. *25*, 84–99.

Kartal, B., van Niftrik, L., Sliekers, O., Schmic, M.C., Schmidt, I., van de Pas-Schoonen, K., Cirpus, I., van der Star, W., van Loosdrecht, M.C.M., Abma, W., Kuenen, G.J., Mulder, J.W., Jetten, M.S.M., Strous, M., and van de Vossenberg, J. (2004). Application, ecophysiology and biodiversity of anaerobic ammonium-oxidizing bacteria. Rev. Environ. Sci. Biotechnol. *3*, 255–264.

Kanagawa, T., Kamagata, Y., Aruga, S., Kohno, T., Horn, M., and Wagner, M. (2000). Phylogenetic analysis of and oligonucleotide probe development for Eikelboom Type 021N filamentous bacteria isolated from bulking activated sludge. Appl. Environ. Microbiol. *66*, 5043–5052.

Kartal, B., Rattray, J., van Niftrik, L.A., van de Vossenberg, J., Schmid, M.C., Webb, R.I., Schouten, S., Fuerst, J.A., Damste, J.S., Jetten, M.S.M., and Strous, M. (2006). Candidatus '*Anammoxoglobus propionicus*' a new propionate oxidizing species of anaerobic ammonium oxidizing bacteria. Syst. Appl. Microbiol. *30*, 39–49.

Kindaichi, T., Ito, T., and Okabe, S. (2004). Eco–physiological interaction between nitrifying bacteria and heterotrophic bacteria in autotrophic nitrifying biofilms as determined by MAR-FISH. Appl. Environ. Microbiol. *70*,1641–1650.

Kindaichi, T., Kawano, Y., Ito, T., Satoh, H., and Okabe S. (2006). Population dynamics and *in situ* kinetics of nitrifying bacteria in autotrophic nitrifying biofilms as determined by real-time quantitative PCR. Biotech. Bioeng. *94*, 1111–1121.

Kindaichi, T., Tsushima, I., Ogasawara, Y., Shimokawa, M., Ozaki, N., Satoh, H., and Okabe, S. (2007). *In situ* activity and spatial organization of anaerobic ammonium-oxidizing (anammox) bacteria in biofilms. Appl. Environ. Microbiol. *76*, 4931–4939.

Konneke, M., Bernhard, A.E., de la Torre, J., Walker, C.B., Waterbury, J.B., and Stahl, D.A. (2005). Isolation of an autotrophic ammonia-oxidizing marine archaeon. Nature. *437*, 543546.

Koops, H.-P., Harms, H., and Wehrmann, H. (1976). Isolation of a moderate halophilic ammonia-oxidizing bacterium, *Nitrosococcus mobilis* nov. sp. Arch. Microbiol. *107*, 277–282.

Kosaka, T., Uchiyama, T., Ishii, S., Enoki, M., Imachi, H., Kamagata, Y., Ohashi, A., Harada, H., Ikenaga, H., and Watanabe, K. (2006). Genomic insights into the central catabolic pathway in a thermophilic syntrophic bacterium, *Pelotomaculum thermopropionicum*. J. Bacteriol., 188, 202–210.

Kuypers, M.M.M., Sliekers, A.O., Lavik, G., Schmid, M., Jørgensen, B.B., Kuenen, J.G., Sinninghe Damsté, J.S., Strous, M., and Jetten, M.S.M. (2003). Anaerobic ammonium oxidation by anammox bacteria in the Black Sea. Nature. *422*, 608–611.

Lee, N., Nielsen, P.H., Andreasen, K.H., Juretschko, S., Nielsen, J.L., Schleifer, K.-H., and Wagner, M. (1999). Combination of fluorescent *in situ* hybridization and microautoradiography-a new tool for structure-function analyses in microbial ecology. Appl. Environ. Microbiol. *65*, 1289–1297.

Lettinga, G. (1995). Anaerobic digestion and wastewater treatment systems. Antonie van Leeuwenhoek. *67*, 3–28.

Liu, W-T., Chan, O.C., and Fang, H.H. (2002). Characterization of microbial community in granular sludge treating brewery wastewater. Wat. Res. *36*, 1767–1775.

Lydmark, P., Lind, M., Sorensson, F., and Hermansson, M. (2006). Vertical distribution of nitrifying populations in bacterial biofilms from a full-scale nitrifying trickling filter. Environ. Microbiol. *8*, 2036–2049.

Mobarry, B.K., Wagner, M., Urbain, V., Rittmann, B.E., and Stahl, D.A. (1996). Phylogenetic probes for analyzing abundance and spatial organization of nitrifying bacteria. Appl. Environ. Microbiol. *62*, 2156–2162.

Nakamura, K., Terada, T., Sekiguchi, Y., Shinzato, N., Meng, X-Y., Enoki, M., and Kamagata, Y. (2006). Application of pseudomurein endoisopeptidase to fluorescence *in situ* hybridization of methanogens within the family Methanobacteriaceae. Appl. Environ. Microbiol. *72*, 6907–6913.

Neef, A., Amann, R., Schlesner, H., and Schleifer, K.-H. (1998). Monitoring a widespread bacteria group: *in situ* detection of planctomycetes with 16S rRNA-targeted probes. Microbiol. *144*, 3257–3266.

Nielsen, J.L., and Nielsen, P.H. (2005). Advances in microscopy: microautoradiography of single cells. Methods Enzymol. *397*, 237–56.

Nielsen, J.L., Christensen, D., Kloppenborg, M., and Nielsen, P.H. (2003). Quantification of cell-specific substrate uptake by probe-defined bacteria under *in situ* conditions by microautoradiography and fluorescence *in situ* hybridization. Environ. Microbiol. *5*, 202–211.

Nielsen, P.H., Andreasen, K., Lee, N., and Wagner, M. (1999). Use of Microautoradiography and fluorescent *in situ* hybridization for characterization of microbial activity in activated sludge. Water Sci. Technol. *39*,Nielsen, M., Bollmann, A., Sliekers, O., Jetten, M., Schmid, M., Strous, M., Schmidt, I., Larsen, L.H., Nielsen, L.P., and Revsbech, N.P. (2005). Kinetics, diffusional limitation and microscale distribution of chemistry and organisms in a CANON reactor. FEMS Microbiol. Ecol. *51*, 247–256.

Okabe, S., Hirata, K., Ozawa, Y., and Watanabe, Y. (1996). Spatial microbial distribution of nitrifiers and heterotrophs in mixed-population biofilms. Biotech. Bioeng. *50*, 24–35.

Okabe, S., Satoh, H., and Watanabe, Y. (1999). *In situ* analysis of nitrifying biofilms as determined by *in situ* hybridization and the use of microelectrodes. Appl. Environ. Microbiol. *65*, 3182–3191.

Okabe, S., Naitoh, H., Satoh, H., and Watanabe, Y. (2002). Structure and function of nitrifying biofilms as determined by molecular techniques and the use of microelectrodes. Water Sci. Technol. *46*, 233–241.

Okabe, S., Kindaichi, T., Ito, T., and Satoh, H. (2004a). Analysis of size distribution and areal cell density of ammonia-oxidizing bacterial microcolonies in relation to substrate microprofiles in biofilms. Biotechnol. Bioeng. *85*, 86–95.

Okabe, S., Kindaichi, T., and Ito T. (2004b). MAR-FISH-An ecophysiological approach to link phylogenetic affiliation and *in situ* metabolic activity of microorganisms at a single-cell resolution. Microbes. Environ. *19*, 83–98.

Okabe, S., Kindaichi, T., and Ito, T. (2005). Fate of ^{14}C-labeled microbial products derived from nitrifying bacteria in autotrophic nitrifying biofilms. Appl. Environ. Microbiol. *71*, 3987–3994.

Ouverney, C.C., and Fuhrman, J.A. (1999). Combined microautoradiography-16S rRNA probe technique for determination of radioisotope uptake by specific microbial cell types *in situ*. Appl. Environ. Microbiol. *65*, 1746–1752.

Park, H.-D., Wells, G.F., Bae H., Criddle, C.S., and Francis, C.S. (2006). Occurrence of ammonia-oxidizing archaea in wastewater treatment plant bioreactors. Appl. Environ. Microbiol. *72*, 5643–5647.

Persson, F., Wik, T., Sorensson, F., and Hermansson, M. (2002). Distribution and activity of ammonia oxidizing bacteria in a large full-scale trickling filter. Water Res. *36*, 1439–1448.

Prosser, J.I. (1989). Autotrophic nitrification in bacteria. Adv. Microb. Physiol. *30*, 125–181.

Purkhold, U., Pommerening-Roser, A., Juretschko, S., Schmid, M.C., Koops, H.-P., and Wagner, M. (2000). Phylogeny of all recognized species of ammonia oxidizers based on comparative 16S rRNA and *amoA* sequence analysis: Implications for molecular diversity surveys. Appl. Environ. Microbiol. *66*, 5368–5382.

Rocheleau, S., Greer, C.W., Lawrence, J.R., Cantin, C., Laramée, L., and Guiot, S.R. (1999). Differentiation of *Methanosaeta concilii* and *Methanosarcina barkeri* in anaerobic mesophilic granular sludge by fluorescent *In situ* hybridization and confocal scanning laser microscopy. Appl. Environ. Microbiol. *65*, 2222–2229.

Rowan, A.K., Snape, J.R., Fearnside, D., Barer, M.R., Curtis, T.P., and Head, I.M. (2003). Composition and diversity of ammonia-oxidising bacterial communities in wastewater treatment reactors of different design treating identical wastewater. FEMS Microbiol. Ecol. *43*, 195–206.

Saiki, Y., Iwabuchi, C., Katami, A., and Kitagawa, Y. (2002). Microbial analyses by fluorescence *in situ* hybridization of well-settled granular sludge in brewery wastewater treatment plants. J. Biosci. Bioeng. *93*, 601–606.

Satoh, H., Okabe, S., Norimatsu, N., and Watanabe, Y. (2000). Significance of substrate C/N ratio on structure and activity of nitrifying biofilms determined by *in situ* hybridization and the use of microelectrodes. Water Sci. Technol. *41*, 317–321.

Satoh, H, Okabe, S, Yamaguchi, Y, and Watanabe, Y. (2003). Evaluation of the impact of bioaugmentation and biostimulation by *in situ* hybridization and microelectrode. Water Res. *37*, 2206–2216.

Satoh, H., Yamakawa, T., Kindaichi, T., Ito, T., and Okabe, S. (2006). Community structures and activities of nitrifying and denitrifying bacteria in industrial wastewater-treating biofilms. Biotech. Bioeng. *94*, 762–772.

Schink, B. (1997). Energetics of syntrophic cooperation in methanogenic degradation. Microbiol. Mol. Biol. Rev. *61*, 262–280.

Schmid, M., Twachtmann, U., Klein, M., Strous, M., Juretschko, S., Jetten, M.S.M., Metzger, J.W., Schleifer, K-.H., and Wagner, M (2000). Molecular evidence for genus level diversity of bacteria capable of catalyzing anaerobic ammonium oxidation. System. Appl. Microbiol. *23*, 93–106.

Schmid, M., Walsh, K., Webb, R., Rijpstra, W.I., van de Pas-Schoonen, K., Verbruggen, M.J., Hill, T., Moffett, B., Fuerst, J., Schouten, S., Damste, J.S., Harris, J., Shaw, P., Jetten, M., and Strous, M. (2003). *Candidatus* 'Scalindua brodae' sp. nov., *Candidatus* 'Scalindua wagneri' sp. nov., two new species of anaerobic ammonium oxidizing bacteria. Syst. Appl. Microbiol. *26*, 529–538.

Schmid, M., Maas, B., Dapena, A., van de Pas-Schoonen, K., van de Vossenberg, J., Kartal, B., van Niftrik, L., Schmidt, I., Cirpus, I., Kuenen, J.G., Wagner, M., Sinninghe Damste, J.S., Kuypers, M., Revsbech, N.P., Mendez, R., Jetten, M., and Strous, M. (2005). Biomarkers for *in situ* detection of anaerobic ammonium-oxidizing (Anammox) bacteria. Appl. Environ. Microbiol. *71*, 1677–1684.

Schmidt, I., Bock, E., and Jetten, M.S.M. (2001). Ammonia oxidation by *Nitrosomonas eutropha* with NO2 as oxidant is not inhibited by acetylene. Microbiol. *147*, 2247–2253.

Schmidt, I., Hermelink, C., van de Pas-Schoonen, K., Strous, M., op den Camp, H., Kuenen, J.G., and Jetten, M.S.M. (2002). Anaerobic ammonia oxidation in the presence of nitrogen oxides (NO_x) by two different lithotrophs. Appl. Environ. Microbiol. *68*, 5351–5357.

Schmidt, J.E., and Ahring, B.K. (1996). Granular sludge formation in upflow anaerobic sludge blanket (UASB) reactors. Biotechnol. Bioeng. *49*, 229–246.

Schramm, A. (2003). *In situ* analysis of structure and activity of the nitrifying community in biofilms, aggregate, and sediments. Geomicrobiol. J. *20*, 313–333.

Schramm, A., Larsen, L.H., Revsbech, N.P., Ramsing, N.B., Amann, R., and Schleifer, K-H. (1996). Structure and function of a nitrifying biofilm as determined by *in situ* hybridization and the use of microelectrodes. Appl. Environ. Microbiol. *62*, 4641–4647.

Schramm, A., de Beer, D., Wagner, M., and Amann, R. (1998). Identification and activity *in situ* of *Nitrosospira* and *Nitrospira* spp. as dominant populations in a nitrifying fluidized bed reactor. Appl. Environ. Microbiol. *64*, 3480–3485.

Schramm, A., de Beer, D., van den Heuvel, J.C., Ottengraf, S., and Amann, R. (1999). Microscale distribution of populations and activities of *Nitrosospira* and *Nitrospira* spp. along a macroscale gradient in a nitrifying bioreactor: quantification by *in situ* hybridization and the use of microsensors. Appl. Environ. Microbiol. *65*, 3690–3696.

Schramm, A., de Beer, D., Gieseke, A., and Amann, R. (2000). Microenvironments and distribution of nitrifying bacteria in a membrane-bound biofilm. Environ. Microbiol. *2*, 680–686.

Sekiguchi, Y. (2006). Yet-to-be cultured microorganisms relevant to methane fermentation processes. Microbes Environ. *21*, 1–15.

Sekiguchi, Y., and Kamagata, Y. (2004). Microbial community structure and functions in methane fermentation technology for wastewater treatment., pp. 361–384. In Zuber, P., and Nakano, M.M. (ed.), Strict and facultative anaerobes: medical and environmental aspects. Horizon Scientific Press, Norfolk, UK.

Sekiguchi, Y., Kamagata, Y., Syutsubo, K., Ohashi, A., Harada, H., and Nakamura, K. (1998). Phylogenetic diversity of mesophilic and thermophilic granular sludges determined by 16S rRNA gene analysis. Microbiology *144*, 2655–2665.

Sekiguchi, Y., Kamagata, Y., Nakamura, K., Ohashi, A., and Harada, H. (1999). Fluorescence *in situ* hybridization using 16S rRNA targeted oligonucleotides reveals localization of methanogens and selected uncultured bacteria in mesophilic and thermophilic sludge granules. Appl. Environ. Microbiol. *65*, 1280–1288.

Sekiguchi, Y., Takahashi, H., Kamagata, Y., Ohashi, A., and Harada, H. (2001). *In situ* detection, isolation and physiological properties of a thin filamentous microorganism abundant in methanogenic granular sludges: a novel isolate affiliated with a clone cluster, the Green Non Sulfur Bacteria subdivision I. Appl. Environ. Microbiol. *67*, 5740–5749.

Sekiguchi, Y., Yamada, T., Hanada, S., Ohashi, A., Harada, H., and Kamagata, Y. (2003). *Anaerolinea thermophila* gen. nov., sp. nov., and *Caldilinea aerophila* gen. nov., sp. nov., two novel filamentous thermophiles that represent a previously uncultured lineage of the domain Bacteria at the subphylum leve l. Int. J. Syst. Evol. Microbiol. *53*, 1843–1851.

Siripong, S., and Rittmann, B.E. (2007). Diversity study of nitrifying bacteria in full-scale municipal wastewater treatment plants. Wat. Res. *41*, 1110–1120.

Strous, M., Heijnen, J.J., Kuenen, J.G., and Jetten, M.S.M. (1998). The sequencing batch reactor as a powerful tool to study very slowly growing micro-organisms. Appl. Microbiol. Biotechnol. *50*, 589–596.

Strous, M., Kuenen, J.G., and Jetten, M.S.M. (1999). Key physiology of anaerobic ammonium oxidation. Appl. Environ. Microbiol. *65*, 3248–3250.

Teske, A., Alm, E., Regan, J.M., Toze, S., Rittmann, B.E., and Stahl, D.A. (1994). Evolutionary relationship among ammonia- and nitrite-oxidizing bacteria. J. Bacteriol. *176*, 6623–6630.

Toh, S.K., Webb, R.I., and Ashbolt, N.J. (2002). Enrichment of autotrophic anaerobic ammonium-oxidizing consortia from various wastewaters. Microb. Ecol. *43*, 154–167.

Treusch, A.H., Leininger, S., Kletzin, A., Schuster, S.C., Klenk, H.-P., and Schleper, C. (2005). Novel genes for nitrite reductase and Amo-related proteins indicate a role of uncultivated mesophilic crenarchaeota in nitrogen cycling. Environ. Microbiol. *7*, 1985–1995.

Tsushima, I., Kindaichi, T., and Okabe, S. (2007). Quantification of anaerobic ammonium-oxidizing bacteria in enrichment cultures by real-time PCR. Water Res. *41*, 785–794.

Tsushima, I., Ogasawara, Y., Shimokawa, M., Kindaichi, T., and Okabe, S. (2007). Development of High-Rate Anaerobic Ammonium-Oxidizing (ANAMMOX) Biofilm Reactors. Water Res. *41*, 1623–1634.

van Dongen, U., Jetten, M.S.M., and van Loosdrecht, M.C.M. (2001). The SHARON-Anammox process for treatment of ammonium rich wastewater. Water Sci. Technol. *44*, 153–160.

Wagner, M., Nielsen, P.H., Loy, A., Nielsen, J.L., and Daims, H. (2006). Linking microbial community structure with function: fluorescence *in situ* hybridization-microautoradiography and isotope arrays. Curr. Opin. Biotechnol. *17*, 83–91.

Weisburg, W.G., Barns, S.M., Pelletier, D.A., and Lane, D.J. (1991). 16S ribosomal DNA amplification for phylogenetic study. J. Bacteriol. *173*, 697–703.

Wintzingerode, V.F., Selent, B., Hegemann, W., and Goebel, U.B. (1999). Phylogenetic analysis of an anaerobic, trichlorobenzene transforming microbial consortium. Appl. Environ. Microbiol. *65*, 283–286.

Wu, J-H., Liu, W-T., Tseng, I-C., and Cheng, S-S. (2001). Characterization of microbial consortia in a terephthalate-degrading anaerobic granular sludge system. Microbiology *147*, 373–382.

Yamada, T., Yamauchi, T., Shiraishi, S., Hugenholtz, P., Ohashi, A., Harada, H., Kamagata, Y., Nakamura, K., and Sekiguchi, Y. (2007). Identification, distribution and diversity of filamentous bacteria, belonging to candidate phylum KSB3, that are associated with bulking in methanogenic granular sludges. The ISME J., *1*, 246–255.

11 The Impact and Molecular Genetics of Bacterial Biofilms

Shuwen An, Yi-Hu Dong and Lian-Hui Zhang

Abstract

Many bacteria can grow and live as biofilms, in which single microbial cells individually interconnect with each other through an extracellular matrix. Biofilm-forming bacteria pose severe problems in the environment, industry and health care sector due to increased bacterial survival competence in the environment and the protective nature of biofilms that prevent effective eradication. Technological progress in microscopy, molecular genetics and genome analysis has significantly advanced our understanding of the structural and molecular aspects of biofilms, especially of extensively studied model organisms such as *Pseudomonas aeruginosa*. Biofilm development can be divided into several key steps including attachment, microcolony formation, biofilm maturation and dispersion; and in each step bacteria may recruit different components and molecules including flagella, type IV pili, DNA and exopolysaccharides. The rapid progress in biofilm research has also unveiled several genetic regulation mechanisms implicated in biofilm regulation such as quorum sensing and the novel secondary messenger cyclic-di-GMP. Understanding the molecular mechanisms of biofilm formation has facilitated the exploration of novel strategies to control bacterial biofilms.

Introduction

A biofilm is a complex aggregation of microbial cells marked by the excretion of a protective and adhesive matrix. In most cases, a biofilm is recognized as a surface attached community with microbial cells interconnected by an extracellular matrix of polymeric substances (Fig. 11.1A), but occasionally non-attached or loosely attached bacterial cell aggregates can also be observed (Fig. 11.1C). Biofilms play important roles in most aspects of microbiology, which can be either beneficial or harmful. For example, bacterial biofilms in bioremediation and fermentation facilities may facilitate stable operation, but medical device- and ship-associated biofilms are serious healthcare or industry problems (Costerton *et al.*, 1999; O'Toole *et al.*, 2000b; Donlan and Costerton, 2002; Parsek and Singh, 2003).

The majority of microorganisms, if not all, form and exist as biofilms under various conditions (Sutherland, 2001). Since the 1970s, tremendous progress has been made in understanding the environmental impact, structural features and developing mechanisms of bacterial biofilm formation, attributing to rapid technological advancements in microscopy, molecular genetics and genome analysis. Understanding the genetic basis and molecular mechanisms involved in biofilm formation are essential for control and prevention of biofilm development, and for solving biofilm-associated medical and environmental problems. Among many biofilm-associated bacterial pathogens, a few medically or environmentally important bacterial species, including *Escherichia coli, Pseudomonas aeruginosa, Pseudomonas fluorescens, Vibrio cholerae* and a few Gram-positive bacterial pathogens, have been characterized extensively at the genetic level. Since many bacterial genes and their encoded functions are conserved in bacteria and archaea, the research findings from these model organisms

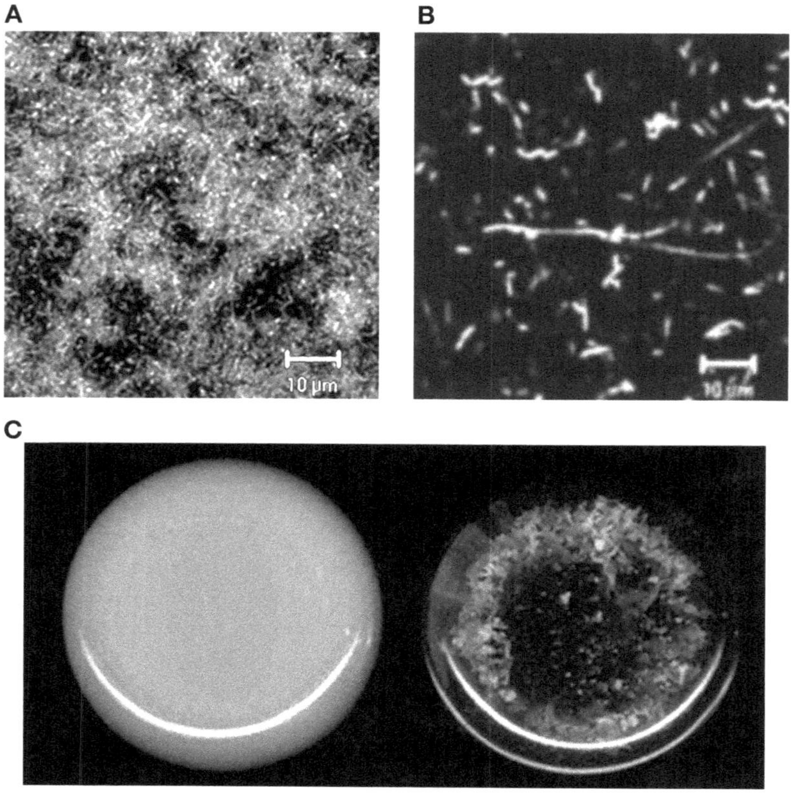

Figure 11.1 Bacterial biofilms and the role of quorum sensing in biofilm formation. **A** Biofilm formed on glass by *Pseudomonas aeruginosa* wild type strain PAO1. **B** Expression of AHL-lactonase that degrades AHL quorum sensing signals in strain PAO1 substantially reduces biofilm formation. **C** Plant bacterial pathogen *Xanthomonas campestris* forms cell aggregation-type biofilm in flask containing liquid medium in the absence of cell–cell communication signal DSF. Left, wild type strain 8004; right, DSF minus mutant 8004dF (He *et al.*, 2006).

may serve as important clues to prevent and control biofilms formed by other bacterial species. In this chapter, we review and discuss the structural features of bacterial biofilms, the molecular basis of biofilm development, the impact of biofilms on health care and the environment, and potential treatment strategies with a focus on *P. aeruginosa*.

The structural characteristics of bacterial biofilms

Biofilm matrix

A biofilm has been defined as a matrix-enclosed bacterial population adherent to each other and/or to a surface or interface (Costerton and Lewandowski, 1995). In the mature biofilm the matrix mass is composed of about 75–90% extracellular polymeric substances, with only 10–25% of the mass comprising organisms (Padera, 2006). The extracellular polymeric substances produced by bacteria are necessary for the irreversible attachment of bacterial cells to environmental surfaces or native tissue surfaces. These extracellular polymeric substances are composed of polysaccharides, extracellular DNA and proteinaceous compounds, which function as matrix components in cell-to-cell interconnections in the *P. aeruginosa* biofilm (Sutherland, 2001).

Exopolysaccharide (EPS) is a key component of the biofilm matrix in many biofilm-forming bacteria and may be composed of various sugar polymers (Watnick and Kolter 2000). These sugars in EPS are often found to be rhamnose, glucose, glucosamine, galactose, and fructose. In addition, glucuronic and guluronic acids are also found in some bacterial EPS. In *P. aeruginosa*, alginate has traditionally been considered as the major EPS in biofilms. This is based on the finding that the isolates from the lungs of cystic fibrosis patients exhibit a mucoid phenotype due to overproduction of alginate (Hoiby, 1993). However, it was later found that the strains with reduced alginate production are also capable of forming mature biofilms *in vitro* (Nivens *et al.*, 2001). Further characterization shows that under common laboratory conditions alginate is not a significant component of the EPS in *P. aeruginosa* biofilms (Hentzer *et al.*, 2001; Nivens *et al.*, 2001; Wozeniak *et al.*, 2003).

Although alginate is not required for biofilm formation by non-mucoid strains *in vitro*, its production (especially the O-acetylated form) contributes significantly to the biofilm architecture. An alginate overproducing *P. aeruginosa* strain forms highly structured biofilms on an abiotic surface as compared with its isogenic, non-mucoid strain (Hentzer *et al.*, 2001; Nivens *et al.*, 2001; Stapper *et al.*, 2004). Matsukawa and Greenberg (2004) showed that alginate may be a minor constituent of the total extracellular material of strain PAO1. The sugars that constitute alginate are hardly detectable but DNA and other sugars including glucose, rhamnose, and mannose are abundant in the extracellular material. Mutations in the polysaccharide synthesis locus (*psl*) result in mutant strains that produce an EPS consisting of predominantly glucose and rhamnose (Matsukawa and Greenberg, 2004; Friedman and Kolter, 2004b). It was suggested that EPS is synthesized early in biofilm formation and serves as a base upon which the subsequent 'mushroom-like' structures are built, or alternatively, EPS functions as a scaffold on which other polymers, such as DNA, become entangled (Matsukawa and Greenberg 2004).

Evidence for extracellular DNA as a cell-to-cell interconnecting compound in *P. aeruginosa* biofilms is accumulating (Whitchurch *et al.*, 2002; Webb *et al.*, 2003; Matsukawa and Greenberg, 2004; Allesen-Holm *et al.*, 2006). Destruction of biofilms by DNase treatment has indicated the importance of extracellular DNA as a biofilm matrix component in *P. aeruginosa* (Whitchurch *et al.*, 2002; Nemoto *et al.*, 2003). Evidence suggests that these biofilm-associated extracellular DNA molecules are released from the dead cells present in the biofilm (Webb *et al.*, 2003). They are localized primarily on the microcolonies in young biofilms and in the stalks of mushroom-shaped multicellular structures (Allesen-Holm *et al.*, 2006).

In addition to EPS and extracellular DNA, proteinaceous polymers are also important matrix components in bacterial biofilms. Recent evidence indicates that cell appendages, termed CupA fimbria, are an important constituent of the extracellular matrix in *P. aeruginosa* biofilms (Vallet *et al.*, 2001; D'Argenio *et al.*, 2002; Friedman and Kolter, 2004b). In conclusion, the available evidence indicates that the matrix of *P. aeruginosa* biofilms consists of exopolysaccharide, extracellular DNA and proteinaceous molecules, whereas alginate plays a key role in forming highly structured biofilms.

Biofilm structures

A structured biofilm is developed by single or mixed bacterial populations through attachment to abiotic or biotic surfaces via exopolysaccharides, extracellular DNA and proteinaceous compounds. Examination using confocal laser scanning microscopy of single-species and complex multispecies biofilms isolated from natural ecosystems has revealed that the basic physical structure of biofilms is universal, but with some minor variations. Most biofilms have a complex three-dimensional structure with internal architecture including channels and pores through which fluid can flow and deliver nutrients deep within this complex structure (Stoodley *et al.*, 2002). The bacterial cells in the mature biofilm generally concentrate themselves away from the underlying material surface and are not uniformly distributed throughout the biofilm.

Based on the structural characteristics, two general types of *P. aeruginosa* biofilms have been described. One is the 'flat' biofilm and other is the 'structured' biofilm. A flat biofilm is characterized by a relatively confluent, uniform community of bacteria on a surface, whereas a structured biofilm consists of a mushroom-shaped structure and large cell aggregates (Kirisits and Parsek, 2006). The mushroom-shaped structure includes a 'stalk' and a 'cap'. The stalk is a non-motile subpopulation and the cap is a motile subpopulation that migrates and resides on the surface of the stalk. Understanding of these structural features facilitates characterization of the molecular mechanisms of biofilm development, which will be discussed in detail below.

The impact of biofilms

Biofilms and bacterial adaptation

Bacteria switch between planktonic growth and a biofilm lifestyle in response to environmental changes. It is evident that bacteria reap a number of benefits from the biofilm mode of growth: protection from harmful conditions in the host; sequestration to a nutrient-rich area (colonization); and utilization of cooperative benefits (community) (Jefferson, 2004). In addition, since bacteria often exist as biofilms in natural environments rather than free-floating single cells, and bacterial cells within biofilms have a reduced growth rate relative to planktonic cells, this suggests that formation of biofilms may confer significant survival fitness advantages.

Bacteria growing in biofilms can withstand nutrient limitation and other stressful conditions, such as osmotic stress, mechanical stress, anoxia, pH changes, and exposure to antibiotics. Bacterial cells in a biofilm are exposed to gradients of nutrients and physicochemical factors, as nutrients reaching bacterial cells through the water channels within the EPS matrix. The latter plays an important role in enabling biofilms to resist the shear forces produced by blood flow and the washing action of saliva in animal hosts (Donlan and Costerton, 2002). In addition, the biofilm structure places bacterial cells in close proximity thus facilitating intercellular communications. This may allow bacterial populations to respond collectively via quorum signalling pathways to cope with adverse environmental conditions (Costerton *et al.*, 1999).

Biofilm components also provide a protective environment for bacterial cells living inside an animal host against the host defense system. Formation of biofilms, in many cases, drastically increases the bacterial resistance to biocides, antibiotic treatments and host immune responses (Costerton and Lewandowski, 1995). The EPS matrix in a biofilm can reduce the phagocytic ability of macrophages and increase resistance to reactive oxygen species produced by phagocytes (Yasuda *et al.*, 1994; Shiau and Wu, 1998). It has been shown that the phagocytosis of alginate-negative biofilm bacteria is significantly increased in the presence of IFN-gamma and exogenous addition of purified alginate to the alginate-negative biofilm restores the resistance to human leukocyte killing (Leid *et al.*, 2005).

The ability to form biofilms is recognized as one of the key determinants associated with successful bacterial pathogenesis. For example, *Pseudomonas aeruginosa,* a Gram-negative, aerobic, rod-shaped bacterium, can survive in diverse environmental conditions, including soil, marine habitats, plant and animal tissues, and non-living surfaces. The basis of this versatility is attributed to its ability to produce a wide array of enzymes that enables the organism to use a variety of substances as nutrients and its ability to form biofilms. These two abilities also benefit *P. aeruginosa* tremendously for surviving in harsh environments and as a pathogen capable of causing diseases in plants, animals and humans. As an opportunistic human pathogen, *P. aeruginosa* causes diseases in patients suffering from cystic fibrosis, burns, cancer and HIV infections and has become a severe problem in intensive care units. Clinical data show that *P. aeruginosa* is the fourth leading cause of nosocomial infection and the foremost cause of hospital-acquired pneumonia (Jarvis and Martone, 1992).

Biofilms and bacterial infection

The formation of biofilms, on one hand, benefits the bacterial population, but, on the other hand, causes serious problems to human health, the environment and industry. The US National Institute of Health reported that biofilms account for 80% of microbial infections in human populations (Davies, 2003). Many persistent and chronic bacterial infections, including periodontitis, otitis media, biliary tract infections and endocarditis, are linked to biofilms (Costerton *et al.*, 1999; Costerton, 2001; Parsek and Singh, 2003). Life-threatening infection caused by *P. aeruginosa* biofilms in cystic fibrosis patients is a well-known example (Govan and Deretic, 1996). During the course of infection on patients with cystic fibrosis, *P. aeruginosa* forms biofilms characterized by cellular aggregates embedded within the mucus layers present in the airways (Lam *et al.*, 1980). Once chronic colonization occurs, the pathogen can hardly be eradicated, even with a high dosage of antibiotics.

Furthermore, the biofilms formed in medical implants and devices often serve as a source of recurrent infections (Donlan and Costerton, 2002; Padera, 2006). These biofilm infections are notoriously difficult to eradicate, even after prolonged antimicrobial therapy due to the protective properties of biofilms formed in the implants and devices. In addition, biofilms provide protection from the host immune response and the bacterial cells living inside biofilms are less susceptible to antimicrobial agents than free-living bacteria.

Biofilms and antibiotic resistance

Biofilms are often highly resistant to antibiotics. The bacterial cells in biofilms can tolerate up to 1000 times higher concentrations of antimicrobial agents than that needed to kill the corresponding planktonic bacterial cells (Hoyle and Costerton, 1991). Several factors have been suggested to contribute to this phenomenon: the physical barrier of the biofilm matrix against antibiotic penetration; the slow growth rates of biofilm bacterial cells; the emergence of resistant strains and phenotypic variants within biofilm populations, and expression of biofilm-specific genes, which will be discussed separately in the following subsections.

Antibiotic penetration

The structural features of a biofilm matrix suggest its role as a physical barrier that may reduce the exposure of bacterial cells within the biofilm to antibiotics. Not surprisingly, poor penetration of antibiotics is suggested as one of the important factors accounting for the observed antimicrobial resistance of biofilms.

Bacteria in biofilms tend to be organized within the EPS matrix rather than on its surface, making them less accessible to therapeutic antibiotics. The physicochemical properties of the biofilm matrix result in slow diffusion rates and delayed penetration of antibiotics within the matrix. As a result, much higher concentrations of antibiotics are needed to kill biofilm-associated bacteria than their planktonic counterparts. Hoyle *et al.* (1992) found that a *P. aeruginosa* biofilm formed on dialysis membranes retards piperacillin diffusion and suggested that the biofilm is a diffusion barrier to piperacillin. In addition, Suci *et al.* (1994) reported that the transport of ciprofloxacin, a fluoroquinolone, to the biofilm–substratum interface is significantly impeded by the biofilm. This delay in penetration was proposed to be caused by binding of the antibiotic to biofilm components.

Different antibiotics have been shown to penetrate biofilms at different rates (Kumon *et al.*, 1994; Shigeta *et al.*, 1997; Vrany *et al.*, 1997; Abdi-Ali *et al.*, 2006). However, the ability to penetrate a biofilm is not necessarily correlated to the bactericidal efficacy. Some antibiotics, however, penetrate biofilms rather well, and still display poor bactericidal activity against biofilm bacteria (Walters *et al.*, 2003). These results suggest that other factors, in addition to the biofilm barrier, may also contribute to antibiotic tolerance in *P. aeruginosa* biofilms.

Slow growth rate of biofilm bacteria

Slow-growing and/or anaerobic conditions have been shown to antagonize antibiotic action against *P. aeruginosa* (Evans *et al.*, 1991; Davies, 2003; Drenkard, 2003). In a mature bacterial biofilm, cells divide at a slow rate, making them less susceptible to certain antibiotics. Oxygen limitation in biofilms is believed to slow bacterial growth rates as it restricts the metabolic activity of bacterial cells (Walters *et al.*, 2003; Borriello *et al.*, 2004). For example, imipenem and beta-lactams show poor bactericidal activity against mature biofilms, but become more effective with younger biofilms. Tobramycin and ciprofloxacin penetrate biofilms but fail to effectively kill the bacteria, and evidence suggests that oxygen limitation and low metabolic activity may be responsible for the antibiotic's resistance (Walters *et al.*, 2003). However, fluoroquinolones have been shown to possess a high bactericidal activity to biofilm bacteria regardless of the growth rate (Shigeta *et al.*, 1997; Tanaka *et al.*, 1999).

Persisters and phenotypic variants

Brooun *et al* (2000) observed that the majority of bacterial cells in *P. aeruginosa* biofilms were killed by a low concentration of antibiotics, but a subpopulation of cells survived, even after further treatment with higher concentrations of antibiotics. These small portions of cells were termed 'persisters'. In addition, exposure of *P. aeruginosa* stationary-phase cultures to very high concentrations of antibiotics also results in a subpopulation of cells with decreased susceptibility to antibiotics (Spoering and Lewis, 2001). These data suggest that the observed antibiotic resistance of stationary-phase cultures and biofilm cells may be largely dependent on the presence of persister cells.

Antibiotic-resistant phenotypic variants of *P. aeruginosa* arise at a high frequency both *in vitro* and in the lungs of cystic fibrosis patients (Drenkard and Ausubel, 2002). Antibiotic treatment in cystic fibrosis patients seems to cause the formation of phenotypic variants. After growing on the antibiotic-free medium for several days, these phenotypic variants revert to normal colony phenotypes and become susceptible to antibiotics. These results suggest that the phenotypic changes observed in the resistance variants are transient. Antibiotic treatment induces not only phenotypic variants but also biofilm formation. It has been reported that low concentrations of aminoglycoside antibiotics could induce biofilm formation in *P. aeruginosa* and the enhanced biofilm formation, in turn, increases the antibiotic's resistance (Hoffman *et al.*, 2005).

Biofilm development

Biofilm formation is a dynamic multi-stage process that includes initial attachment, microcolony formation, biofilm maturation, and ultimately dispersion (detachment) (Costerton *et al.*, 1999; O'Toole *et al.*, 2000b; Stoodley *et al.*, 2002; Padera, 2006; Klausen *et al.*, 2006) (Fig. 11.2A). As summarized in Section 2, a number of molecules are involved in biofilm formation. These components appear to be recruited for biofilm formation in response to prevailing environmental conditions, and their roles in various biofilm developmental stages have now become clear.

Initial attachment

Initial attachment is the first step of biofilm formation (Fig. 11.2A). At this stage, planktonic cells randomly attach to a conditioned surface nonspecifically via electrostatic, van der Waals, and/or hydrophobic forces. This process is reversible, as the bacteria have not yet committed to the biofilm lifestyle. There are no significant changes in gene expression at this stage and the bacteria can freely return to their planktonic form.

Early attachment involves bacterial motility. Flagella driven swimming and type IV pili-mediated surface-associated twitching motility have been shown to contribute to *P. aeruginosa* early attachment as the mutations in genes encoding flagella, type IV pili and cup fimbria were shown, in most cases, to affect the bacterial ability to attach to surfaces (Klausen *et al.*, 2006) (Fig. 11.2B). It was speculated that swimming motility might enable bacteria to overcome repulsive forces at the surface–water interface to facilitate settlement on the surface (O'Toole and Kolter, 1998). Intriguingly, the requirement of flagella for initial adherence to a surface appears to be dependent on the carbon source on which *P. aeruginosa* is grown. For example, mutation of the *fliM* and *pilA* genes of *P. aeruginosa* strain PAO1, which encode the flagellar and type IV pili biosynthesis, respectively, attenuates or even abolishes biofilm formation when grown in glucose-casamino acids minimal medium, whereas in citrate minimal medium the mutants form biofilms equally well as the wild-type strain (Klausen *et al.*, 2003b).

Microcolony formation

Following initial attachment, the surface-associated cells become irreversibly attached as a monolayer of cells. Subsequently, microcolonies are formed (Fig. 11.2A). This irreversible attachment is a crucial step in biofilm formation because the initial surface colonizers are likely the foundation upon which the mature biofilm will be built. O'Toole and Kolter (1998) found that microcolony formation is due largely to the aggregation of *P. aeruginosa* dispersed cells. However, when *P. aeruginosa* is grown with citrate as the sole carbon source in a flow chamber, the formation of initial microcolonies was found through clonal growth from single cells attached to the substratum, and

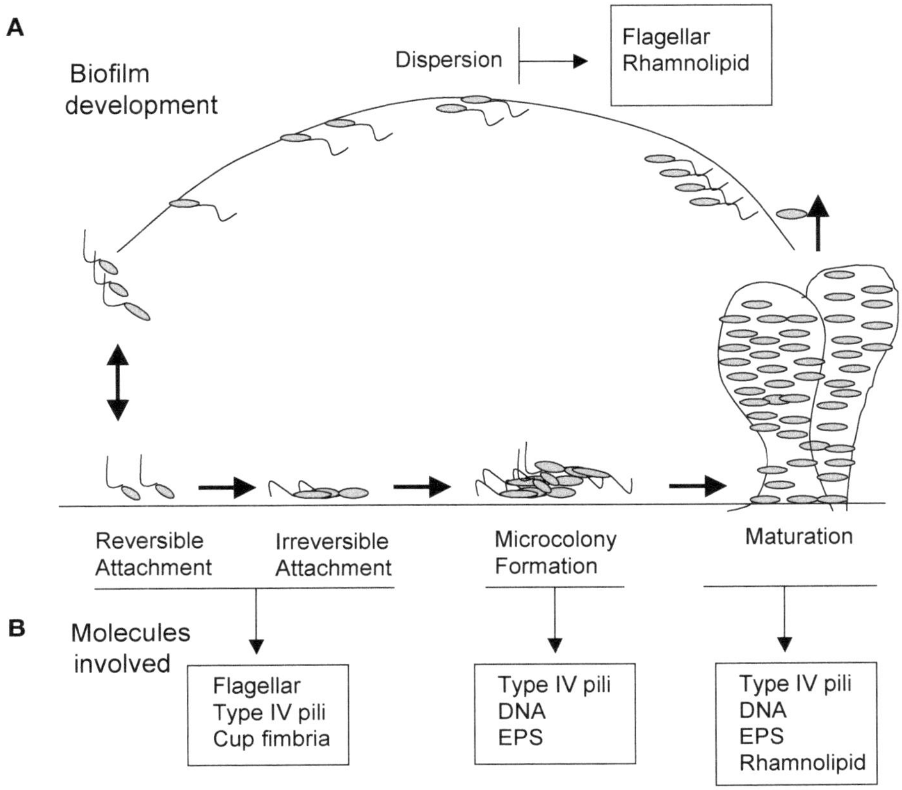

Figure 11.2 A schematic model of bacterial biofilm formation. **A** *P. aeruginosa* biofilm development processes. **B** The key molecules (boxed) that are involved in biofilm development at different stages of biofilm development.

cell aggregation did not play a significant role in *P. aeruginosa* biofilm formation (Klausen *et al.*, 2003a). It appears that biofilm development may occur differently under different nutritional and environmental conditions.

Microcolony formation requires the type IV pili-driven motility which allows bacterial cells to move along the surface, thereby leading to cellular aggregation and microcolony formation (Fig. 11.2B). Type IV pili biogenesis mutants could attach to the surface to form a monolayer similarly to wild type cells, but they were unable to form microcolonies (O'Toole and Kolter, 1998). It is known that type IV pili are required for a surface-associated movement known as twitching motility. Twitching motility is thought to be a consequence of extension and retraction of type IV pili, which propels the bacteria across a surface by an uncharacterized mechanism. The *P. aeruginosa* mutants lacking functional pili could eventually form biofilms in flowing conditions, but the structure of biofilm formed was markedly different from that observed for the wild-type strain (Klausen *et al.*, 2003b). In addition, DNA molecules are present in the microcolonies of young biofilms (Allesen-Holm *et al.*, 2006) (Fig. 11.2B).

Biofilm maturation

After establishment of microcolonies, the biofilm undergoes a maturation process that leads to formation of larger microcolonial structures (Fig. 11.2A). Depending on the experimental conditions, mature *P. aeruginosa* biofilms appear either as a mushroom-shaped or a flat structure. The mushroom-shaped structure is formed by a motile subpopulation of biofilms migrating to the surface of a non-motile subpopulation (stalk), forming the caps of the mushroom-like structure. Time-lapse confocal laser scanning microscopy shows that the shift from non-motile to migrating cells occurs when the initial microcolonies reach a certain size (Klausen *et al.*, 2003b). It appears that the formation of the mushroom caps is caused by high concentrations of extracellular DNA presenting between the stalk-forming and the cap-forming subpopulations in mature biofilms as the type IV pili can bind to DNA (van Schaik *et al.*, 2005; Allesen-Holm *et al.*, 2006; Klausen *et al.*, 2006) (Fig. 11.2B). It is believed that the stalk population produces the surfactant rhamnolipid, which may 'grease' the surface of the stalk and facilitate the movement of a motile subpopulation onto the stalk (Lequette and Greenberg, 2005).

A flat biofilm is formed as the motile bacterial cells multiply and continue to move on the surface thus forming a confluent mat of cells. In this case, bacterial attachment is independent of flagella and type IV pili. By means of twitching motility bacteria spread on the substratum and finally form flat mature biofilms (Singh *et al.*, 2002; Klausen *et al.*, 2003a; Shrout *et al.*, 2006).

The formation of a structured biofilm or flat biofilm in *P. aeruginosa* is influenced by nutritional and environmental conditions. *P. aeruginosa* PAO1 forms heterogeneous biofilms containing mushroom-shaped multicellular structures separated by water-filled channels when glucose is used as a carbon source. However, when citrate, benzoate or casamino acids are used as a carbon source, *P. aeruginosa* PAO1 forms flat and very dynamic biofilms (Klausen *et al.*, 2003a). Overproduction of the exopolysaccharide alginate under certain conditions has also been shown to contribute to the formation of structured biofilms (Hentzer *et al.*, 2001; Nivens *et al.*, 2001).

Biofilm dispersion

Dispersion is the last stage of biofilm development, where a subpopulation of biofilm cells detach and swim away from the mature biofilm and revert to a planktonic lifestyle (Fig. 11.2A). Biofilm dispersion is often observed in aged biofilms. This process involves the formation of internal cavities that are eventually fractured to release motile bacterial cells and is termed as 'central hollowing' (Sauer *et al.*, 2002; Hunt *et al.*, 2004). Two-dimensional-gel analysis shows that the protein profiles of the dispersing cells of *P. aeruginosa* are more similar to planktonic cells than the mature biofilm cells (Sauer *et al.*, 2002).

Dispersion of *P. aeruginosa* biofilms has been demonstrated to be influenced by environmental factors such as availability of organic nutrients (Bollinger *et al.*, 2001; Singh *et al.*, 2002; Sauer *et al.*, 2004; Hunt *et al.*, 2004). The sudden increase in carbon substrate availability has been reported to induce *P. aeruginosa* biofilm dispersal. The extent of dispersion is dependent on the carbon source used. It is associated with the increased expression of flagellar genes and down regulation of the genes involved in twitching motility (Sauer *et al.*, 2004) (Fig. 11.2B). Apparently, available nutrients attract microcolonies to move out of biofilm. On the other hand, nutrient depletion in biofilms was hypothesized to force the bacterial cells to seek new 'food' outside the biofilm, which leads to initiation of the dispersion process (Hunt *et al.*, 2004). In addition, the production of rhamnolipid biosurfactants has been found to be associated with dispersal of cells from a *P. aeruginosa* biofilm (Boles *et al.*, 2005) (Fig. 11.2B).

Interestingly, in several other bacterial species, including *X. campestris* and *Burdholderia cenocepatia*, biofilm dispersion is modulated by DSF-like quorum sensing signals (Dow *et al.*, 2003; He *et al.*, 2006; Boon *et al.*, 2008). Mutation of the *rpfF* gene encoding DSF biosynthesis results in formation of cell aggregates (Fig. 11.1C), which can be dispersed by exogenous addition of DSF (He *et al.*, 2006).

Genetic basis of biofilm formation

In *P. aeruginosa*, a number of genes implicated in biofilm formation have been identified from characterization of the mutants defective in biofilm formation. Increasing evidence suggests that the biofilm development and formation are dependent on expression of certain specific genes. These genes can be catalogued into functional groups for cellular motility, adhesiveness, EPS production, and genetic regulation.

Cellular motility

Genetic analysis reveals that the genes encoding swimming and twitching motility are essential for biofilm formation. Mutations in the genes encoding flagella, type IV pili and cup fimbria were shown to affect the ability of *P. aeruginosa* to attach to surfaces in a microtiter tray and in flow-chambers. A surface attachment defective (*sad*) mutant was found by disruption of a structural gene, *flgK* required for the synthesis of flagella, demonstrating that flagella function is required in the initial attachment of cells (O'Toole and Kolter, 1998). Genetic analysis further showed that type IV pili and related twitching motility are required for the assembly of a monolayer of *P. aeruginosa* cells into microcolonies (O'Toole and Kolter, 1998).

Characterization of the *cup* genes in *P. aeruginosa* has revealed the important role of their encoded products at early stages of biofilm formation (Vallet *et al.*, 2001). The *cup* (*cupABC*) genes encode components of a chaperone-usher pathway associated with the assembly of fimbrial subunits in *P. aeruginosa*. The *cup* mutants have normal twitching and swimming motility, but are defective in biofilm formation. In addition to their role in initial biofilm formation, the cup fimbria were also reported to play a role as cell-to-cell interconnecting compounds in mature biofilms (Klausen *et al.*, 2006).

Cellular adhesiveness

Cell–cell adhesion could be important for the formation of structured biofilms. Two lectins encoded by *lecA* and *lecB* respectively were shown to play a role in this process. LecA and LacB are cytotoxic lectins and adhesins produced by *P. aeruginosa,* and they exert different cytotoxic effects on respiratory epithelial cells and presumably facilitate bacterial adhesion to the airway mucosa (Giboa-Garber, 1982). LecB was found to be present in the surface of *P. aeruginosa* cells and the *lecB*-deficient mutant is impaired in biofilm formation (Tielker *et al.*, 2005). Another study showed that *lecA* is expressed in biofilm cells and that the mutation in *lecA* reduced the biofilm depth and surface coverage. Consistently, over expression of *lecA* enhanced biofilm formation (Diggle *et al.*, 2006). Since Lectins are known as carbohydrate-binding proteins (Giboa-Garber, 1982), it is thought that LecA and LecB could contribute to cell adhesion within biofilms.

EPS production

As exopolysaccharide (EPS) is an essential component in biofilms, the genes involved in EPS production play a vital role in biofilm formation and development. Four gene clusters encoding functions or putative enzymes in EPS biosynthesis have been found in *P. aeruginosa*.

The *pel* (PA3058–PA3064) cluster contains seven *pel* genes that are involved in pellicle formation. The *pel* mutants are defective in formation of solid surface-associated biofilms and display altered colony morphologies (Friedman and Kolter, 2004a), suggesting that the *pel* genes are required not only for pellicle but also for biofilm formation. Carbohydrate composition analysis showed that the *pel* mutants lack a glucose-rich component in their extracellular matrix, suggesting that the *pel* genes are responsible for the production of glucose-rich matrix materials required for biofilm formation by *P. aeruginosa* (Friedman and Kolter, 2004a).

The *psl* (for polysaccharide synthesis locus) cluster contains 15 open reading frames and evidence suggests that this cluster encodes an important determinant in *P. aeruginosa* biofilm development. Disruption of the *psl* locus results in a biofilm defective phenotype in several assays. Mutation of *pslA* delays microcolony formation in a flow chamber setup (Matsukawa and Greenberg, 2004) and the *psl* locus was also found to be important for early biofilm formation in a microtiter tray assay (Friedman and Kolter, 2004b; Jackson *et al.*, 2004). Most of the *psl* genes are predicted to encode proteins with sequence homologies to enzymes involved in the synthesis of a mannose-rich EPS using glucose, mannose and rhamnose as substrates (Wozeniak *et al.*, 2003).

The *alg* cluster is involved in alginate biosynthesis, a major component of EPS that is important for biofilm development and maintenance in cystic fibrosis-derived mucoid isolates of *P. aeruginosa*. The mucoid phenotype is due to overproduction of alginate. The *alg* operon does not appear to be expressed in *P. aeruginosa* biofilms formed in flow-chambers (Wozniak *et al.*, 2003), but seems to play a role in biofilm formation by *P. aeruginosa* in other conditions such as in the lungs of cystic fibrosis patients (Govan and Deretic, 1996).

Quorum sensing and genetic regulation

Quorum sensing is a community genetic mechanism by which bacteria monitor their cell population density through extracellular accumulation of signalling molecules to synchronize expression of a set of genes. Quorum sensing has been shown to play a role in *P. aeruginosa* biofilm maturation. Mutation of the *lasI* gene that encodes a long-chain *N*-acyl-*L*-homoserine lactone (AHL) synthase, leads to formation of flat, undifferentiated biofilms. However, mutation in *rhlI* that encodes a short-chain AHL synthase does not seem to affect the formation of structured biofilms (Davies *et al.*, 1998). In addition to AHLs, the quinolone signalling compound (PQS), also plays a role in biofilm formation. PQS is an additional quorum-sensing signal linking between the *las* and *rhl* quorum-sensing systems in *P. aeruginosa*. PQS signalling has been shown to promote *P. aeruginosa* biofilm formation. Addition of exogenous PQS to a growing culture of *P. aeruginosa* significantly enhanced adherence and biofilm formation (Diggle *et al.*, 2003). Although the molecular basis for this phenomenon has not been fully understood, an increase in production of lectins and rhamnolipids may at least partially account for the increase in biofilm formation (Diggle *et al.*, 2006).

Many quorum-sensing regulated genes were previously shown to associate with biofilm formation. The examples include *rhlAB*, *pvdA* and *pvdS*, and *lecA* and *lacB*, which control production of surfactant rhamnolipids, siderophore pyoverdine, and lectins, respectively. The mutants of *rhlAB* produce architecturally flat and undifferentiated biofilms (Davey *et al.*, 2003; Lequette and Greenberg, 2005). The *pvdA* and *pvdS* mutants, that are unable to produce pyoverdine, form thin uniform flat biofilms. Consistently, complementation of the mutants with the wild-type *pvdA* and *pvdS* restores the normal biofilm architecture (Banin *et al.*, 2005).

A two-component response regulator GacA was found to play a critical role in biofilm formation. Disruption of *gacA* in *P. aeruginosa* PA14 resulted in a 10-fold reduction in biofilm formation and the mutant was unable to aggregate into microcolonies on the surface (Parkins *et al.*, 2001). GacA was reported to regulate the production of short-chain AHL in *P. aeruginosa* (Reimmann *et al.*, 1997), but motility and AHL assays revealed no differences between the wild type and the *gacA* mutant, thus suggesting that GacA is a new and independent regulatory element in *P. aeruginosa* biofilm formation (Parkins *et al.*, 2001).

Cyclic-di-GMP (c-di-GMP) was shown to regulate the transition between sessile and motile phenotypes in a range of bacterial species, including *V. cholerae* (Tischler and Camilli, 2004), *Salmonella enterica* and *E. coli* (Simm *et al.*, 2004). The steady-state level of c-di-GMP in the cell is controlled by the relative rate of synthesis and degradation of this molecule. In *P. aeruginosa*, several findings indicate the important role of c-di-GMP in biofilm formation. An early study shows that a membrane-associated novel sensor-regulator protein MorA, which contains domains implicated in c-di-GMP turnover, is required for biofilm formation by *P. aeruginosa* and *Pseudomonas putida* (Choy *et al.*, 2004). Mutation of *morA* in both bacterial species leads to reduced biofilm formation, suggesting that the cellular content of c-di-GMP is important in modulation of biofilm formation.

Global gene expression profiles in biofilm development

Recent advances in genome sequencing and DNA microarray technologies allow a holistic approach to explore the global gene expression profiles between bacterial cells in biofilm and planktonic forms under different growth conditions and developmental stages (Whiteley *et al.*, 2001; Sauer *et al.*, 2004; Waite *et al.*, 2005, 2006). These gene expression data should accelerate our understanding of the biofilm physiology and corresponding regulatory mechanisms.

Early DNA microarray analyses identified 73 genes (1% of the genome) with differential expression profiles in biofilm and planktonic cells of *P. aeruginosa*. Among these genes 34 were activated and 39 were repressed in biofilm populations (Whiteley *et al.*, 2001). About 34% of the 73 biofilm-regulated genes code for hypothetical proteins of unknown function. Seven genes for synthesis of pili and flagella are repressed and nine bacteriophage genes are highly expressed in *P. aeruginosa* biofilm.

The gene expression profiles between biofilm and newly dispersed cells are also markedly different. In dispersed cells, genes for flagellar, pili and ribosomal proteins, kinases, and phage PF1 are upregulated (Sauer, *et al.*, 2004). The global transcriptional profiles for two planktonic growth phases (logarithmic and stationary growth phase) and two biofilm structures (microcolony and confluent biofilms) of *P. aeruginosa* have shown differential gene expression at different stages of biofilm development (Waite, *et al.*, 2005). Noticeably, a range of genes encoding transportation of small molecules are significantly induced in developing biofilms.

More recently, a comprehensive cluster analysis of microarray data provided an overview of *P. aeruginosa* gene expression in planktonic, developing biofilm and confluent biofilm states (Waite, *et al.*, 2006). The results provide further evidence to the participation of flagella, type IV pili and quorum sensing in biofilm formation, and useful clues for understanding the regulatory mechanisms associated with biofilm development.

Biofilm control

As mentioned above, biofilms have become a serious problem in human health, the environment and industry due to their resistance to antibiotics and difficulty to eradicate. Based on our understanding of the underlying molecular mechanisms and physiology of biofilms, several anti-biofilm strategies have been explored.

Preventing biofilm formation

In theory, any molecules that can effectively interfere with the key step(s) in biofilm development could be potentially used to prevent biofilm formation. One promising approach is to target the regulatory mechanisms involved in biofilm formation, for example, quorum sensing which regulates many genes essential for biofilm formation as discussed above. Several classes of enzymes and small molecules were shown to be able to quench bacterial quorum sensing effectively (Dong *et al.*, 2001; Zhang, 2003; Hentzer and Givskov 2003; Zhang and Dong 2004), and their potentials in control and prevention of bacterial virulence and biofilm have been demonstrated in recent years (Dong *et al.*, 2001; Reimmann *et al.*, 2002; Lin *et al.*, 2003; Hentzer *et al.*, 2003) (Fig. 11.1B).

An alternative approach to control biofilm formation is to inhibit the production of the biofilm matrix material. Bismuth-dimercaprol blocks EPS synthesis and consequently reduces biofilm formation. In addition, it inhibits adherence of *P. aeruginosa* to human cell lines and collagen matrix (Huang and Stewart, 1999; Wu *et al.*, 2002). These results highlight the importance of developing potent agents that block polysaccharide production in our battle to control biofilm formation.

Biofilm control could also be achieved by targeting a critical juncture in biofilm development. Lactoferrin, a component of innate immunity has been reported to be able to prevent biofilm formation by the bacteria that have escaped initial killing by the host immune response. In the presence of subinhibitory concentrations of lactoferrin, bacteria attach and multiply, but they fail to form microcolonies or differentiate biofilm structures. Lactoferrin appears to function by stimulating cellular twitching motility and thus mobilizing the bacterial cells to wander across the surface instead of forming cell clusters and biofilms (Singh *et al.*, 2002; Singh, 2004).

Disrupting existing biofilms

Disruption of already existing biofilms has also been considered over the last few years. Enzymes have been used to remove bacterial biofilms from surfaces, such as stainless steel, polypropylene, and soft contact lenses (Johansen *et al.*, 1997). Alginate lyase is one of such enzymes which can degrades EPS produced by the mucoid strains of *P. aeruginosa*. Alginate lyase and gentamicin when administered together eliminate biofilms formed by mucoid bacteria, whereas biofilms treated only with gentamicin persist (Alkawash *et al.*, 2006). These findings suggest that co-administration of antibiotics with alginate lyase might benefit cystic fibrosis patients by increasing the efficacy of antibiotic in the respiratory tract.

Like EPS, extracellular DNA is also an important component in a biofilm. The treatment by DNases, either DNase I or Varidase which breaks down DNA, results in effective destruction of early and mature *P. aeruginosa* PAO1 biofilms (Whitchurch *et al.*, 2002; Nemoto *et al.*, 2003). These results are consistent with the notion that extracellular DNA is one of the major components in the biofilm matix of *P. aeruginosa*.

Conclusion

Research over the last few decades has demonstrated that biofilms are a widely conserved phenomenon and that they have significant implications in disease control and environmental protection. Extensive investigations on biofilm formation by the bacterial pathogen *P. aeruginosa* have established that biofilm formation can be viewed as a highly regulated developmental process that results in the formation of a complex community of bacterial cells with distinct structures. To form these structured communities, *P. aeruginosa* seems to be able to respond to and integrate various environmental cues and internal signals, and coordinates a timed series of multicellular behaviours that lead to morphological changes. The investigations have also unveiled a range of structural and regulatory genes involved in biofilm formation. Several promising strategies for biofilm control have been explored, including interference of bacterial quorum sensing, inhibiting production of key biofilm matrix molecules, and disruption of existing biofilms using enzymes to degrade DNA and exopolysaccharides that are the key components of biofilm matrix.

Future trends

For better understanding of the molecular mechanisms and control of biofilms, it is essential to have a thorough characterization of the genetic and signalling networks that regulate biofilm formation. Several groups of molecules including proteins, DNA and exopolysaccharides implicated in biofilm formation have been identified, but how their production is regulated and how they are recruited in biofilm formation are far from clear. In this regard, the knowledge of how quorum sensing mechanisms and the sensors/regulators associated with c-di-GMP metabolism respond to various environmental conditions and thus modulate the expression of corresponding genes would be particularly valuable for designing and developing novel strategies against bacterial biofilm formation.

In addition, research on biofilm community behaviour, the associated gene functions, as well as the timing of protein expression and turnover will continue to be an active area of investigation. A detailed knowledge of the developmental process from single cells scattered on a surface to complex multicellular biofilm structures would help to formulate strategies to control biofilm development and to provide novel avenues for therapeutic intervention of biofilm related infections. It should be noted that most previous research findings on bacterial biofilms were based on a single bacterial species model under laboratory conditions. In natural environments, however, a biofilm complex commonly consists of multiple bacterial species even including fungi (Hu *et al.*, 2003). As not all the bacterial species share the same genetic regulatory mechanisms, for example, there are only about 80 bacterial species that produce AHL-type quorum sensing signals; certain control strategies that are effective on single-species biofilms may not be useful against the multiple-species biofilms. Therefore, determination of the dominant microbial species of biofilm samples in a particular environment and identification of the key molecules essential for mixed-species biofilm development may have significant implications in designing and developing practical means to control biofilm formation.

References

Abdi-Ali, A., Mohammadi-Mehr, M., and Agha Alaei, Y. (2006). Bactericidal activity of various antibiotics against biofilm-producing *Pseudomonas aeruginosa*. Int. J. Antimicrob. Agents *27*, 196–200.

Alkawash, M.A., Soothill, J.S., and Schiller, N.L. (2006). Alginate lyase enhances antibiotic killing of mucoid *Pseudomonas aeruginosa* in biofilms. APMIS *114*, 131–138.

Allesen-Holm, A., Barker, K.B., Yang, L., Klausen, M,. Webb, J.S., Kjelleberg, S., Molin, S,. Givskov, M., and Tolker-Nielsen, T. (2006). A characterization of DNA release in *Pseudomonas aeruginosa* cultures and biofilms. Mol. Microbiol. *59*, 1114–1128.

Banin, E., Vasil, M.L., and Greenberg, E.P. (2005). Iron and *Pseudomonas aeruginosa* biofilm formation. Proc. Natl. Acad. Sci. U.S.A. *102*, 11076–11081.

Boles, B.R., Thoendel, M., and Singh, P.K. (2005). Rhamnolipids mediate detachment of *Pseudomonas aeruginosa* from biofilms. Mol. Microbiol. *57*, 1210–1223.

Bollinger, N., Hassett, D.J., Iglewski, B.H., Costerton, J.W., and McDermott, T.R. (2001). Gene expression in *Pseudomonas aeruginosa*: evidence of iron override effects on quorum sensing and biofilm-specific gene regulation. J. Bacteriol. *183*, 1990–1996.

Boon, C., Deng, Y., Wang, L.H., He, Y., Xu, J.L., Fan, Y., Pan, S.Q., and Zhang, L.H. (2008). A novel DSF-like signal from *Burkholderia cenocepacia* interferes with *Candida albicans* morphological transition. ISME J. *2*, 27–36.

Borriello, G., Werner, E., Roe, F., Kim, A.M., Ehrlich, G.D., and Stewart, P.S. (2004). Oxygen limitation contributes to antibiotic tolerance of *Pseudomonas aeruginosa* in biofilms. Antimicrob. Agents Chemother. *48*, 2659–2664.

Brooun, A., Liu, S., and Lewis, K. (2000). A dose-response study of antibiotic resistance in *Pseudomonas aeruginosa* biofilm. Antimicrob. Agents Chemother. *44*, 640–646.

Caiazz,a N.C., and O'Toole, G.A. (2004). SadB is required for the transition from reversible to irreversible attachment during biofilm formation by *Pseudomonas aeruginosa* PA14. J. Bacteriol. *186*, 4476–4485.

Choy, W.K., Zhou, L., Syn, C.K.C., Zhang, L.H., and Swarup, S. (2004). MorA defines a new class of regulators affecting flagellar development and biofilm formation in diverse *Pseudomonas* species. J. Bacteriol. *186*, 7221–7228.

Christen, M., Christen, B., Folcher, M., Schauerte, A., and Jenal, U. (2005). Identification and characterization of a c-di-GMP-specific phosphodiesterase and its allosteric control by GTP.J. Biol. Chem. *280*, 30829–30837.

Costerton, J.W. (2001). Cystic fibrosis pathogenesis and the role of biofilms in persistent infection. Trends Microbiol. *9*, 50–52.

Costerton, J.W., and Lewandowski, Z. (1995). Microbial biofilms. Ann. Rev. Microbiol. *49*, 711–745.

Costerton, J.W., Stewart, P.S., and Greenberg, E.P. (1999). Bacterial biofilms: a common cause of persistent infections. Science *284*, 1318–1322.

D'Argenio, D.A., Calfee, M.W., Rainey, P.B., and Pesci, E.C. (2002). Autolysis and autoaggregation in *Pseudomonas aeruginosa* colony morphology mutants. J. Bacteriol. *184*, 6481–6489.

Davey, M.E., Caiazza, N.C., and O'Toole, G.A. (2003). Rhamnolipid surfactant production affects biofilm architecture in *Pseudomonas aeruginosa* PAO1. J. Bacteriol. *185*, 1027–1036.

Davies, D. (2003). Understanding biofilm resistance to antibacterial agents. Nature Rev. Drug Discov. *2*, 114–122.

Davies, D.G., and Geesey, G.G. (1995). Regulation of the alginate biosynthesis gene *algC* in *Pseudomonas aeruginosa* during biofilm development in continuous culture. Appl. Environ. Microbiol. *61*, 860–867.

Davies, D.G., Parsek, M.R., Pearson, J.P., Iglewsky, B.H., Costerton, J.W., and Greenberg, E.P. (1998). The involvement of cell-to-cell signals in the development of a bacterial biofilm. Science *280*, 295–298.

Diggle, S.P., Winzer, K., Chhabra, S.R., Worrall, K.E., Camara, M., and Williams, P. (2003). The *Pseudomonas aeruginosa* quinolone signal molecule overcomes the cell density-dependency of the quorum sensing hierarchy, regulates *rhl*-dependent genes at the onset of stationary phase and can be produced in the absence of LasR. Mol. Microbiol. *50*, 29–43.

Diggle, S.P., Stacey, R.E., Dodd, C., Cámara, M., Williams, P., and Winzer, K. (2006). The galactophilic lectin, LecA, contributes to biofilm development in *Pseudomonas aeruginosa*. Environ. Microbiol. *8*, 1095–1104.

Dong, Y.H., Wang, L.H., Xu, J.L., Zhang, H.B., Zhang, X.F., and Zhang, L.H. (2001). Quenching quorum sensing-dependent bacterial infection by an *N*-acyl homoserine lactonase. Nature *411*, 813–817.

Donlan, R.M., and Costerton, J.W. (2002). Biofilms: survival mechanisms of clinically relevant microorganisms. Clin. Microbiol. Rev. *15*, 167–93.

Dow, J.M., Crossman, L., Findlay, K., He, Y.Q., Feng, J.X., and Tang, J.L. (2003). Biofilm dispersal in *Xanthomonas campestris* is controlled by cell–cell signaling and is required for full virulence to plants. Proc. Natl. Acad. Sci. U.S.A. *100*, 10995–11000.

Drenkard, E. (2003). Antimicrobial resistance of *Pseudomonas aeruginosa* biofilms. Microb. Infect. *5*, 1213–1219.

Drenkard, E., and Ausubel, F.M. (2002). *Pseudomonas* biofilm formation and antibiotic resistance are linked to phenotypic variation. Nature *416*, 740–743.

Evans, D.J., Allison, D.G., Brown, M.R., and Gilbert, P. (1991). Susceptibility of *Pseudomonas aeruginosa* and *Escherichia coli* biofilms towards ciprofloxacin: effect of specific growth rate. J. Antimicrob. Chemother. *27*, 177–184.

Friedman, L., and Kolter, R. (2004a). Genes involved in matrix formation in *Pseudomonas aeruginosa* PA14 biofilms. Mol. Microbiol. *51*, 675–690.

Friedman, L., and Kolter, R. (2004b). Two genetic loci produce distinct carbohydrate-rich structural components of the *Pseudomonas aeruginosa* biofilm matrix. J. Bacteriol. *186*, 4457–4465.

Gilboa-Garber, N. (1982). *Pseudomonas aeruginosa* lectins. Methods Enzymol. *83*, 378–385.

Govan, J.R., and Deretic, V. (1996). Microbial pathogenesis in cystic fibrosis: mucoid *Pseudomonas aeruginosa* and *Burkolderia cepacia*. Microbiol. Rev. *60*, 539–574.

He, Y.W., Xu, M., Lin, K., Ng, Y.J.A., Wen, C.M., Wang, L.H., Liu, Z.D., Zhang, H.B., Dong, Y.H., Dow, J.M., and Zhang, L.H. (2006). Genome scale analysis of DSF regulon in *Xanthomonas campestris* pv. *campestris:* identification of novel cell–cell communication-dependent genes and functions. Mol. Microbiol. *59*, 610–622.

Hentzer, M., and Givskov, M. (2003). Pharmacological inhibition of quorum sensing for the treatment of chronic bacterial infections. J. Clin. Invest. *112*, 1300–1307.

Hentzer, M., Teitzel, G.M., Balzer, G.J., Heydorn, A., Molin, S., Givskov, M., and Parsek, M.R. (2001). Alginate overproduction affects *Pseudomonas aeruginosa* biofilm structure and function. J. Bacteriol. *183*, 5395–5401.

Hentzer, M., Wu, H., Andersen, J.B., Riedel, K., Rasmussen, T.B., Bagge, N., Kumar, N., Schembri, M.A., Song, Z., Kristoffersen, P., Manefield, M., Costerton, J.W., Molin, S., Eberl, L., Steinberg, P., Kjelleberg, S., Hoiby, N., and Givskov, M. (2003). Attenuation of *Pseudomonas aeruginosa* virulence by quorum sensing inhibitors. EMBO J. *22*, 3803–3815.

Hoffman, L.R., D'Argenio, D.A., MacCoss, M.J., Zhang, Z.Y., Jones, R.A., and Miller, S.I. (2005). Aminoglycoside antibiotics induce bacterial biofilm formation. Nature *436*, 1171–1175.

Hoiby, N. (1993). Antibiotic therapy for chronic infection of *Pseudomonas* in the lung. Annu. Rev. Med. *44*, 1–10.

Hoyle, B.D., and Costerton, W.J. (1991). Bacteria resistance to antibiotics: the role of biofilms. Prog. Drug Res. *37*, 91–105.

Hoyle, B.D., Alcantara, J., and Costerton, W.J. (1992). *Pseudomonas aeruginosa* biofilm as a diffusion barrier to piperacillin. Antimicrob. Agent Chemother. *36*, 2054–2056.

Hu, J.Y., Yang, F., Lin, Y.H., Zhang, H.B., Ong, S.L., Dong, N., Xu, J.L., Ng, W.J., and Zhang, L.H. (2003). Microbial diversity and prevalence of virulent pathogens in biofilms developed in a water reclamation system. Res. Microbiol. *154*, 623–629.

Huang, C.T., and Stewart, P.S. (1999). Reduction of polysaccharide production in Pseudomonas aeruginosa biofilms by bismuth dimercaprol (BisBAL). treatment. J. Antimicrob. Chemother. *44*, 601–605.

Hunt, S.M., Werner, E.M., Huang, B., Hamilton, M.A., and Stewart, P.S. (2004). Hypothesis for the role of nutrient starvation in biofilm detachment. Appl. Environ. Microbiol. *70*, 7418–7425.

Jackson, K.D., Starkey, M., Kremer, S., Parsek, M.R., and Wozniak, D.J. (2004). Identification of psl, a locus encoding a potential exopolysaccharide that is essential for *Pseudomonas aeruginosa* PAO1 biofilm formation. J. Bacteriol. *186*, 4466–4475.

Jarvis, W.R., and Martone, W.J. (1992). Predominant pathogens in hospital infections. J. Antimicrob. Chemother. *29*, S19–S24.

Jefferson, K.K. (2004). What drives bacteria to produce a biofilm? FEMS Microbiol. Lett. *236*, 163–173.

Johansen, C., Falholt, P., and Gram, L. (1997). Enzymatic removal and disinfection of bacterial biofilms. Appl. Environ. Microbiol. *63*, 3724–3728.

Kirisits, M.J., and Parsek, M.R. (2006). Does *Pseudomonas aeruginosa* use intercellular signalling to build biofilm communities? Cell Microbiol. *8*, 1841–1849.

Klausen, M., Aaes-Jorgensen, A., Molin, S., and Tolker-Nielsen, T. (2003a). Involvement of bacterial migration in the development of complex multicellular structures in *Pseudomonas aeruginosa* biofilms. Mol. Microbiol. *50*, 61–68.

Klausen, M., Heydorn, A., Ragas, P., Lambertsen, L., Aaes-Jorgensen, A., Molin, S., and Tolker-Nielsen, T. (2003b). Biofilm formation by *Pseudomonas aeruginosa* wild type, flagella and type IV pili mutants. Mol. Microbiol. *48*, 1511–1524.

Klausen, M., Gjermansen, M., Kreft, J.U., and Tolker-Nielsen, T. (2006). Dynamics of development and dispersal in sessile microbial communities: examples from *Pseudomonas aeruginosa* and *Pseudomonas putida* model biofilms. FEMS Microbiol. Lett. *261*, 1–11.

Kuchma, S.L., Connolly, J.P., and O'Toole, G.A. (2005). A three-component regulatory system regulates biofilm maturation and type III secretion in *Pseudomonas aeruginosa* J. Bacteriol. *187*, 1441–1454.

Kulesekara, H., Lee, V., Brencic, A., Liberati, N., Urbach, J., Miyata, S., Lee, D.G., Neely, A.N., Hyodo, M., Hayakawa, Y., Ausubel, F.M., and Lory, S. (2006). Analysis of *P. aeruginosa* diguanylate cyclases and phosphodiesterases reveals a role for bis-(3'-5')-cyclic-GMP in virulence. Proc. Natl. Acad. Sci. U.S.A. *103*, 2839–2844.

Kumon, H., Tomochika, K., Matunaga, T., Ogawa, M., and Ohmori, H. (1994). A sandwich cup method for the penetration assay of antimicrobial agents through *Pseudomonas* exopolysaccharides. Microbiol. Immunol. *38*, 615–619.

Lam, J., Chan, R., Lam, K., and Costerton, J.W. (1980). Production of mucoid microcolonies by *Pseudomonas aeruginosa* within infected lungs in cystic fibrosis. Infect. Immun. *28*, 546–556.

Leid, J.G., Willson, C.J., Shirtliff, M.E., Hassett, D.J., Parsek, M.R., and Jeffers, A.K. (2005). The exopolysaccharide alginate protects *Pseudomonas aeruginosa* biofilm bacteria from IFN-gamma-mediated macrophage killing. J. Immunol. *175*, 7512–7518.

Lequette, Y., and Greenberg, E.P. (2005). Timing and localization of rhamnolipid synthesis gene expression in *Pseudomonas aeruginosa* biofilms. J. Bacteriol. *187*, 37–44.

Lin, Y.H., Xu, J.L., Hu, J., Wang, L.H., Ong, S.L., Leadbetter, J.R., and Zhang, L.H. (2003). Acyl-homoserine lactone acylase from *Ralstonia* strain XJ12B represents a novel and potent class of quorum-quenching enzymes. Mol. Microbiol. *47*, 849–860.

Matsukawa, M., and Greenberg, E.P. (2004). Putative exopolysaccharide synthesis genes influence *Pseudomonas aeruginosa* biofilm development. J. Bacteriol. *186*, 4449–4456.

Nemoto, K., Hirota, K., Murakami, K., Taniguti, K., Murata, H., Viducic, D., and Miyake, Y. (2003). Effect of Varidase (streptodornase) on biofilm formed by *Pseudomonas aeruginosa*. Chemother. *49*, 121–125.

Nivens, D.E., Ohman, D.E., Williams, J., and Franklin, J. (2001). Role of alginate and its O acetylation in formation of *Pseudomonas aeruginosa* microcolonies and biofilm. J. Bacteriol. *183*, 1047–1057.

O'Toole, G.A., and Kolter, R. (1998). Flagellar and twitching motility are necessary for *Pseudomonas aeruginosa* biofilm development. Mol. Microbiol. *30*, 295–304.

O'Toole, G.A., Gibbs, K.A., Hager, P.W., Phibbs, P.V. Jr, and Kolter, R. (2000a). the global carbon metabolism regulator Crc is a component of a signal transduction pathway required for biofilm development by *Pseudomonas aeruginosa*. J. Bacteriol. *182*, 425–431.

O'Toole, G.A., Kaplan, H.B., and Kolter, R. (2000b). Biofilm formation as microbial development. Annu. Rev. Microbiol. *54*, 49–79.

Padera, R.F. (2006). Infection in ventricular assist devices: the role of biofilm. Cardiovasc. Pathol. *15*, 264–270.

Parkins, M.D., Ceri, H., and Storey, D.G. (2001). *Pseudomonas aeruginosa* GacA, a factor in multihost virulence, is also essential for biofilm formation. Mol. Microbiol. *40*, 1215–1226.

Parsek, M.R., and Singh, P.K. (2003). Bacterial biofilms: an emerging link to disease pathogenesis. Annu. Rev. Microbiol. *57*, 677–701.

Paul, R., Weiser, S., Amiot, N.C., Chan, C., Schirmer, T., Giese, B., and Jenal, U. (2004). Cell cycle-dependent dynamic localization of a bacterial response regulator with a novel di-guanylate cyclase output domain. Genes Dev. *18*, 715–727.

Reimmann, C., Beyeler, M., Latifi, A., Winteler, H., Foglino, M., Lazdunski, A., and Haas, D. (1997). The global activator GacA of *Pseudomonas aeruginosa* PAO positively controls the production of the autoinducer N-butyryl-homoserine lactone and the formation of the virulence factors pyocyanin, cyanide and lipase. Mol. Microbiol. *24*, 309–319.

Reimmann, C., Ginet, N., Michel, L., Keel, C., Michaux, P., Krishnapillai, V., Zala, M., Heurlier, K., Triandafillu, K., Harms, H., Defago, G., and Haas, D. (2002). Genetically programmed autoinducer destruction reduces virulence gene expression and swarming motility in *Pseudomonas aeruginosa* PAO1. Microbiology *148*, 923–932.

Sauer, K., Camper, A.K., Ehrlich, G.D., Costerton, J.W., and Davies, D.G. (2002). *Pseudomonas aeruginosa* displays multiple phenotypes during development as a biofilm. J. Bacteriol. *184*, 1140–1154.

Sauer, K., Cullen, M.C., Richard, A.H., Zeef, L.A.H., Davies, D.G., and Gilbert, P. (2004). Characterization of nutrient induced dispersion in *Pseudomonas aeruginosa* PAO1 biofilm. J. Bacteriol. *186*, 7312–7326.

Schmidt, A.J., Ryjenkov, D.A., and Gomelsky, M. (2005). Ubiquitous protein domain EAL encodes cyclic diguanylate-specific phosphodiesterase: enzymatically active and inactive EAL domains. J. Bacteriol. 187, 4774–4781.

Shiau, A.L., and Wu, C.L. (1998). The inhibitory effect of *Staphylococcus epidermidis* slime on the phagocytosis of murine peritoneal macrophages is interferon-independent. Microbiol. Immunol. *42*, 33–40.

Shigeta, M, Komatsuzawa, H., Suga,i M., Suginaka, H., and Usui, T. (1997). Effect of the growth rate of *Pseudomonas aeruginosa* biofilms on the susceptibility to antimicrobial agents. Chemother. *43*, 137–141.

Shrout, J.D., Chopp, D.L., Just, C.L., Hentzer, M., Givskov, M., and Parsek, M.R. (2006). The impact of quorum sensing and swarming motility on *Pseudomonas aeruginosa* biofilm formation is nutritionally conditional. Mol. Microbiol. *62*, 1264–1277.

Simm, R., Morr, M., Kader, A., Nimtz, M., and Römling, U. (2004). GGDEF and EAL domains inversely regulate cyclic di-GMP levels and transition from sessility to motility. Mol. Microbiol. *53*, 1123–1134.

Singh, P.K. (2004). Iron sequestration by human lactoferrin stimulates *P. aeruginosa* surface motility and blocks biofilm formation. Biometals. *17*, 267–270.

Singh, P.K., Parsek, M.R., Greenberg, E.P., and Weish, M.J. (2002). A component of innate immunity prevents bacterial biofilm development. Nature *417*, 552–555.

Stapper, A.P., Narasimhan, G., Ohman, D.E., Barakat, J., Hentzer, M., Molin, S., Kharazmi, A., Hoiby, N., and Mathee, K. (2004). Alginate production affects *Pseudomonas aeruginosa* biofilm development and architecture, but is not essential for biofilm formation. J. Med. Microbiol. *53*, 679–690.

Sutherland, L.W. (2001). The biofilm matrix – An immobilized but dynamic microbial environment. Trends Microbiol. *9*, 222–227.

Spoering, A.L., and Lewis, K. (2001). Biofilms and planktonic cells of *Pseudomonas aeruginosa* have similar resistance to killing by antimicrobials. J. Bacterial. *183*, 6746–6751.

Stoodley, P., Sauer, K., Davies, D.G., and Costerton, J.W. (2002). Biofilms as complex differentiated communities. Annu. Rev. Microbiol. *56*, 187–209.

Suci, P.A., Mittelman, M.W., Yu, F.P., and Geesev, G.G. (1994). investigation of ciprofloxacin penetration into *Pseudomonas aeruginosa* biofilms. Antimicrob. Agents Chemother. *38*, 2125–2133.

Tal, R., Wong, H.C., Calhoon, R., Gelfand, D., Fear, A.L., Volman, G., Mayer, R., Ross, P., Amikam, D., Weinhouse, H., Cohen, A., Sapir, S., Ohana, P., and Benziman, M. (1998). Three *cdg* operons control cellular turnover of c-di-GMP in *Acetobacter xylinum*: genetic organization and occurrence of conserved domains in isoenymes. J. Bacteriol. *180*, 4416–4425.

Tanaka, G., Shigeta, H., Komatsuzawa, M., Sugai, M., Suginaka, H., and Usui, T. (1999). Effect of growth rate of Pseudomonas aeruginosa biofilms on the susceptibility to antimicrobial agents: beta-lactams and fluoroquinolones. Chemotherapy *45*, 28–36.

Tielker, D., Hacker, S., Loris, R., Strathmann, M., Wingender, J., Wilhelm, S., Rosenau, F., and Jaeger, K.E. (2005). *Pseudomonas aeruginosa* lectin LecB is located in the outer membrane and is involved in biofilm formation. Microbiology *151*, 1313–1323.

Tischler, A.D., and Camilli, A. (2004). Cyclic diguanylate (c-di-GMP) regulates Vibrio cholerae biofilm formation. Mol. Microbiol. *53*, 857–869.

Vallet, I., Slson, J.W., Lory, S., Lazdunsk,i A., and Filloux, A. (2001). The chaperone/usher pathways of *Pseudomonas aeruginosa*: identification of fimbrial gene clusters (*cup*) and their involvement in biofilm formation. Proc. Natl. Acad. Sci. U.S.A. *98*, 6911–6916.

Vallet, I., Diggle, S.P., Stacey, R.E., Camara, M., Ventre, I., Lory, S., Lazdunsk,i A., Williams, P., and Filloux, A. (2004). Biofilm formation in *Pseudomonas aeruginosa*: fimbrial *cup* gene clusters are controlled by the transcriptional regulator MvaT. J. Bacteriol. *186*, 2880–2890.

van Schaik, E.J., Giltner, C.L., Audette, G.F., Keizer, D.W., Bautista, D.L., Slupsky, C.M., Sykes, B.D., and Irvin, R.T. (2005). DNA binding: a novel function of *Pseudomonas aeruginosa* type IV pili. J. Bacteriol. *187*, 1455–1464.

Vrany, J.D., Stewart, P.S., and Suci, P.A. (1997). Comparison of recalcitrance to ciprofloxacin and levofloxacin exhibited by *Pseudomonas aeruginosa* bofilms displaying rapid-transport characteristics. Antimicrob. Agents Chemother. *41*, 1352–1358.

Waite, R.D., Papakonstantinopoulou, A., Littler, E., and Curtis, M.A. (2005). Transcriptome analysis of *Pseudomonas aeruginosa* growth: comparison of gene expression in planktonic cultures and developing and mature biofilms. J. Bacteriol. *187*, 6571–6576.

Waite, R.D., Paccanaro, A., Papakonstantinopoulou, A., Hurst, J.M., Saqi, M., Littler, E., and Curtis, M.A. (2006). Clustering of *Pseudomonas aeruginosa* transcriptomes from planktonic cultures, developing and mature biofilms reveals distinct expression profiles. BMC Genomics 7, 162–175.

Walters, M.C., Roe, F., Bugnicourt, A., Franklin, M.J., and Stewart, P.S. (2003). Contributions of antibiotic penetration, oxygen limitation, and low metabolic activity to tolerance of *Pseudomonas aeruginosa* biofilms to ciprofloxacin and tobramycin. Antimicrob. Agents Chemother. *47*, 317–323.

Watnick, P., and Kolter, R. (2000). Biofilm, city of microbes. J. Bacteriol. *182*, 2675–2679.

Webb, J., Thompson, L.S., James, S., Charlton, T., Tolker-Nielsen, T., Koch, B., Givskov, M., and Kjelleberg, S. (2003). Cell death in *Pseudomonas aeruginosa* biofilm development. J. Bacteriol. *185*, 4585–4592.

Whitchurch, C.B., Tolker-Nielsen, T., Ragas, P.C., and Mattick, J.S. (2002). Extracellular DNA required for bacterial biofilm formation. Science *295*, 1487.

Whiteley, M., Bangera, M.G., Bumgarner, R.E., Parsek, M.R., Teirzel, G.M., Lory, S., and Greenberg, E.P. (2001). Gene expression in *Pseudomonas aeruginosa* biofilms. Nature *413*, 860–864.

Wozniak, D.J., Wyckoff, T.J.O., Starkey, M., Keyser, R., Azadi, P., O'Toole, G.A., and Parsek, M.R. (2003). Alginate is not a significant component of the extracellular polysaccharide matrix of PA14 and PAO1 *Pseudomonas aeruginosa* biofilms. Proc. Natl. Acad. Sci. U.S.A. *100*, 7907–7912.

Wu, C.L., Domenico, P., Hassett, D.J., Beveridge, T.J., Hauser, A.R., and Kazzaz, J.A. (2002). Subinhibitory Bismuth-Thiols Reduce Virulence of *Pseudomonas aeruginosa*. Am. J. Respir. Cell Mol. Biol. *26*, 731–738.

Yasuda, H., Ajiki, Y., Aoyama, J., and Yokota, T. (1994). Interaction between human polymorphonuclear leukocytes and bacteria released from *in vitro* bacterial biofilm models. J. Med. Microbiol. *41*, 359–367.

Zhang, L.H. (2003). Quorum quenching and proactive host defense. Trends Plant Sci. *8*, 238–244.

Zhang, L.H., and Dong Y.H. (2004). Quorum sensing and signal interference: diverse implications. Mol. Microbiol. *53*, 1563–1571.

Index

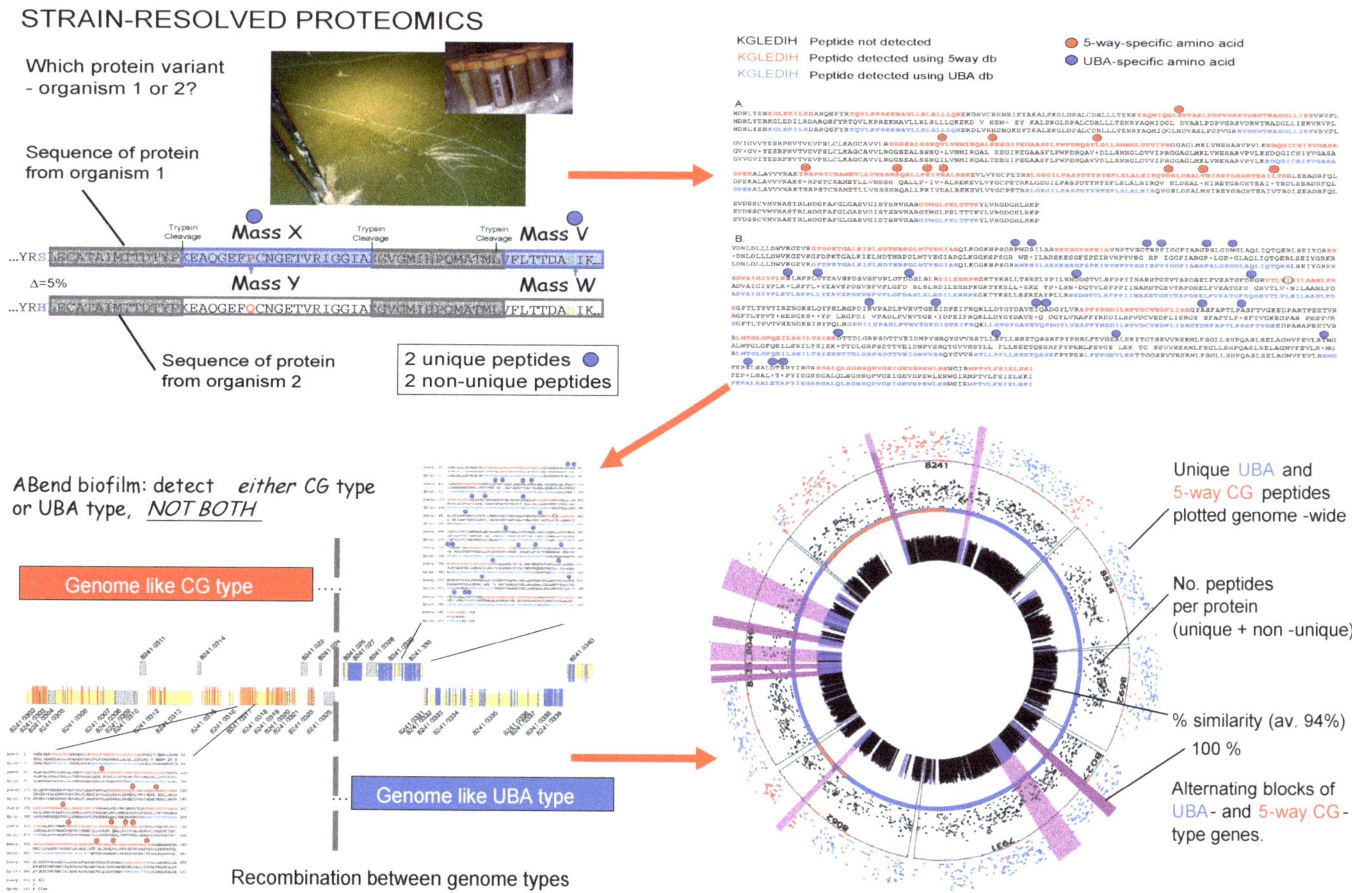

Plate 3.5 The PIGT (Proteomics-inferred genome typing) method. Proteomics is capable of strain-typing sampled bacteria by matching detected peptides to those belonging to a given variant or variants. Additionally, the genetic alignment of these variant peptides onto a genetic map can give information on genetic exchange including recombination between strains and phage-derived integration. First, peptides with unique sequences are identified between two strains from genomics-predicted sequences (top left panel). Next, mass spectra for peptides unique to one strain are matched to a database from one of the two strains (top right panel). Typing of unique peptides is related to parent proteins, which can then be mapped to the genome (lower left panel). Evidence for recombination events can be inferred from alternating blocks of proteins belonging to one strain type (bottom right panel). Adapted from Lo *et al.*, 2007.

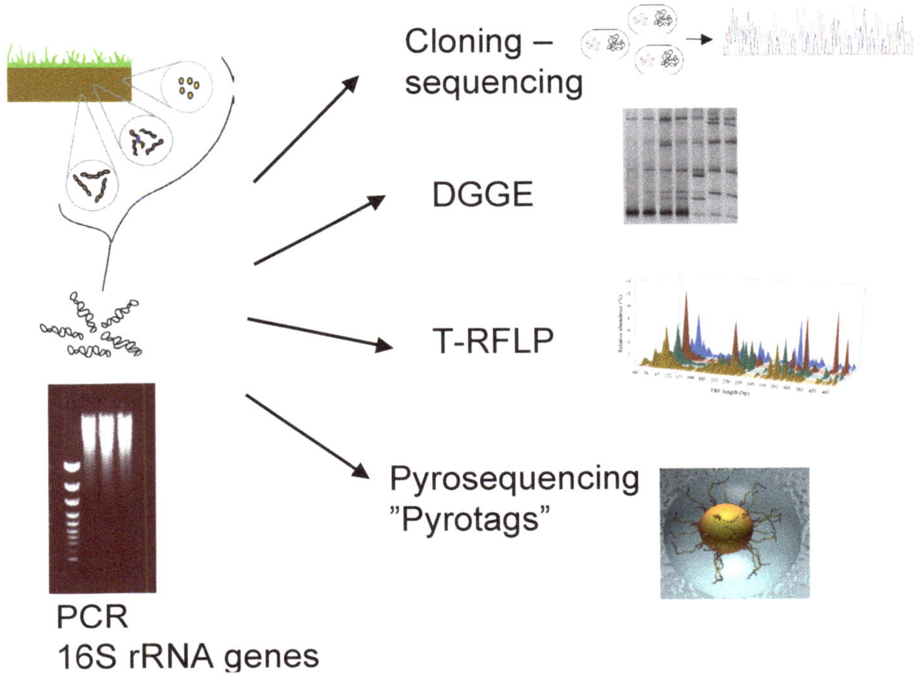

Plate 4.1 An overview of fingerprinting techniques to study complex microbial community structures in environmental samples 'Microbiomics' based on differentiation and/or identification of 16S rRNA genes amplified from the community.

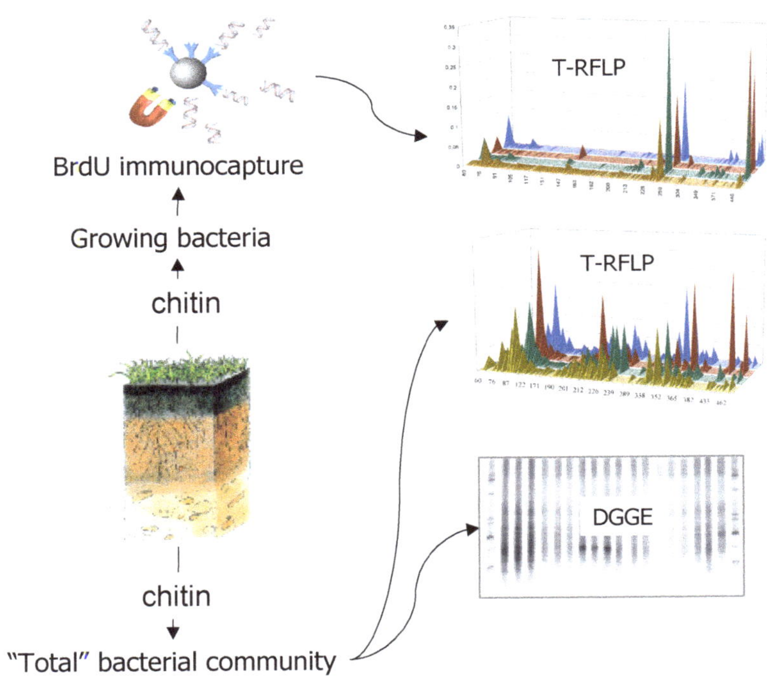

Plate 4.4 Illustration of the use of bromodeoxyuridine (BrdU) immunocapture to fingerprint actively growing bacterial populations upon chitin addition, in comparison to the 'total' microbial community represented in directly extracted DNA (adapted from results described in Hjort *et al*., 2007).

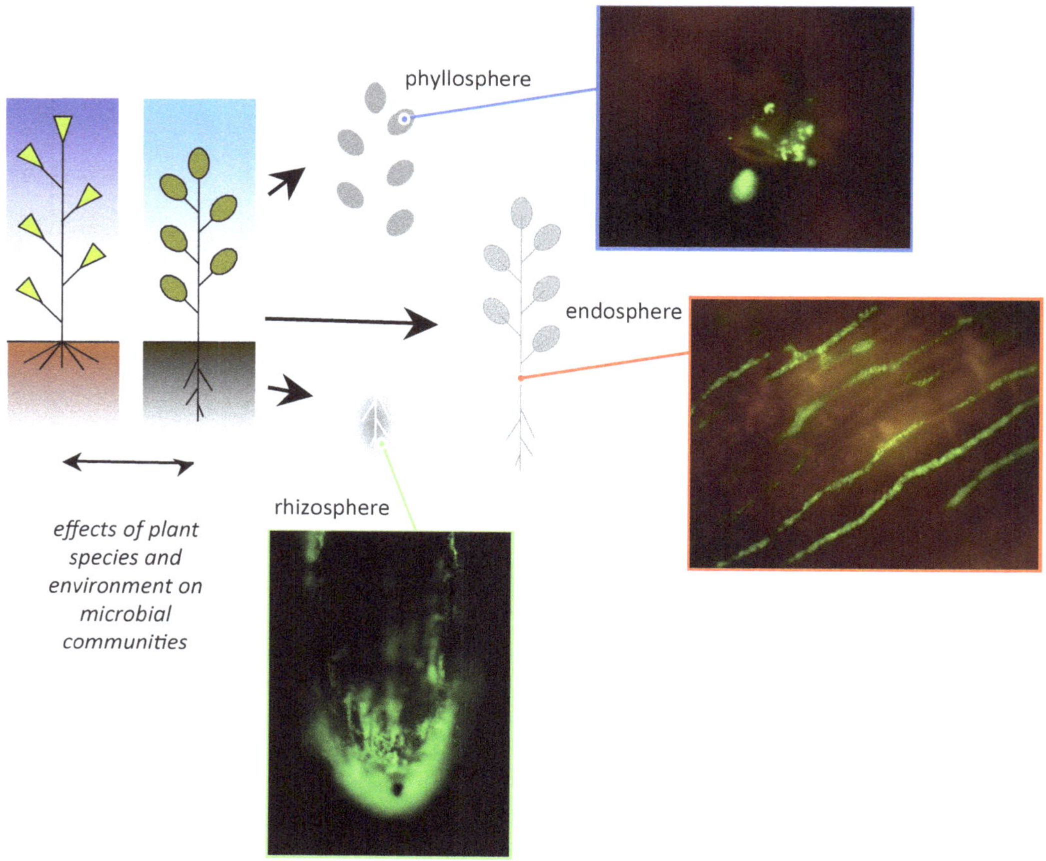

Plate 7.1 Schematic of plant-associated environments colonized by bacteria, highlighting the phyllosphere, endosphere and rhizosphere. Images depict bacterial colonization by GFP-labelled cells of *Burkholderia phytofirmans* strain PsJN, from top to bottom, on a leaf surface (surrounding a stomate), between epidermal cells, and at the root tip. Images courtesy of Stephane Compant.

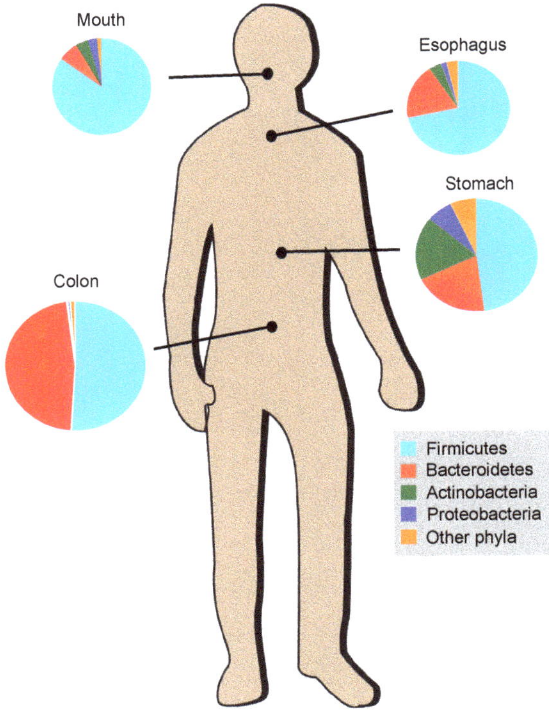

Plate 9.2 Distributions of bacterial phyla at different anatomical sites of the body. Modified from Dethlefsen *et al*. (2007).

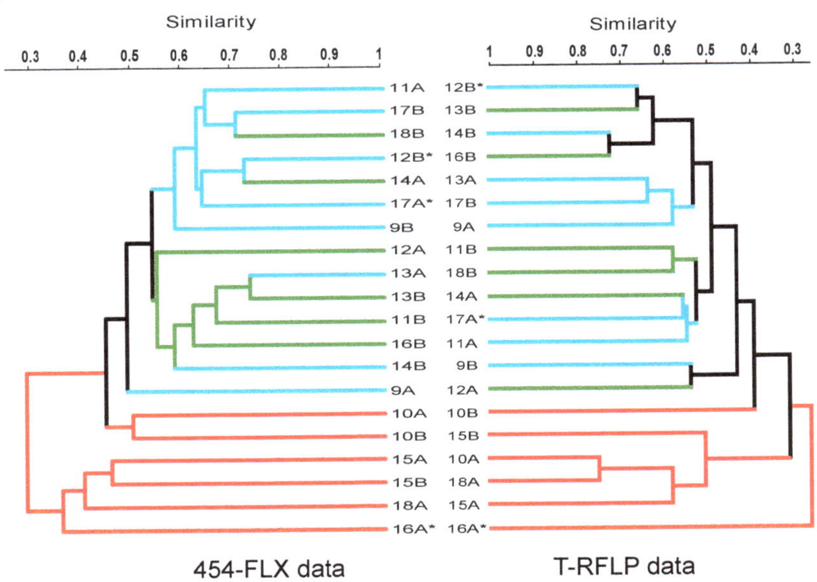

Plate 9.4 Clustering of bacterial community profiles in faecal samples originated from twin pairs that were either healthy, condordant or discordant for Crohn's disease (samples derived from Dicksved *et al*., 2008; Willing *et al*., 2009; see Fig. 9.3, which represents same sample cohort). Bray Curtis index was used as a distance matrix and the dendrogram was based on UPGMA clustering representing data from T-RFLP analysis and 454-FLX pyrosequencing of 16S rRNA genes. The numbers represent the twin pairs (Dicksved *et al*., 2008) and the letters (A or B) represent the individuals within a pair. Colouring of branches refers to the disease phenotype: red, Ileal Crohn's disease (ICD); blue, Colonic Crohn's disease (CCD); and green, healthy individuals. *Represent individuals that originally were classified as ileocolonic disease phenotype, but were reclassified as predominantly having ICD or CCD (Willing *et al*., 2009).

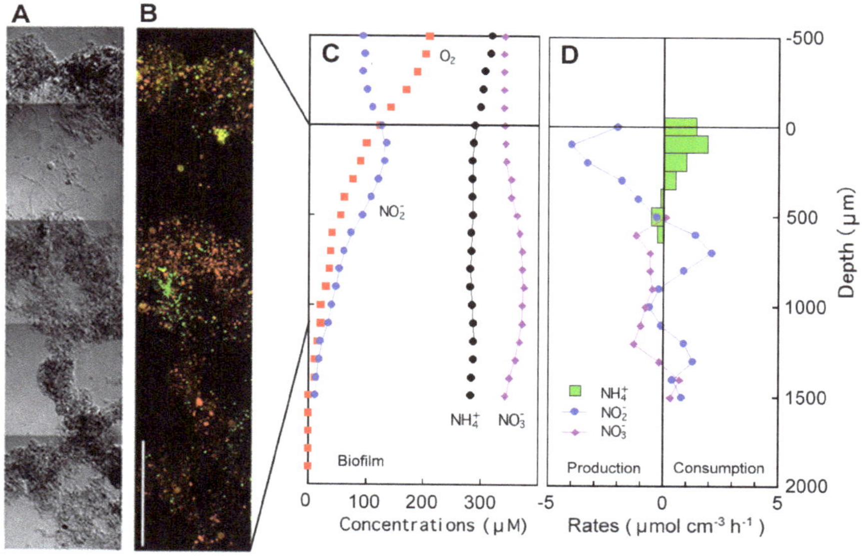

Figure 10.1 Simultaneous *in situ* hybridization of vertical sections of a domestic wastewater biofilm with TRITC-labelled Nso190 probe and FITC-labelled Ntsp454 probe. A composite DIC image (A) and corresponding CSLM projection image (B) of the entire biofilm vertical section (scale bar = 200 μm). The biofilm thickness was about 950 μm. Steady-state concentration profiles of O_2, NH_4^+, NO_2^-, and NO_3^- in a domestic wastewater biofilm incubated in air-saturated medium with 0 μM acetate, 300 μM NH_4^+, 300 μM NO_3^-, and 100 μM NO_2^- (C) and the spatial distribution of the estimated volumetric consumption and production rates of NH4+, NO_2^-, and NO_3^- (D). The points are measured mean values ± standard deviations, the lines are calculated profiles. The biofilm surface is at a depth of 0 μm. Figures are reproduced from Okabe *et al*. (1999).

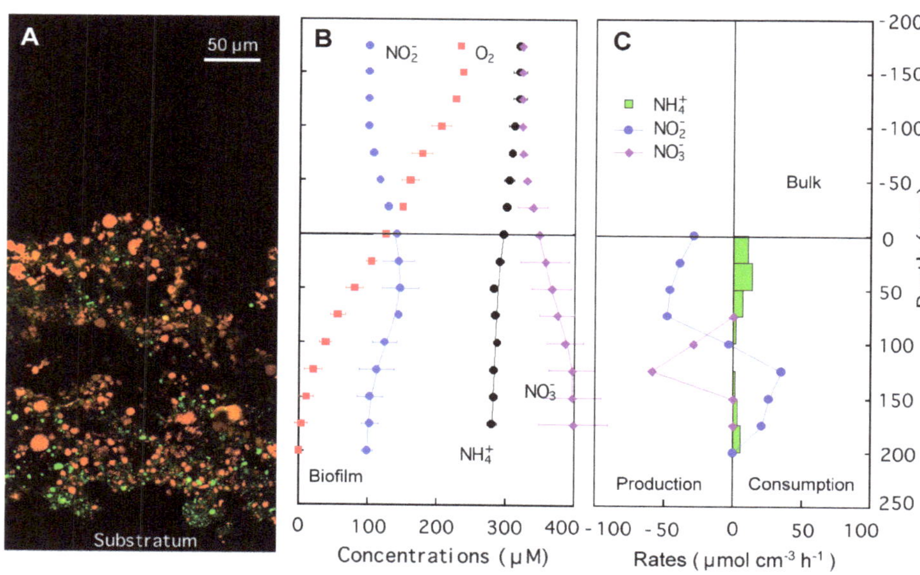

Plate 10.2 Simultaneous *in situ* hybridization of vertical sections of an autotrophic nitrifying biofilm with TRITC-labelled NEU probe and FITC-labelled Ntsp454 probe. A CSLM projection image of the entire biofilm vertical section (scale bar = 100 μm) (A). The biofilm thickness was about 350 μm. Steady-state concentration profiles of O_2, NH_4^+, NO_2^-, and NO_3^- in an autotrophic nitrifying biofilm incubated in air-saturated medium with 0 M acetate, 300 μM NH_4^+, 300 μM NO_3^-, and 100 μM NO_2^- (B) and the spatial distribution of the estimated volumetric consumption and production rates of NH_4^+, NO_2^-, and NO_3^- (C). The points are measured mean values ± standard deviations, the lines are calculated profiles. The biofilm surface is at a depth of 0 μm. Figures are reproduced from Okabe *et al*., (1999).

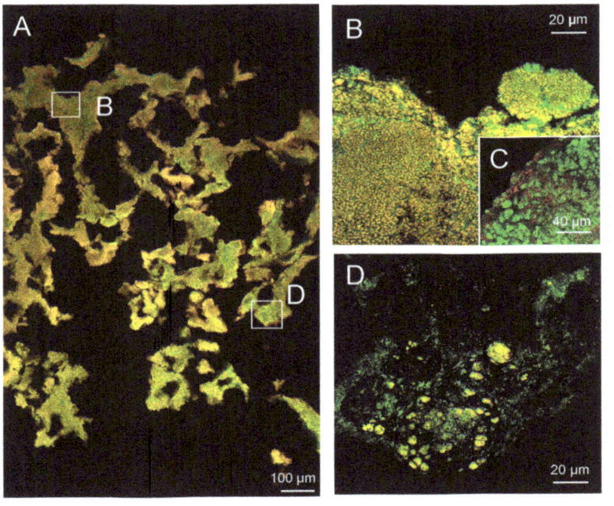

Figure 10.3 CLSM images of vertical sections (30 µm thick) of the anammox biofilm. FISH with TRITC-labelled EUB338 probe mixture (green) and FITC-labelled Amx820 probe (red) (specific to anammox bacteria; Candidatus *Brocadia anammoxidans* and Candidatus *Kuenenia stuttgartiensis*). Yellow signals result from binding both probes to one cell (**A**, **B** and **D**). FISH with FITC-labelled Amx820 probe (green) and TRITC-labelled Nse1472 probe (red) (specific to *Nitrosomonas eutropha*, *N. europaea* and *N. Halophila*) (**C**). The biofilm surfaces are at the top of all images. Figures are reproduced from Tsushima *et al*. (2007).

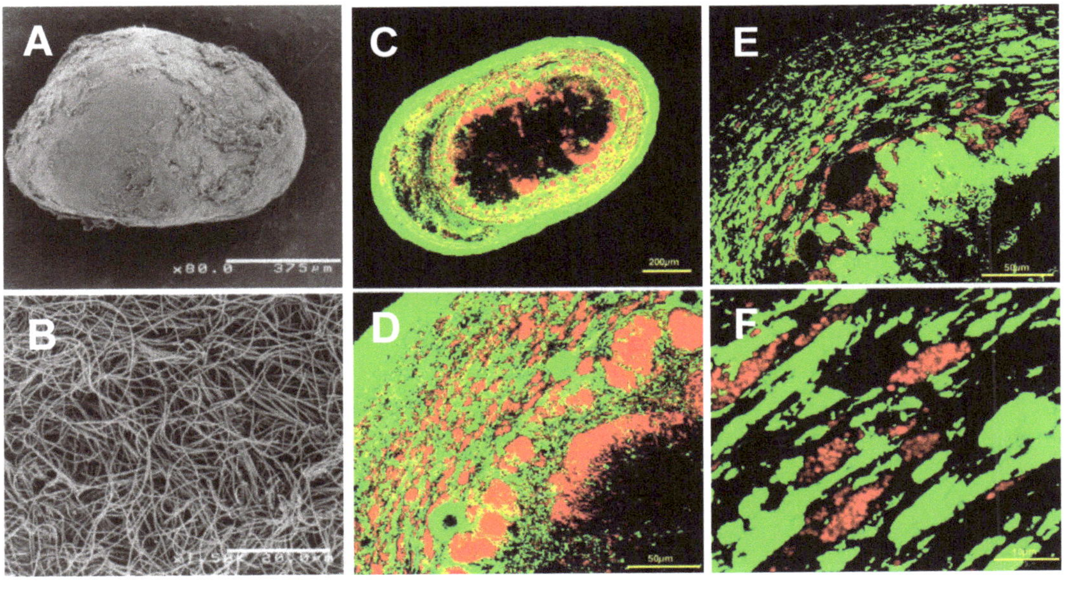

Plate 10.7 Micrographs of mesophilic methanogenic granular sludge. Scanning electron micrographs at **A** high and **B** low magnifications. **C–F** *In situ* hybridization of sections from a mesophilic granule viewed by confocal laser scanning microscopy. **C** and **D** Sections simultaneously hybridized with Cy-5-labelled bacterial-domain probe (EUB338) (green) and rhodamine-labelled archaeal-domain probe (ARC915) (red). **C** and **D** are at low and high magnifications, respectively. **E** and **F** Sections simultaneously hybridized with Cy-5-labelled archaeal-domain probe (ARC915) (green) and rhodamine-labelled *Syntrophobacter* specific probe (SYB701) (red). **E** and **F** are at low and high magnifications, respectively. Figures are reproduced from Sekiguchi *et al.* (1999).

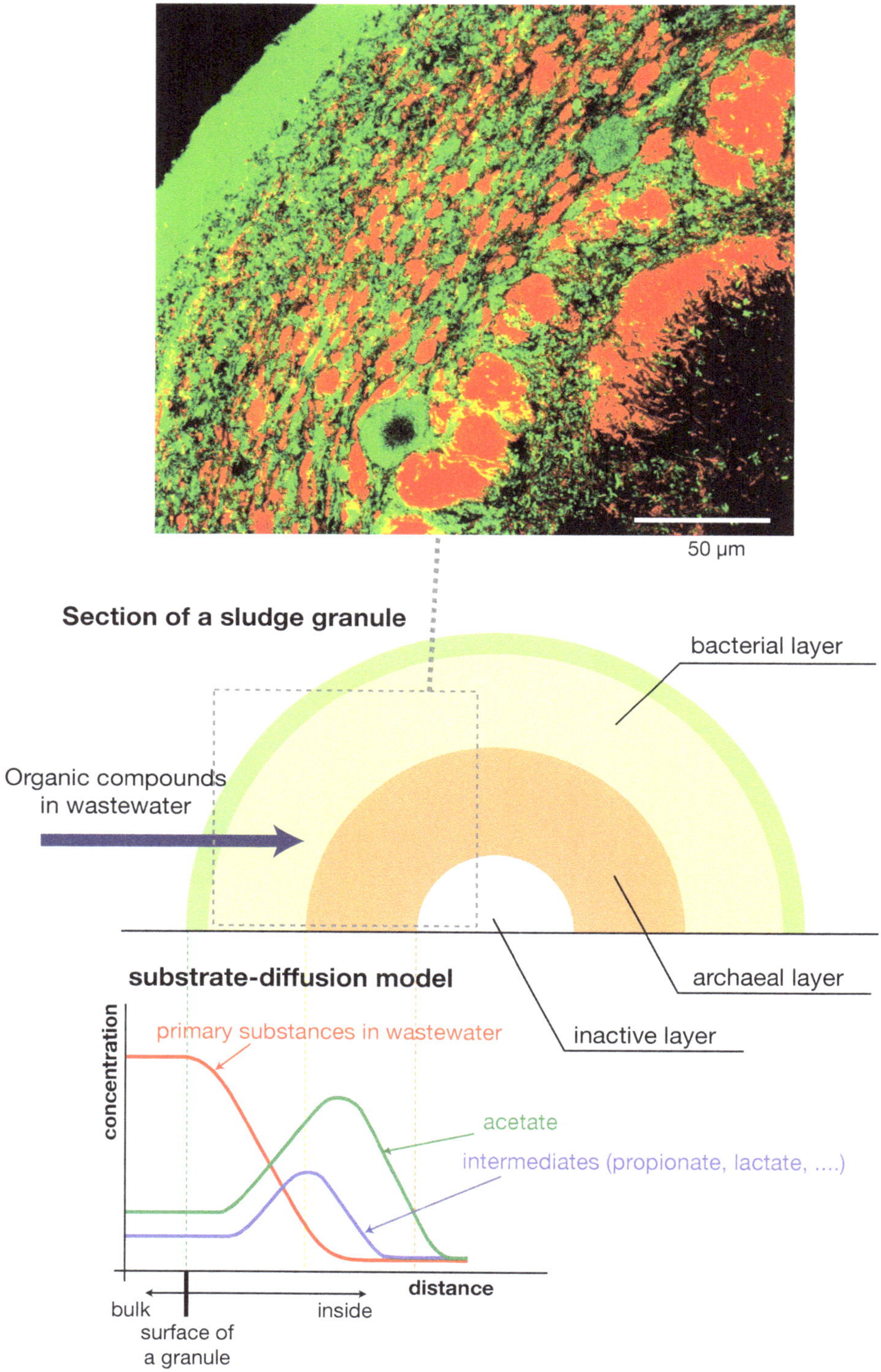

Plate 10.8 **A** *In situ* detection of bacteria (shown as green) and archael (indicated as red) cells cells within a section of a mesophilic methanogenic granule viewed by CSLM (see Fig. 10.7). **B** Schematic diagram of the typical internal structure of a sludge granule, indicating a microbial layered structure and flux profile of organic substrates and intermediates. Figures are reproduced from Sekiguchi (2006).

www.ingramcontent.com/pod-product-compliance
Lightning Source LLC
LaVergne TN
LVHW071337210926
848486LV00003B/167

* 9 7 8 1 9 0 4 4 5 5 5 2 3 *